Life Cycle Costing for Facilities

Alphonse J. Dell'Isola, PE, CVS
and Stephen J. Kirk, FAIA, CVS

RSMeans

Life Cycle Costing for Facilities

Economic Analysis for Owners and Professionals in:

- *Planning, Programming, and Real Estate Development*
- *Designing, Specifying, and Construction*
- *Maintenance, Operations, and Procurement*

**Alphonse J. Dell'Isola, PE, CVS
and Stephen J. Kirk, FAIA, CVS**

 Reed Construction Data

Copyright © 2003
Construction Publishers & Consultants
63 Smiths Lane
Kingston, MA 02364-0800
781-422-5000

Reed Construction Data, Inc., and its authors, editors and engineers, endeavor to apply diligence and judgment in locating and using reliable sources for the information published herein. **However, Reed Construction Data makes no express or implied representation, warranty or guarantee in connection with the content of the information contained herein, including the accuracy, correctness, value, sufficiency, or completeness of the data, methods and other information contained herein. Reed Construction Data makes no express or implied warranty of merchantability or fitness for a particular purpose.** Reed Construction Data shall have no liability to any customer or other party for any loss, expense, or damages, whether direct or indirect, including consequential, incidental, special or punitive damages, including lost profits or lost revenue, caused directly or indirectly by any error or omission, or arising out of, or in connection with, the information contained herein.

No part of this publication may be reproduced, stored in a retrieval system, or transmitted in any form or by any means without prior written permission of Reed Construction Data.

The editor for this book was Danielle Georges. The managing editor was Mary Greene. The production manager was Michael Kokernak. The production coordinator was Marion Schofield. Composition was supervised by Sheryl Rose. The proofreader was Robin Richardson. The book and cover were designed by Norman R. Forgit.

Printed in the United States of America

10 9 8 7 6 5 4 3 2 1

Library of Congress Cataloging in Publication Data

ISBN 0-87629-702-5

Table of Contents

Preface and Acknowledgments — ix
About the Authors — xi
How to Use This Book — xiii
Management Briefing — xv

PART 1: Life Cycle Costing — 1

Chapter 1: Introduction to LLC — 3
    Recent Trends that Support the Need for Life Cycle Costing — 3
    LCC for Design Professionals — 7
    What is Life Cycle Costing? — 12
    Life Cycle Costing Logic — 15
    Application in the Design Process — 17
    Relationship of LCC to Value Engineering — 22
    Program Requirements — 24

Chapter 2: Life Cycle Costing Fundamentals — 27
    Time Value of Money — 28
    Life Cycle — 41
    Discount Rate — 42
    Dealing with Inflation and Cost Growth — 43
    Analysis Period — 47
    Present Time — 49
    Costs — 49

Chapter 3: Economic Analysis — 55
    Terminology — 55
    Office Economic Analysis Example — 57

Types of Economic Analysis	57
Benefits and Costs	60
Economic Analysis Approaches	61

Chapter 4: Life Cycle Cost Analysis — 65
- Life Cycle Cost Analysis — 65
- Present Worth Method Calculations — 65
- Annualized Method Calculations — 67
- LCCA Calculation Worksheets — 68
- LCCA Spreadsheet — 72
- More Complex LCC Analyses — 76

Chapter 5: Estimating Life Cycle Costs — 79
- Estimating Framework — 80
- Initial Costs — 83
- Energy Costs — 92
- Maintenance, Repair, and Custodial Costs — 103
- Alteration and Replacement Costs — 115
- Associated Costs — 119

Chapter 6: Economic Risk Assessment — 125
- Range of Probable Values — 127
- Confidence Index (CI) Approach — 129

Chapter 7: Conducting an LCCA Study — 153
- Identifying Economic Criteria — 154
- Generating Design Alternatives — 155
- Preliminary Evaluation — 157
- Evaluating Life Cycle Costs and Benefits — 158
- Selecting the Design Alternative — 162
- Alternative Ranking/Selection — 172
- Example LCCA Study — 172

Chapter 8: Case Studies — 185
- LCCA Case Study 1: Office to Museum Renovation — 186
- LCCA Case Study 2: District Court Consolidation — 188
- LCCA Case Study 3: Branch Bank Prototype Layout — 189
- LCCA Case Study 4: Health Care Facility Layout — 189
- LCCA Case Study 5: Daylighting — 192
- LCCA Case Study 6: Glass/HVAC Replacement — 193
- LCCA Case Study 7: Lighting System — 194

LCCA Case Study 8: Elevator Selection	194
LCCA Case Study 9: Equipment Purchase Examples	195
LCCA Case Study 10: Construction and Service Contracts	197
LCCA Case Study 11: Chemical Stabilization Plant—Dryers vs. Windows	199
LCCA Case Study 12: Campus University Planning Using LCC and Choosing by Advantages	200
LCCA Case Study 13: High School Trade-Off Analysis of Initial vs. Staffing Costs	203
LCCA Case Study 14: Highway for Regional Transportation Authority	203
LCCA Case Study 15: Life Cycle Cost Assessment —HVAC System for a High School	203
LCCA Case Study 16: Lease vs. Build LCC Analysis	204
Case Study Figures	205-236

Chapter 9: Management Considerations — 237
- Strengths and Weaknesses of Life Cycle Costing — 238
- Level of Effort — 239
- Planning the Overall Study Effort — 239
- Contingency Planning — 247

PART 2: Life Cycle Cost Data — 253

Introduction to Cost Data — 255
Maintenance & Replacement Estimating Data
- Structural — 257-260
- Architectural — 260-268
- Mechanical — 268-288
- Electrical — 289-296
- Equipment — 297
- Sitework — 297-301

PART 3: Appendix — 303

- Appendix A: Economic Tables — 305
- Appendix B: Energy-Estimating Data — 319
- Appendix C: Elevator and Escalator LCC Considerations — 329
- Appendix D: Sample LCC Scope of Services — 337
- Appendix E: Life Cycle Costing Forms — 341
- Appendix F: Selected Government Requirements for LCC — 357
- Appendix G: Historical Development of LCC — 369

PART 4: Bibliography, Glossary & Index	375
Bibliography	377
Glossary	384
Index	389

Preface and Acknowledgments

The cost of ownership for facilities continues to rise. Owners are having difficulty in both obtaining initial capital to construct facilities, and receiving adequate annual funding for maintenance, energy, replacements, and other costs associated with properly operating the facilities. Smaller government budgets and worldwide competition in the marketplace have created a situation in which owners are seeking out design, construction, and facilities professionals who are capable of giving them the most effective life cycle cost design solutions. In addition, recent trends to minimize facility obsolescence, achieve environmental sustainability, and improve operational effectiveness have forced decision-makers to examine, in greater detail, the impact of various alternative design courses of action.

New initiatives with LEED™ (Leadership in Energy and Environmental Design) by the U.S. Green building Council have encouraged both government and private industry clients to become more "green" in their building solutions. Being green also means producing solutions that are cost effective over the facility's life cycle.

The problem professionals and owners face is not a lack of ideas. Ideas are historically what they have been best at providing. The problem is a lack of data and methodology. The challenge is dealing with limited economic resources in such a way that optimum design alternative proposals are selected and implemented.

The emphasis of this book is to make proven life cycle costing (LCC) guidelines available to owners and professionals. Currently-used methodology and data are provided to illustrate the evaluation of alternatives that best meet owners' needs—both economic and non-monetary considerations. The end result is a valuable assessment of an owner's various courses of action.

This book has been organized to serve as a resource for professionals as they prepare life cycle cost analyses for facilities. It takes into account each step of analysis: planning, design, construction, and operation processes.

Examples and illustrations are provided.

The authors would like to acknowledge Michael Dell'Isola for his contribution of the initial management briefing. They would also like to acknowledge the early confidence provided by the American Consulting Engineers Council and the American Institute of Architects for allowing them to conduct their life cycle costing (LCC) training courses for many years, beginning in 1978. The questions from the design professionals in attendance and discussions that ensued at each seminar allowed constant testing and evaluation of the techniques included in both the first and second editions of this book's predecessor *Life Cycle Costing for the Design Professional*. Government agencies such as the National Aeronautics and Space Administration (NASA), the Naval Facilities Engineering Command, the Corps of Engineers, and the General Services Administration are acknowledged for their early enthusiasm and support with regard to LCC. The Worldwide Facilities Group of the General Motors Corporation is also recognized for its use of life cycle costing as documented in this book.

Acknowledgement is also due to Mary Greene of RSMeans for her patience in gaining approval for publication of this book. Danielle Georges is gratefully recognized for her review and helpful comments during publication.

Finally, the authors' families are thanked for their love and support during the time required to bring this book to a successful completion. This book is dedicated to Jan, Gina, Stephen, and Ashley Kirk and Rosie, Mark, David, Paul, Elizabeth, and Marie Dell'Isola, and grandchildren.

Stephen J. Kirk

Alphonse J. Dell'Isola

About the Authors

Alphonse J. Dell'Isola, PE, HRICS, FSAVE, is a Senior Value Engineering Consultant located in Melbourne, Florida. He offers consultant services for project management, value engineering, life cycle costing (LCC) and project cost control. He has worked since 1963 in value engineering and in construction management. Mr. Dell'Isola has conducted more than 1,000 contracts for various organizations and agencies—with projects totaling more than $200 billion dollars in construction that have resulted in implemented savings of some $15 billion.

For twenty years he was Director of the Value Management Division in Washington, D.C. for the SmithGroup, a large design firm headquartered in Detroit, MI. Serving as Director of Value Engineering for the Naval Facilities Engineering Command and for the Army Corps of Engineers, he introduced value engineering in 40 government agencies, and in an equal number of corporations in the United States and abroad. A major portion of these efforts was spent in the Middle East. Between 1950 and 1963, he worked principally on airfields overseas as a materials and cost engineer.

Engineering News-Record cited him in 1964 for outstanding achievement in value engineering. In 1980, he received a Presidential Citation from the Society of Japanese Value Engineers, and in 1993 an Exceptional Service Award received from the World Trade Center for his active role in the disaster reconstruction. In 1994 The Royal Institute of Chartered Surveyors (U.K.) elected him an Honorary Associate. In 1996 SAVE International recognized his achievements by establishing a new honor and award for outstanding achievement—the "Alphonse J. Dell'Isola Award for Construction," and in 1998 he received their highest honor the "Larry Miles Award."

He has given expert testimony to several U.S. Senate and House Committees and was a consultant to the Presidential (President Ford) Advisory Council on Management Improvement. His testimonies were instrumental in leading to the adoption of VE in the government construction agencies.

The author's publications include more than 100 articles on VE, LCC and cost control, as well several professional texts: *Value Engineering in the Construction Industry (Third Edition)* (Smith, Hinchman & Grylls, 1988); *Life Cycle Costing for Design Professionals (Second Edition)* (McGraw-Hill, Inc., 1995) with Dr. Stephen J. Kirk; *Life Cycle Cost Data* (McGraw-Hill, Inc., 1983); *Project Budgeting for Buildings* (Van Nostrand Reinhold, 1991) with Donald E. Parker, and *Value Engineering: Practical Applications* (R.S. Means, 1997).

Dr. Stephen J. Kirk, FAIA, FSAVE, LEED™AP, CVS, is President of Kirk Associates, LLC, an international multi-discipline consulting firm that specializes in facility economics, project planning, and value management services. He is a registered architect, a LEED-certified professional, and a certified value specialist. He received his doctorate in architecture at the University of Michigan. Recognized as a Fellow of the American Institute of Architects for his work in life cycle costing and value management, Dr. Kirk was also selected as a Senior Fulbright Scholar in architecture. SAVE International (Value Management Society) committee on technology advancement awarded special recognition to two unique techniques that he developed. Dr. Kirk is past president of SAVE International and is also a Fellow of SAVE. He has personally led more than 250 studies of offices, research facilities, universities, hospitals, and manufacturing facilities; as well as federal, state, and city government projects and environmental and transportation projects.

Dr. Kirk has more than 30 years of professional design experience, including 25 years devoted exclusively to life cycle costing, estimating, project planning, and value management (VM). He established VM programs for General Motors Worldwide Facilities; Leighton Contractors in Australia; and the General Directorate of Military Works, Saudi Arabia; and has managed VM programs for the SmithGroup and United Technologies Corporation. Dr. Kirk has also conducted more than 200 training seminars for national and international clients. He currently teaches value-related short courses at Harvard University.

Dr. Kirk is the author of several books related to facility economics and value management: *Life Cycle Costing for Design Professionals,* (2nd Edition) (McGraw Hill, 1995) with Alphonse Dell'Isola, *Enhancing Value in Design Decisions* (Meseraull Company, 1993) with Kent Spreckelmeyer, Korean Edition (Ki Moon Dang Publishing Company, 1997); *Creative Design Decisions, A Systematic Approach to Problem Solving in Architecture,* (Van Nostrand Reinhold, 1988); *Life Cycle Costing for Design Professionals* with Al Dell'Isola, (McGraw-Hill,1981); Japanese Edition, 1985; *Life Cycle Cost Data,* (McGraw-Hill,1983); and *Economic Studies for Military Construction: Design Applications* (U.S. Army Tech. Manual 5-802-1: 1986).

How To Use This Book

In this book, the authors show how life cycle costing can be applied to every aspect of construction—from all types of buildings—to roads and bridges—to HVAC equipment and electrical systems upgrades—to materials and equipment procurement.

At the beginning of this book, prior to Chapter 1 is a Management Briefing that is designed to facilitate the process of obtaining approval to use the outlined LCC methodology. The first six chapters, "Introduction," "Life Cycle Costing Fundamentals," "Economic Analysis," "Life Cycle Cost Analysis," "Estimating Life Cycle Costs," and "Economic Risk Assessment," describe the various elements of the total process. Chapter 7, "Conducting an LCCA Study," describes the step-by-step procedures for conducting an LCC study. "Case Studies," Chapter 8, shows how LCC can be applied to particular types—and components—of facilities, in new construction and remodeling projects. The final chapter discusses a variety of management considerations of importance to professionals who use, or work with others to employ this technique.

The appendices contain tools such as economic tables, format sheets, and selected life cycle data, to assist the user in actual applications. A sample scope of work for life cycle cost analysis is included to aid the professional in negotiating with an owner for additional fees to perform the analysis. Excerpts from various government regulations and LCC requirements further emphasize the importance of performing life cycle cost analysis. A glossary provides quick definitions for key words used in life cycle costing. Finally, appropriate sources for additional information, organized by topic, are listed in the bibliography.

This book also has a dedicated Web site with an Excel-based LCC program, including several examples. The Web site, which is directly accessible at http://www.rsmeans.com/supplement/67341.asp, provides the reader with a worksheet that organizes the process and seeks out the required economic factors.

In addition to its use as a professional reference, this book serves as a university text on the process of life cycle costing as it is used in architectural, engineering, and business courses. Finally, *Life Cycle Costing for Facilities* can be used in conjunction with related problem-solving topics, including value engineering, engineering economics, cost estimating, and project management.

Management Briefing

Life cycle costing (LCC) is an economic assessment of an item, system, or facility over its life, expressed in terms of equivalent cost, using baselines identical to those used for initial cost. Life cycle costing is used to compare various options by identifying and assessing economic impacts over the life of each option. In facilities, future costs will often match or exceed the initial cost of procurement. If staffing and other costs are factored into the analysis, the initial procurement costs may be less than 20% of the total costs. Owners are recognizing the relationships between costs and looking for ways to reduce total life cycle cost without adding undue burdens to initial procurements. Designers need to find ways to better respond to owners' needs for improving decisions on both initial and follow-on life cycle costs.

This Briefing will discuss methods for defining and estimating life cycle costs. It will also outline some approaches that owners and design and construction professionals can utilize to optimize design decisions or procurements that are sensitive to life cycle costs.

Economic Principle Analysis

In making decisions, both present and future costs should be taken into account and related to one another. Today's dollar is not equal to tomorrow's dollar. Money invested in any form earns, or has the capacity to earn, interest. For example, $100 invested at 10% annual interest, compounded annually, will grow to $673 in 20 years. A current dollar is worth more than the prospect of a dollar at some future time. The exact amount of future worth depends on the investment rate (cost of money) and the length of time of the investment. The terms "interest rate" and "discount rate" are generally used synonymously, and refer to the annual growth rate for the time value of money.

Figure B.1, "Investment Return on One Dollar," demonstrates the time value of money for various discount rates. Note that for a 5% discount rate, a dollar will grow in value by a factor of approximately 3.5 over the

25-year period, while at a 15% discount rate, the factor is nearly 35. Even though the discount rates are a factor of 3, the resulting relationship is nearly 10.

The *discount rate* means either the minimum acceptable rate of return for the client for investment purposes, or the current prime or borrowing rate of interest. In establishing this rate, several factors must be considered, including the source of finance (borrowed money or capital assets), the client (government agency or private industry), and the rate of return for the industry (before or after income taxes).

At times, the owner may establish the minimum attractive rate of return based only on the cost of borrowed money. Although this approach is particularly common in government projects and in personal economic studies, the same approach may not be applicable to a project in a competitive industry.

Also, inflation affects economic analysis because it reduces the purchasing power of currency over time. This effect, more correctly termed "deflation" means that in the future more currency is required to purchase the same goods. Figure B.2, "Purchasing Power Deflation for Various Inflation Rates," presents this effect over time and, as with discounting, demonstrates the non-linear relationship between the rate and the end result.

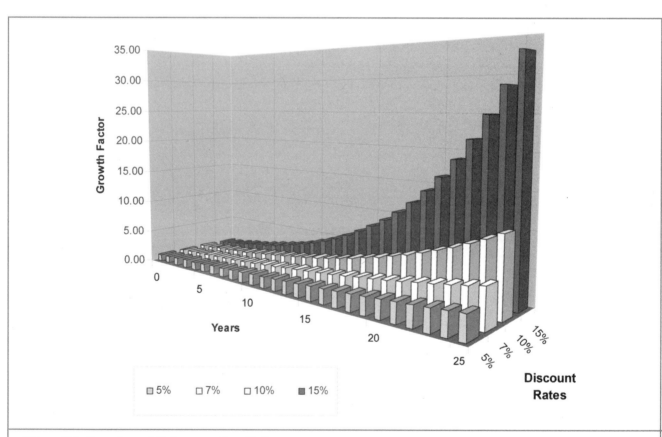

Figure B.1: **Investment Return on One Dollar**

Inflation does not directly affect the actual time value of money, since money has a time value under any circumstance. Inflation, however, does affect how the time value is calculated and must be added to the calculation. In other words, if the real time value of money is 7% to a particular owner, and inflation is predicted to be 3%, then any cash flow analysis would need to use 10% as an interest rate, and inflate all future costs by 3%. This is called a *current dollars* analysis.

As a simplification (especially in comparative analyses such as LLC, which is not used for cash flow calculations), *constant dollars* should be used. In a comparative analysis, all dollars used must be equal-buying-power dollars at the time of the analysis. Future costs that follow the annual inflation factor do not require adjustment. However, cash flows expected not to follow general inflation require adjustment. For example, energy costs have tended to increase at 1–2% above inflation over the past 10 years. In this case, future energy costs would be inflated *differentially* by 1–2%. This effect is referred to as *escalation*.

Note: The appendices at the end of this book contain charts, economic tables, and spreadsheets that provide discounting formulas, including escalated values used for comparative LCC analyses. For more detailed analysis see the book's Web site: **www.rsmeans.com/supplement/67341.asp**

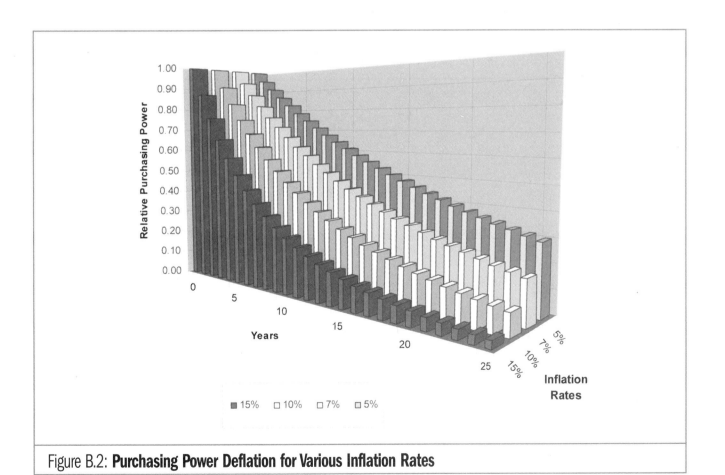

Figure B.2: **Purchasing Power Deflation for Various Inflation Rates**

Escalation can be seen when we look at a $10,000 item that will need to be replaced in year 20. We should bear in mind that the current interest rate is 7%, and the annual dollar inflation is estimated at 3% per year. However, this item is energy-sensitive and estimated to inflate annually at 6%. In a life cycle analysis, the constant dollar replacement cost in year 20 will be calculated using $10,000 escalated at 3% over 20 years. The formula is:

$$A_1 = A \times (1+i)^n$$

where i is the interest rate in decimals, and A is the amount and n the number of years.

A_1 is just over $18,000. When this amount is discounted back to today, the present value is $2,680 ($18,000 × 0.148). *

*Use 7% and 3% Differential escalation – See file – LCC sheet PW 7% on Web site: www.rsmeans.com/supplement/67341.asp

Stated another way, $2,680 placed in the bank today at 10% (7% + 3%) interest rate would grow to provide $18,000 at year 20. This growth would reflect the real discount rate of 7% and the differential escalation rate of 3%. In terms of equivalent costs, $2,680 today (baseline year) is *equivalent* to $18,000 at year 20 at a 7% (today's) interest plus 3% differential inflation rate.

Another example of equivalence (buying power) is presented in Figure B.3 and demonstrates that $1,000 per year for five years is equivalent to $4,123 today at a 10% interest rate, and 3% differential inflation rate. See Text Table A-9 in Appendix A.

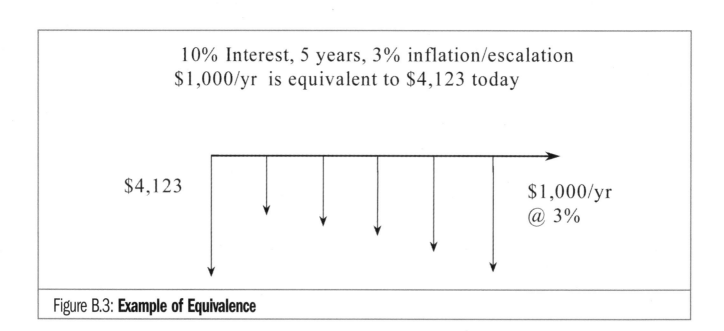

Figure B.3: **Example of Equivalence**

Economic Analysis Period

The economic or study period used in comparing alternatives is an important consideration. Generally, 25–40 years is long enough to predict future costs for economic purposes to capture the most significant costs. Figure B.4 illustrates and plots accumulated annual costs for 100 years discounted to present worth at a 10% interest rate. Note that 90% of the total equivalent cost is consumed in the first 25 years. This is because early year dollars are worth much more than later year dollars. As a result, periods longer than 40 years will generally have no significant impact to the analysis.

A time frame must also be used for each system under analysis. The useful life of each system, component, or item under study may be the physical, technological, or economic life. The useful life of any item depends on such factors as the frequency with which it is used, its age when acquired, the policy for repairs and replacements, the climate in which it is used, the state of the art, economic changes, technological advances, and other developments within the industry. There may be several periods for component replacement in an overall facility cycle.

Categories of Cost

Over the life of a facility, costs will be expended on a broad range of components and for numerous purposes. Because a life cycle cost analysis is a *comparative analysis*, it is important that costs be properly identified

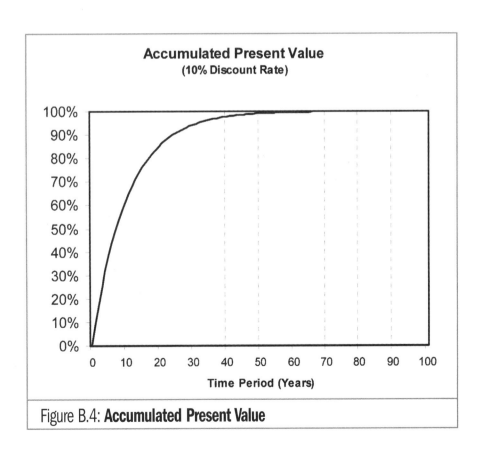

Figure B.4: **Accumulated Present Value**

and categorized so that common items can be eliminated from the analysis, and sufficient effort can be spent on critical items.

Costs can be categorized as listed below:

A. Initial Costs
   1. Construction
   2. Fees
   3. Other Initial Costs
B. Future Facility One-time Costs
   1. Replacements
   2. Alterations
   3. Salvage
   4. Other One-time Costs
C. Future Facility Annual Costs
   1. Operations
   2. Maintenance
   3. Financing
   4. Taxes
   5. Insurance
   6. Security
   7. Other Annual Costs
D. Functional Use Costs
   1. Staffing
   2. Materials
   3. Denial of Use
   4. Other Functional Use Costs

Initial costs include construction fees and other costs such as land acquisition and moving. These represent "up-front" costs associated with facility development.

Future one-time costs represent major expenditures that are not annual (although they may be periodic) and include replacement, elective alterations, and salvage.

Facility annual costs include all costs to run the facility itself, exclusive of what the facility produces. These costs include operations, maintenance, and other built environment costs.

Functional use costs are those associated with the production of the facility and include staffing, materials, and any other non-facility costs. Other items, such as denial-of-use costs, may be necessary during construction because of the specific option selected and may include temporary space, operations, and added security.

Figure B.5 presents the total owning costs of a typical high school. Note that initial costs are only 25% of the total owning costs of the facility. Other facility types such as hospitals, research laboratories, and judicial facilities may be even more weighted toward future costs. In nearly all cases, the initial costs are 25% or less when compared to total owning costs.

Life Cycle Costing Procedures

Life cycle costing focuses on comparing competing alternatives. To compare alternatives, both present and future costs for each alternative must be brought to a common point of time. One of two methods can be used: costs may be converted into today's cost by the *present worth method*, or they may be converted to an annual series of payments by the *annualized method*. Either method will allow proper comparison between alternatives.

Present Worth Method

The present worth method requires conversion of all present and future expenditures to a baseline using today's cost. Initial (present) costs are already expressed in present worth. Future costs are converted to present value by applying the factors presented earlier in this chapter.

A key issue in this process is for the owner to determine the rate of return or the discount rate. The federal government (through OMB Circular A-94) has established 7% as the interest rate to be used in facility studies, excluding the lease or purchase of real property. Normally, a life cycle between 25 and 40 years is considered adequate for estimating future expenses. Differential escalation (that rate of inflation above the general economy) is taken into account for recurring costs, as necessary.

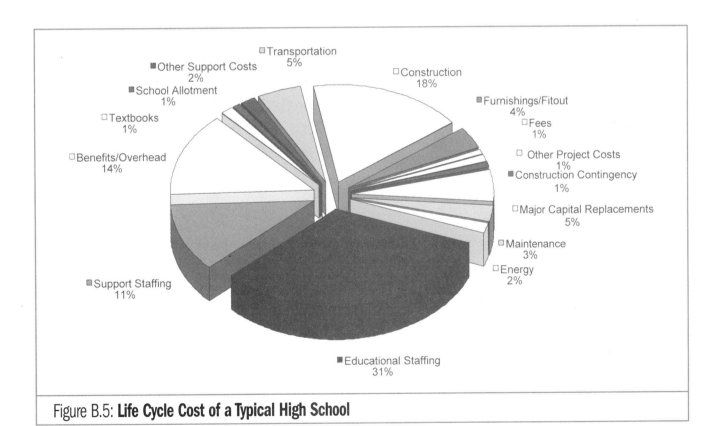

Figure B.5: **Life Cycle Cost of a Typical High School**

Annualized Method

The second method converts initial, recurring, and nonrecurring costs to an annual series of payments and may be used to express all life cycle costs as an annual expenditure. Home payments are an example of this procedure. With mortage a buyer opts to purchase a home for $1,050 a month (360 equal monthly payments at 10% yearly interest) rather than pay $150,000 all at once. Recurring costs are already expressed as annual costs; therefore no adjustment is necessary. Initial costs, however, require equivalent cost conversion. Nonrecurring costs (future expenditures) must be converted to current cost (present worth) and then to an annual expenditure (annualized cost).

Other Economic Analysis Methods

Other methods of economic analysis can be used in a life cycle study. With additional rules and mechanics, it is possible to:

- Perform a sensitivity analysis;
- Determine the payback period;
- Establish a break-even point between alternatives;
- Determine the rate of return and extra-investment;
- Determine rate-of-return alternatives;
- Perform a cash flow analysis;
- Review the benefits and costs.

Note: Also see Chapter 3.

All methods, correctly applied, will yield results pointing to the same conclusion. In other words, the alternative with superior economic performance will be selected. However, since the construction industry is capital cost-intensive, the present worth method is recommended. Furthermore, the present worth method tends to be easier to use and produces results that can be easily understood.

The Relationship Between Quality and Cost

Life cycle costing provides a methodology to compare competing alternatives over the life of the facility. These alternatives will generally reflect differing performances of the systems relative to their useful life, required maintenance and operating costs. The performance of systems will tend to reflect their quality in the sense that higher "quality" systems usually have high initial cost and low future costs. Low "quality" systems tend to be the opposite, with low initial cost, but high future cost.

Figure B.6 presents the diagrammatic relationship between quality and cost, and demonstrates the fact that there generally exists a "lowest life cycle cost" design choice. If the owner's quality expectations are well-founded, this point will occur within the range of the specified quality.

It is interesting to note that in traditional design, the tendency has been for systems to "creep" up the quality curve under the premise that better quality is always desirable. However, quality that exceeds actual needs is not necessarily a good investment if the added quality is obtained at a high price. Conversely, "cost cutting," which is often done to maintain budget requirements, may reduce quality below what is in fact needed, simply to achieve budget.

Owners can optimize the quality-to-cost relationships by requiring their decision-makers to use life cycle costing.

Life Cycle Costing Applications

The debate in the construction industry continues as to which is the "best" delivery system. There is no easy answer, but it becomes apparent that life cycle costing can be a critical component in analyzing design solutions and in ultimately providing *improved quality* facilities.

There appear to be three basic opportunities for life cycle costing to impact delivery:

1. By owners emphasizing performance and quality in systems, facility design, and operations and maintenance procurements, and potentially requiring bids based on life cycle cost analyses for selection. The industry initiative to move to "performance specifying" will assist this outcome in the preparation of bidding documents.

2. By owners including life cycle costing impact in initial design selections. The owner can require the designers to indicate life cycle costs for the

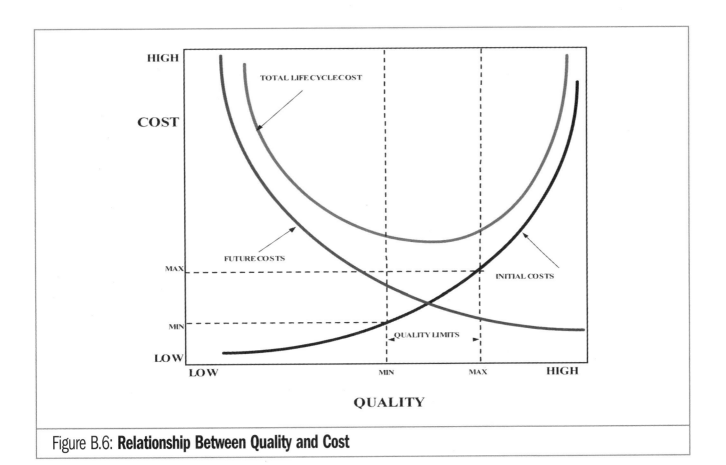

Figure B.6: **Relationship Between Quality and Cost**

entire facility, for only deviations from owner specifications, or for selected systems. The owner may also require a performance "guarantee" tied to future incentives or penalties. Owners are also including requirements for sustainable design and workspace productivity enhancement.
3. By designers emphasizing life cycle performance when presenting competing design solutions. This can be in response to owner requirements or elective on the part of the designers to gain a competitive advantage. This action will also provide owners with an opportunity to optimize total costs, reducing initial costs, providing funds that can be applied to follow-on expenses.

Life cycle costing has a unique opportunity to "connect" initial cost and ownership costs to optimize total costs. The benefit of this connection to an owner who is responsible for operating a facility is improved decisions and facilities.

Part 1 Life Cycle Costing

Chapter 1: Introduction to LCC

When we mean to build, we first survey the plot, then draw the model; and when we see the figure of the house, then must we rate the cost of the erection; which if we find outweighs ability, what do we then but draw anew the model in fewer offices, or at least desist to build at all?
William Shakespeare
Henry IV, Part 2, I. iii, 1598

A number of recent trends have emerged as issues of concern for the design professional, including: facility obsolescence, environmental sustainability, operational staff effectiveness (re-engineering), total quality management (TQM), and value engineering (VE). Life cycle costing (LCC) permits the economic assessment of alternatives being considered in response to these issues.

Recent Trends That Support the Need For Life Cycle Costing

Facility Obsolescence

Building or facility obsolescence (especially premature) results from such factors as changing facility users and their new demands; new materials, technology, and procedures of construction and operation; new air pollutants; and new laws and regulations. Each factor requires design professionals to consider innovative alternatives that accommodate change in order to minimize premature obsolescence. Response to a fourth dimension—time—is critical to overall project success. LCC is particularly useful for designers in assessing the life cycle consequences of alternatives being considered in order to minimize facility obsolescence. As Admiral Donald Iselin (former commander of the Naval Facilities Engineering Command) pointed out in his book, *The Fourth Dimension in Building: Strategies for Minimizing Obsolescence*, "We cannot afford to build in ways that become obsolete quickly in a changing world."

Environmental Sustainability

Ensuring environmental sustainability involves seeking materials and methods of construction that will not harm the environment or use excess natural resources. Reusing asphalt pavement and existing buildings, and energy conservation techniques such as daylighting are examples of seeking alternatives that are friendly to the earth. Unfortunately, many of these alternatives carry a high initial price tag that must be factored into the LCC considerations in selecting the best choice for a given owner or user.

Operational Effectiveness (Including Re-engineering)

Operational staff effectiveness is particularly important in buildings such as hospitals and correctional facilities where the staffing costs typically exceed the capital cost in as little as three years. Design professionals must seek ways to organize space in order to minimize staff travel distance and otherwise create efficient work flows to enhance work productivity. Another consideration is re-engineering the management organization of the business departments in order to maximize staff interaction. Above all, design professionals, through effective space layout, can assist owners in minimizing the number of staff members required.

Total Quality Management

The potential for improvements in product quality through effective use of total quality management has forced every business to change its management philosophies. Key TQM concepts such as customer-focused, process-driven, people-oriented, management-led, and continuously improved operations, have revolutionized the way businesses approach their work. Design professionals are now required to rethink their approach to design in order to become more TQM-based. This requirement causes the design professional to address owner challenges of international competition, costs of operation and maintenance, and overall business profitability. The use of life cycle costing helps evaluate options for solving these problems.

Value Engineering

Value engineering has been an effective management tool for seeking the best value for the money in facilities design for more than 40 years. VE continues to be mandated by the U.S. federal government and many state, county, and city agencies. It is also used in many businesses because of the savings it achieves for the owner—often 5–10% of the construction cost. Life cycle costing is a tool that is used in VE to help evaluate various alternatives for the purpose of selecting the optimum solutions. VE can result in another 5–10% savings in follow-on costs of ownership. The authors routinely use LCC in their VE studies.

Owners' Rising Expectations

Rising expectations about the services and amenities that a facility should provide are the result of changing users or owners and the differences in their requirements from those the facility was initially intended to fulfill.

The technologies of modern facilities, including hospitals, research laboratories, educational facilities, and manufacturing facilities, have changed substantially in recent decades, and are continuing to change. These changes have also led to rising expectations in buildings. Accommodating those expectations is costly, but failing to accommodate change is more costly because obsolete facilities impose a heavy burden on their owners and users.

The obsolescence burdens carried by owners result in significantly higher life cycle costs for their facilities because of lost productivity of people, increased operating costs to overcome the mismatch of needs and facility capability, increased worker absenteeism, health care costs related to on-the-job-stress, and reduced ability to attract and retain employees. Obsolescence should be considered within the context of a facility's entire life cycle, from initial planning through operations and maintenance.

LCC Assistance for Recent Trends

Life cycle costing can assist the designer in assessing the economic consequences of continuing to use an existing building, system, or component, in comparison with the expense of substituting an alternative that may offer better staffing efficiency, improved performance, new technology, or changing organizational structure. Facilities must accommodate anticipated new communications, building automation, and energy-saving technology. Consideration must be given to changing patterns of space use. For example, in hospitals, same-day surgery has drastically altered the demand for surgical and supporting laboratory facilities. New technologies—such as positron emission tomography (PET)—have added to hospitals' new, large, and heavy equipment that cannot be easily housed in or moved into older buildings. A good example of social and political cause for obsolescence is the Americans with Disabilities Act of 1990. This law requires that new and remodeled buildings be fully accessible and safe for people with disabilities. A more recent social and political cause is the need for increased security as a result terrorism, and green building requirements for new construction in some states and municipalities.

Strategies for addressing these recent trends include identifying changes that may foster obsolescence, performing pre-design analysis, responding to future requirements through innovative and flexible design solutions, and modifying current approaches during construction, and operation and maintenance.

Before they begin to design, architects and engineers should screen published literature in order to spot emerging issues. For example, the Environmental Protection Agency (EPA) provides professional forums to alert designers of emerging environmental issues. The American Institute of Architects (AIA) has an environmental committee that focuses on the environmental implications of various building materials on facilities.

During pre-design, post-occupancy evaluations are helpful in gathering information regarding the best and worst building features, potential cost-saving modifications, and operational effectiveness of the facility that may be used as ideas for improved life cycle cost-effectiveness and minimized obsolescence. Consideration of integrated building systems may be an effective tool for achieving flexibility. The Veterans Administration (now the Department of Veterans Affairs) has developed an integrated hospital building system (VAHBS) for managing obsolescence and minimizing the remodeling costs of its hospitals. The U.S. Postal Service has developed a "kit-of-parts" to respond to a diverse range of functional needs. Open-plan office building designs and modular furniture respond to the search for increased ability to avoid or delay obsolescence. Design approaches for other facility types can be found through the creative imagination of design architects and engineers and the use of life cycle costing to assess the design alternatives. Such ideas as larger bay sizes, raised flooring, use of interstitial ceiling space, increased floor-to-floor heights, modularity of mechanical and electrical systems, and increased shell space permit operation flexibility for future, as yet unknown, owner and user needs.

During construction, attention should be given to achieving the quality envisioned in the design. Failure to do so will result in a more rapid decline in facility performance and operational effectiveness, increased energy use, and an earlier onset of obsolescence. For those items that are particularly sensitive to new technology, such as electronic control components, medical equipment, and data transmission and networking devices, delaying specification and procurement until immediately prior to installation can help avoid obsolescence. Separate "fit-out" construction packages and owner-furnished equipment (independent from the prime construction contract) will help reduce obsolescence and improve the life cycle cost of the facility.

Management actions during the maintenance and operation stage of the life cycle are particularly important to avoiding or delaying facility obsolescence. No less important is continuing to seek ways of improving staffing operational effectiveness. Good maintenance practices have an effect similar to that of quality assurance during construction; they enhance the likelihood that performance will indeed conform to design intent. This responsibility for good practices rests primarily with the facility manager and maintenance staff. Training of maintenance staff, preparation and updating of maintenance manuals, and use of appropriate materials in maintenance activities contribute to avoiding the costs of obsolescence. Use of new computerized facility management systems that support condition monitoring, documentation management, and maintenance scheduling can be linked with other building systems.

Figure 1.1 summarizes trends and related LCC concerns, including delaying facility obsolescence, environmental sustainability, improving operational effectiveness through re-engineering, and practicing TQM and VE as they relate to LCC concerns including energy, maintenance, flexibility, staffing, and capital cost. This chapter is intended to help design professionals assess the alternatives to address these concerns.

LCC for Design Professionals

Life cycle costing for design professionals can be defined as the economic assessment of competing design alternatives, considering all significant costs of ownership over the economic life of each alternative, and expressed in equivalent dollars. P.A. Stone, a British economist, applies the terminology *cost in use* and suggests that the technique is concerned with "the choice of means to a given end with the problem of obtaining the best value for money for the resources spent."[1] In 1972 the U.S. Department of Health, Education, and Welfare summarized life cycle analysis as the systematic consideration of "cost, time, and quality."[2] Life cycle costing most certainly addresses these, as well as several other issues related to decision processes, analytic methods, databases, and component performance.

Federal, state, and institutional owners have each issued LCC directives to the designers of their facilities. In 1977, the U.S. federal government established a goal of reducing energy consumption by 45% for all federally owned new buildings over their prior 1975 counterparts. The U.S.

RECENT TRENDS: LCC CONCERNS:	Total Quality Management (TQM)	Obsolescence	Environmental Sustainability	Operational Effectiveness	Value Engineering
Initial Project Cost					■
Energy/Fuel Costs			■		■
Maintenance & Repair		■			■
Alterations & Replacements		■	■		■
Administrative Costs	■			■	■
Staffing Costs	■				
Safety/Security Systems		■		■	■
Real Estate Taxes					
Water & Sewer Costs			■		
Fire Insurance Costs					
Flexible Furniture Systems	■		■	■	
Air/Water Quality		■	■		
Healthful Environment		■	■		
Sustainable Materials			■		
New Business Technology	■	■		■	■
Communication Systems	■	■		■	■
Automation Equipment	■	■		■	■
Site Environment			■		
Occupant Comfort/Control	■	■	■	■	■
Business Profitability	■	■	■	■	■
Bay Size/Floor Height		■		■	■

Figure 1.1: **Recent Trends and Their LCC Concerns**

Congress, in November 1978, established the National Energy Conservation Policy Act, which mandates that all new federal buildings be *life cycle cost-effective* as determined by LCC methods prescribed by the legislation. Existing buildings were to be reviewed to improve their energy efficiency in general, and to minimize their life cycle cost. Nebraska passed legislation in 1978 requiring a life cycle cost analysis for every state facility with a project cost of more than $50,000. Alaska, Florida, Massachusetts, Wisconsin, Texas, North Carolina, New Mexico, Washington, Maryland, Wyoming, Colorado, Illinois, and Idaho, among others, have since established similar legislation. The federal government's General Services Administration (GSA) has developed elaborate procedures for predicting a facility's total life cycle cost. Cities including Atlanta, Phoenix, and Chicago, also require life cycle cost analysis studies from designers of municipal facilities.

Owner Demands

Owners are feeling the international economic squeeze, as well as social and economic pressures, and are reacting to it. Keen international market competition is forcing owners to be extremely cost-effective and environmentally responsive. These factors, coupled with economic incentives to minimize facility obsolescence, *mean that owners are looking for design professionals with enhanced economic skills.* They want facilities with the lowest possible initial project cost, as well as minimized annual energy consumption, maintenance cost, replacement cost, and staffing costs; together with the longest serviceable building life attainable within the other parameters, the highest possible quality, the best appearance, and the least taxes. How can proper decisions be made with all these parameters to consider? How has the profession historically made these decisions? Many design professionals are not completely fulfilling their responsibility to their clients to adequately consider all these factors in the design process. Also, owners have not been forceful enough in requiring these services, nor have they shown much willingness to pay for them. *Owners must provide the required fees and data and be prepared to become involved in the trade-off life cycle cost decisions necessary for final design selections.*

To ensure the best value for least life cycle cost, design professionals must use a clearly defined methodology and have the tools necessary to perform the economic analysis. Of the typical client demands, all are measurable to some extent in dollars except quality and aesthetics, and even these can be evaluated given the functions required of the facility and a proper methodology. (Chapter 7, "Conducting an LCCA Study," outlines an approach for considering these non-monetary criteria in the selection process.)

Design and LCC Considerations

Design professionals must respond to owners' needs and demands. The designer today requires a methodology to analyze the initial project cost, energy consumption, maintenance cost, replacement cost, financing, staffing cost, and so on—in short, a tool to look at *total cost*. This approach is necessary to determine if owners can afford not only the initial costs, but also the follow-on costs. How many design professionals typically provide the owner with both an estimate of the initial project cost and the annual follow-on costs? What training and expertise does the design professional have to address these client needs?

There is a hierarchy of objectives or goal priorities, from the design professional's point of view, to be solved in any design. These can be related to Maslow's needs-priority model[3] (Figure 1.2), which illustrates certain *basic physiological needs* that the designer must first satisfy for the client. These basic needs include such things as program and user requirements and cost and time constraints of the project. Once these have been fulfilled, the designer concentrates on the client's security and safety needs. The requirements of documented building codes and regulations are reviewed and compared with the initial layout. Next, the designer must establish a design that has *social belonging*. In other words, will the community appreciate and understand the design, and will the solution receive peer designer acceptance?

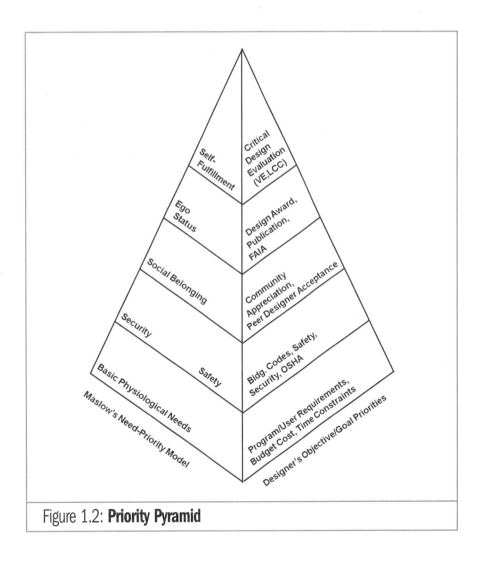

Figure 1.2: **Priority Pyramid**

These needs having been satisfied, the next block on the priority model relates to *professional status or ego*. For an architect, this might be a design award or election to the College of Fellows with the right to use the designation FAIA (Fellow of the American Institute of Architects). For an engineer, it might be a fellowship in an engineering society or recognition through publications and awards. However, there is one more need—that of *self-fulfillment*. That design professionals are beginning to recognize this is shown in their efforts to perform post-occupancy evaluations of their designs.

Can the design stand the test of time? This highest level of the pyramid, from the designer's point of view, can be described as the *critical design-evaluation phase*. To achieve success in this category, such factors as LCC, value engineering, and post-occupancy evaluation need to be considered. How many design professionals are prepared to address these elements, which many of today's facility owners are requiring? In planning of a number of proposed large facilities, owners want to know not only what it will cost to build these facilities, but also what will it cost to maintain and operate them. What are the annual operation and maintenance budgets for the facilities now being designed? *Getting involved in LCC is the first step toward finding the answers.*

Figure 1.3 represents the breakdown of the significant life cycle costs for a typical hospital. It is interesting to note the distribution: initial (capital) costs, 6%; fuel and utilities, 6%; maintenance and contracted costs, 12%; medical supplies and food, 7%; drugs and pharmaceutical, 5%; and staffing costs, 64%. This illustrates the importance of finding design solutions that minimize hospital staffing costs. When staffing costs are not included, the largest block of ownership cost is normally the initial cost. Therefore, the first requirement is to have a standard costing procedure and accounting method to properly define and collect initial costs. In addition, there must be a procedure to define and assemble costs for maintenance and other annual expenses. These accounting procedures must be compatible with the standard costing system for initial costs. The effect of time on the value of money must also be considered, since today's dollar is not equal to a dollar spent in the future. Money invested in any form earns, or has the capacity to earn, interest. *(This concept is discussed at length in Chapter 2, "Life Cycle Costing Fundamentals.")*

What are some of the reasons owners have not always received optimum decisions from their agents? And, in particular, why has there been less-than-optimum decision-making in some of the LCC analyses performed to date? The major reasons are lack of management and owner commitment, lack of innovative ideas, lack of life cycle cost information, recurring circumstances, honest wrong beliefs, and habits and attitudes. If owners do not provide designers the opportunity or incentive to perform LCC, it will seldom be done effectively. *Owners must require specific tasks using the methodology of LCC and be willing to provide additional fees to cover the efforts required from the design professional.*

Who is providing designers with new ideas and better information to ensure optimum decisions? The majority of new ideas come from manufacturers and suppliers. Yet many owners and designers actually discourage vendors' participation in the design process, to the point of not allowing them in the door. At a recent value engineering and LCC workshop held at an engineers' club, club rules did not allow vendors to participate. In addition, very few design firms have a person or a department appointed to assist manufacturers' representatives with new ideas. Seldom do firms conduct in-house seminars concerning the latest uses of new products. Moreover, little effort is spent to seek out top professionals from outside to bring in new ideas and more information. *Yet, if decisions are to be optimized and alternatives properly assessed, designers must continually seek out innovations and new product information.*

Design professionals must overcome recurring circumstances, such as lack of time and lack of redesign funding, that almost always are present and cause poor decisions and unnecessary costs. Since these circumstances seem to follow a standard pattern, designers should either plan ahead, knowing that the chance of their recurrence is great, or offset their impact by generating new solutions on existing projects for later application to follow-on designs.

Designers have to look for, question, challenge, and change honest wrong beliefs that tend to creep into any organization. Many designers are making decisions today that are based on what they think are the best

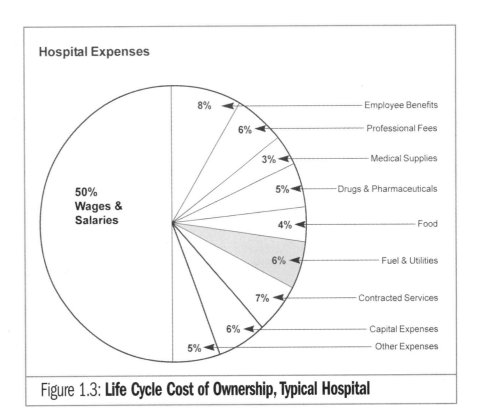

Figure 1.3: **Life Cycle Cost of Ownership, Typical Hospital**

alternatives. These alternatives may not, in fact, be the best ones. For example, many electrical engineers believe that only circuit breakers should be used for main switchgear. On the other hand, there is another generation of engineers who are switch- and fuse-oriented. Whose decision is optimal for a given project? One or the other has some honest wrong beliefs. Where are the quality assurance procedures in facility design similar to those used in the manufacturing process?

Designers must review the habits and attitudes in their organizations to see which ones must be changed to improve decision making. A prime concern is trying to achieve a positive approach to problem-solving. A positive approach enhances optimum decisions, while a negative approach contributes to poor solutions. It can be guaranteed that at the end of each year, the designers with positive attitudes toward new ideas will have made better decisions than designers who express only reasons why an idea will not work. *Remember, positive attitudes achieve positive results!*

What Is Life Cycle Costing?

What is life cycle costing? Basically, LCC is an economic assessment of an item, area, system, or facility that considers all the significant costs of ownership over its economic life, expressed in terms of equivalent dollars. LCC is a technique that satisfies the requirements of owners for adequate analyses of total costs.

A key element in LCC is an economic assessment using equivalent dollars. For example, assume one person has $1,000 on hand, another has $1,000 promised 10 years from now, and a third is collecting $100 a month for 10 months. Each has assets of $1,000. However, are the assets equivalent in terms of today's purchasing power? The answer is not simple because the assets are spread across different points in time. To determine whose assets are worth more, a baseline time reference must first be established. All monies are then brought back to the baseline, using proper economic procedures to develop equivalent costs. Design professionals normally choose between competing design alternatives. So, for design professionals, *life cycle costing is an economic assessment of design alternatives, considering all the significant costs of ownership over an economic life expressed in equivalent dollars.* LCC may also be used to assess the consequences of decisions already made, as well as to estimate the annual operation and maintenance (O&M) costs for budgeting purposes.

For designers to perform a life cycle cost analysis, the owner must provide them with information regarding such things as the facility's economic life, the anticipated return on investment, and financing costs, as well as non-monetary requirements.

From owner to owner, this information will vary greatly. As an example, assume an owner is planning to build a speculative apartment house. The federal government has set up an investment tax credit for this type of facility. However, it requires retention of property by the owner for a minimum of ten years. The owner, in hiring a consultant to perform LCC, would set a ten-year economic life, and would also state the minimum acceptable rate of return for the project to be economically attractive.

Suppose, on the other hand, the owner is a telephone company that will own and operate the facility from 40–100 years. In this case, the owner would require a permanent type of construction, normally with a 40-year economic life. These examples represent the extremes, but they illustrate that the economic information for LCC will vary from one owner to the next.

The LCC definition states that all "significant cost of ownership" should be included. Figure 1.4 illustrates the types of costs that may be considered significant by the designer and owner for an LCC study. These costs are organized so that they may be structured easily into an automated approach as experience is gained in the procedures. For clarification, a brief discussion regarding these blocks of costs follows.

Initial costs include the owner's costs associated with initial development of a facility, such as project costs (fees, real estate, site, and so on) and construction costs. *Financing costs* include the costs of any debt associated with the facility's capital costs. The category of *operation (including energy)* costs is used to keep track of such items as fuel and salaries required to operate the facility. *Maintenance costs* include the regular custodial care and repair, annual maintenance contracts, and salaries of facility staff performing maintenance tasks. Usually, replacement items less than $5,000 in value or having a life of less than 5 years are also included in maintenance costs.

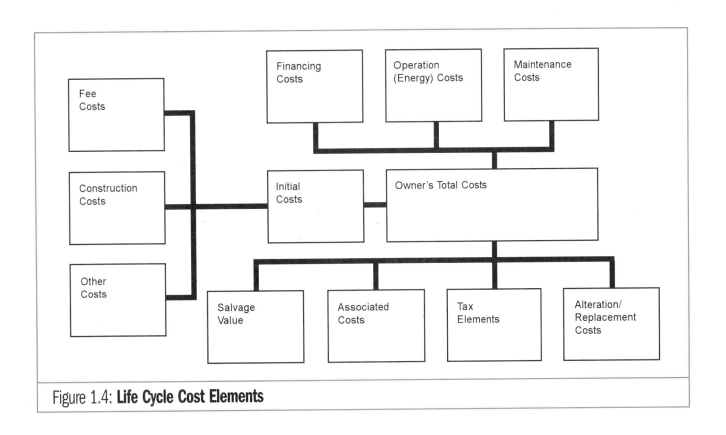

Figure 1.4: **Life Cycle Cost Elements**

Alteration and replacement costs require a more detailed discussion. Alteration costs are those involved in changing the function of the space. For example, a facility is built initially as an office building, but after ten years, the first floor area will function as a bank. The ten-year cost for conversion would be an alteration cost because of the change in the functional use of the space.

A replacement cost is a one-time cost to be incurred in the future in order to maintain the original function of the facility or item. For example, assume the owner defines the LCC period as 40 years, and the design calls for a cooling tower on the roof. The useful life of a cooling tower is approximately 20 years, and the tower would have to be replaced at the end of that period. The cost involved is defined as a replacement cost—a future cost required to maintain the same function for which the facility was intended.

Assignable costs associated with *taxes, credits, and depreciation* must be continually reviewed as tax laws change.

Associated costs include other identifiable costs not covered previously, but related to a facility decision. They may include functional use, denial of use, security, and insurance. Since the functional-use costs, when considered in an LCC, may be the most significant of all follow-on figures, they are discussed first. These costs include the staff, materials, insurance premiums, and taxes required to perform the function of the organization using the facility or installation. As an example, suppose an LCC analysis is required for a branch bank. What is the function of the bank? It is to serve customers. Suppose there are two banks that have exactly the same initial costs. One bank can process 200 clients a day; the other can process only 150. Which is more cost-effective? It is the one that processes more customers. *Functional-use costs* for a branch bank are those of serving customers. In life cycle cost analysis, a cost difference would have to be considered to take into account the difference in capacity to provide the basic function of a facility. A hospital is another good example. The basic function of a hospital is to treat patients. Suppose two hospitals are identical in costs and area, but in one hospital a doctor can treat a patient in half the time it would take in the other hospital. The LCC would be tempered for the difference in functional-use cost.

Denial-of-use costs include the extra costs or lost income during the life cycle because occupancy or production is delayed for some reason. As an example of denial-of-use cost, suppose that in making an alteration there are two approaches whose construction costs are the same. One alternative would require moving people out of the space for six months; the other alternative could be accomplished during non-working hours. In the LCC, the cost of not being able to use the space would have to be recognized.

The associated cost effect of *insurance* was illustrated by a recent study of a food distribution warehouse. When different fire protection system alternatives were examined, all costs were comparable, except that one system had a lower annual insurance premium. The annual difference in cost was accounted for in the LCC analysis.

Salvage value is the value (positive if it has residual economic value, and negative if demolition is required) of competing alternatives at the end of the life cycle period. This cost can become quite important if one alternative requires a major replacement toward the end of its economic life. For example, two automobiles are being analyzed for purchase. The plan is to keep these cars for four years. It is estimated that one car will reach the end of four years without requiring any major repairs, but it will then need a major overhaul. The other car will require a major engine overhaul at the end of the third year. As a result, the salvage value of the car with the more recently overhauled engine at the end of four years will be significantly higher. The same concept applies to building equipment.

Life Cycle Costing Logic

Within its more general definitions, LCC is not new. Many specific studies of facilities have had extremely large numbers of work hours devoted to answering the same life cycle cost questions. The difference between the majority of previous efforts and the approach outlined in this book is the establishment of a formal, consistent methodology. This concept is central to LCC. It not only determines the ultimate validity of the answers developed, but it also lays the groundwork for future refinement of life cycle cost decision-making.

This relatively simple methodology was developed by the authors a number of years ago and continues to be valid today. The best way to illustrate its simplicity is to use a hypothetical facility. For example, a hospital and its design team are considering two alternative nursing-station designs for each in-patient wing. One will cost far more to construct than the other because it relies more heavily on automated devices for patient monitoring and recordkeeping. However, the question is: does the savings in nursing salaries justify the increased facility cost?

Using life cycle cost methodology to obtain the answer is a process of several steps. First—to reduce the time and complexity of the analysis—those facility elements that will be the same in any of the options under consideration are identified and removed, or fixed, during the comparative analysis.

Next, the decision-making team isolates the significant costs associated with each alternative. The hypothetical, automated solution in the above example has higher capital investment, operation, and maintenance costs, but lower functional-use (nursing salary) costs. The costs isolated for each alternative must be grouped by year over a number of years equal to the economic life of the facility. Or, if more appropriate, they must be grouped by time spans equal to the mode of user operation. In either case, probable replacement and alteration costs should be considered. A salvage value, if relevant, is also added for the end of the life cycle period.

All costs are converted to today's dollars by *present worth* techniques (see Chapter 2) using a reasonable discount factor (7% is used by federal agencies, but many private owners use a higher rate). This discounting is done because a cost incurred in the tenth year of a facility's life is, of course, not the same present value as one incurred in the first year.

Finally, the team adds up the discounted costs and identifies the lowest-cost alternative. It may be necessary to make a sensitivity analysis of each of the assumptions to see if a reasonable modification in any of the cost assumptions would change the conclusions. If it would, the probability of such an occurrence must be carefully weighed. If it appears that two or more events have roughly the same probability of occurrence, the option will normally be based principally on cost. However, the effect on total cost of any non-monetary factor, such as aesthetics or safety, should be factored in by the decision-maker at the time of the decision.

The methodology consists of identifying the significant costs associated with each alternative, adding each group of costs by year, discounting them back to a common base, selecting the lowest-cost alternative, and tempering final selection with non-economic considerations. This process is illustrated in detail in the following chapters.

In summary, most major facility decisions have life cycle cost implications. The question is—which ones have the greatest impact? All projects are different, but a lesson learned from past construction cost control programs applies: in most cases, the early decisions are the most important in terms of the savings potential. For example, the decision on how many operating suites there will be in a hospital usually has far greater cost implications than the decisions regarding the detailed design for each suite. The selection of the type of HVAC system is more important than the choice of a manufacturer of the system selected. In both cases, the final choice, theoretically, could be made late in a project's development, but the cost penalty for making changes after the early sequences increases geometrically. This is illustrated in Figure 1.5, which indicates that the earlier the LCC, the greater the potential for cost reduction, and the lower the cost to implement. When major document revisions are required, their costs tend to negate the cost reduction potential.

Therefore, it is never too early to begin facility analysis in life cycle cost terms. There are many decisions with life cycle cost implications at every phase or decision point in the life of a project. Chapter 9, "Management Considerations," further discusses these potential study areas.

How would the flow of life cycle cost activities proceed in a typical project? Figure 1.6 illustrates a recommended LCC logic flow. The first requirement is the input data, which would normally consist of three types:

1. Specific Project Information: program requirements, criteria and standards, operational mode, quantities and economic data such as time value of money and life cycle period.
2. Facility Components: initial cost, useful life, and maintenance and operational costs.
3. Site Data: climatic and environmental conditions.

With these data, alternatives would be generated. Then the life cycle costs are predicted—and tempered by non-economic comparisons—before a final recommendation is made.

Of the input data required, the specific project information and site data are usually easily accessible. It is a different story for the facility components data. Where does a design professional go to get data regarding useful life, maintenance, and operations costs needed to calculate roughly one-half of the overall costs? Few designers have had access to comprehensive data in these areas. The real problem is that there is no readily available storage and retrieval format containing these data. (This problem will be discussed in Chapter 5, "Estimating Life Cycle Costs.")

Application in the Design Process

As discussed previously, the greatest savings potential for a life cycle cost analysis occurs in the earliest stages of a project. Not only are the potential savings more significant, but the costs of making changes in the plans and specifications are much less. As the project moves into the construction and occupancy phases, the LCC value tends to become the feedback data for use in other projects. These data are useful in three phases: concept development, preliminary design, and design development. During later phases, LCC analysis data should be more or less used as a review and a validation of the assumptions made during the earlier phases.

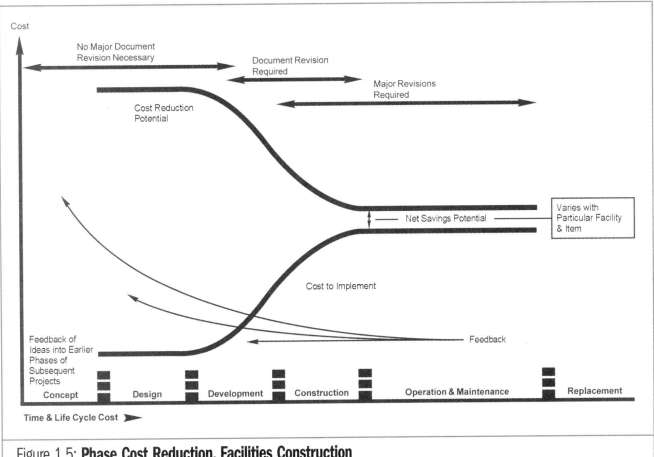

Figure 1.5: **Phase Cost Reduction, Facilities Construction**

Study areas may include:

- Facility versus other economic investment,
- New facility versus retrofit existing structure,
- High-rise versus low-rise construction,
- Active/passive solar energy versus conventional HVAC,
- Design layouts versus staffing efficiencies,
- Space flexibility versus interior partitioning,
- Natural landscaping versus conventional landscaping,
- Fire sprinkler systems versus insurance premiums,
- Fixed partitions versus demountable partitions,
- Interstitial space versus floor-to-floor height,
- Insulation and glazing versus energy requirements,
- Fenestration and shading versus lighting requirements,
- And other key areas, such as the selection of alternative designs and major building systems.

In many cases, LCC is applied to a single discipline. This has led to restrictive solutions. Figure 1.7 illustrates the design decision-makers' influence on total building costs. It portrays the design process as a team effort in which there are various disciplines making decisions in a

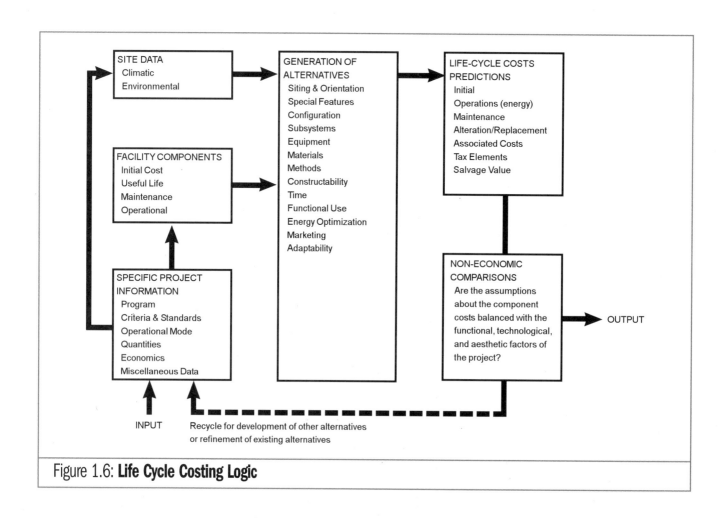

Figure 1.6: **Life Cycle Costing Logic**

discipline-oriented environment. One of the principal reasons for unnecessary initial costs has been the uni-disciplinary approach used by designers. In too many cases, the architect dictates the design and the other disciplines merely respond to it. A multi-disciplinary approach to optimize the building as a system has produced more desirable results.

LCC must avoid the same problem. The building industry overall seems to be moving toward a discipline-oriented solution to environmental problems. Mechanical and electrical engineers, in some cases, design their systems with the expectation that others will have to design around those systems. The application of LCC has centered on energy, with the mechanical engineers providing the lead. The result is an exaggerated energy orientation. It almost appears that the primary function of a facility in some instances is to conserve energy rather than to house people. *The best solutions will be developed only when all participants collaborate as a team and seek an optimum solution for the total problem.*

Concept and Design Development

Figure 1.8 shows the recommended timing during the design process for various activities, such as life cycle cost analysis and value engineering. This diagram formalizes the integration of cost and quality activities during the design process. The concept design portion is further broken

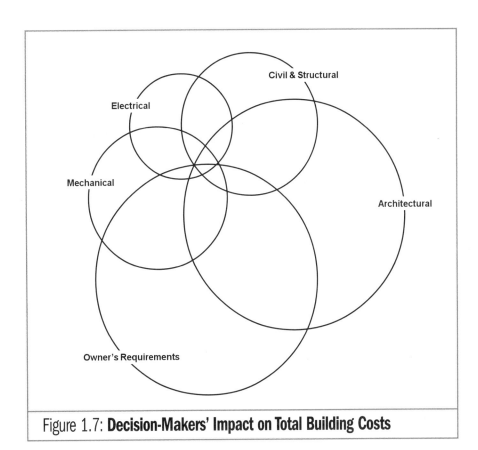

Figure 1.7: **Decision-Makers' Impact on Total Building Costs**

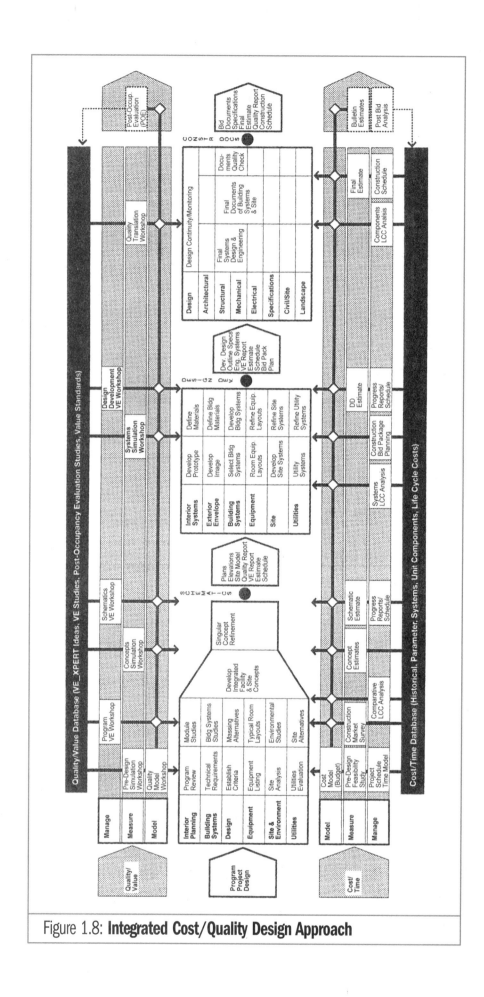

Figure 1.8: **Integrated Cost/Quality Design Approach**

down into a sequence of several events, which are presented in Figure 1.9. At the end of each event (starting with concept schematics), formal, concurrent reviews are held using the procedures outlined in this text. This series of reviews culminates in a formal five-day workshop, which might study in depth such pre-selected items as exterior closure, computer room HVAC, emergency generators, fire protection, structural systems, and floor-to-floor height. Techniques of energy conservation, value engineering, LCC, and cost control are also used during the workshop. Participants might include representatives of the owner, users of the facility, facility manager [operation and maintenance (O&M)], architects and engineers, construction manager, and a value engineering and LCC consultant. This process is repeated during the design development phase, focusing on more discrete building elements. *Implemented savings on a recent project budgeted at $60 million included over $4 million in first costs and $10 million in follow-on costs.*

Planning and Feasibility

LCC may also be used even sooner in the decision-making process, for activities such as project planning, feasibility, and programming. Examples of LCC applications at this phase include: site analysis (given use); use analysis (given site); impact analysis (given use and site); and building programming (given basic requirements). See Case Studies included in Chapter 8.

Use of LCC by the owner or builder of residential properties is a good example of the potential benefit of an effective early-stage LCC study. In most housing financial feasibility analyses, $1 taken out of operating expenses through better planning or design is equivalent to a $10–$12 reduction in construction costs.

The size of most new facilities depends on the results of the project's financial feasibility analysis. The feasibility depends on a combination of the facility income, capital investment (including cost of debt service), and operating costs. A project becomes financially feasible when the owner can find the right balance between these three factors. In the equation, however, a reduction in capital investment cost of $1 for a facility, borrowing at a rate of 8%, will increase a project's projected net annual income by only 8 cents. [See Appendix A, Table 3, Periodic Payment (PP): 40 years at 8% interest, PP = 0.0839.] A reduction of $1 in operating costs, however, will increase net income $1. The difference is 12.5 to 1. This type of analysis is very common in private real estate development.

It is often less difficult to take a dollar out of annual operating costs through better planning and design than it is to take it out of construction costs. When, in addition, the effect of every dollar of savings can potentially be multiplied by 10 or 12, annual costs are many times more important than a project's construction costs. With the large savings that can be achieved from life cycle cost reductions, the question is clearly not whether to, but *how to* use LCC.

Relationship of LCC to Value Engineering

Value engineering is a systematic procedure directed at analyzing the function of facilities, systems, processes and equipment for the purpose of achieving required functions at the lowest total cost of ownership. VE is directed toward analysis of functions. It is concerned with elimination or modification of anything that adds costs without contributing to required functions. VE is a team approach that analyzes a function by systematically developing the answers to such questions as:

- What is it?
- What does it do?
- What must it do?
- What does it cost?
- What other material or method could be used to do the same job without sacrificing required performance or degradation to safety, reliability, maintainability, etc? What would the alternative material or method cost?

Value engineering may utilize life cycle costing in developing the lowest cost of ownership, except in those cases where initial cost reductions are the owner's major consideration. The primary focus of VE is analysis of function and elimination or modification of unneeded functions in seeking out the optimum course of action that satisfies the true functional requirement. The

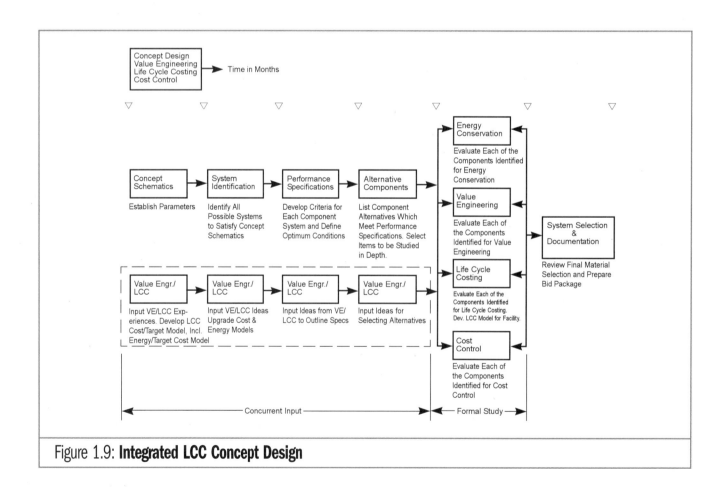

Figure 1.9: **Integrated LCC Concept Design**

primary focus of life cycle costing is to determine which of several courses of action would be least costly over a specified period of time.

Value engineering can be used to complement a life cycle cost analysis when selected LCC alternatives cannot be adopted without exceeding the project budget. VE can be utilized to reduce initial costs of design features other than those under study in an LCC analysis. If the VE effort results in sufficient reduction in initial cost, then the selected LCC alternatives can be adopted, thus optimizing the long-term cost-effectiveness of the project as a whole.

Figure 1.10 graphically portrays the interrelationship of VE and LCC. At one end is the start of the design process, and at the other end is the ultimate design selection; between is the life cycle cost analysis process. In developing the design selection, the basic site and facility program information is combined with the subsystem information to generate alternative solutions. It is in the generation of the alternative solutions that the VE methodology should be utilized. Because of its team approach to understanding the problem using function analysis, and later creativity and brainstorming techniques, VE is a powerful tool to assist in developing alternatives objectively. After developing alternatives, the process focuses

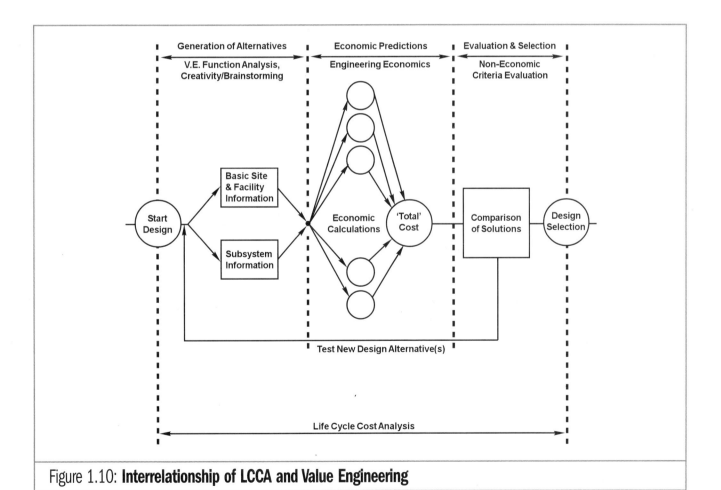

Figure 1.10: **Interrelationship of LCCA and Value Engineering**

on economic analysis. At this point, LCC is performed on the alternatives using the principles of engineering economics. The results are expressed as the total life cycle cost of each competing alternative.

In making an economic analysis, are designers required to take the lowest-cost alternative? Is the lowest-cost alternative always the optimum solution financially for the owner? There normally are non-economic considerations such as politics, aesthetics, safety, and the environment that have to be taken into account. Therefore, after the life cycle costs are estimated, there must be an evaluation of non-economic criteria. From this evaluation, a design selection is made. The objective of LCC analysis is an optimum solution—not a solution based strictly on costs. For this book, the whole process (including VE) is defined as an LCC analysis study.

Program Requirements

How should an organization start LCC planning? If owners and design professionals are going to be serious about LCC, they must work together in setting up a comprehensive approach to life cycle costing. The approach should be designed so that each year, the activities progress one step further into a more refined LCC process. Eventually, a system should be created that will complete the circle, recycling LCC cost data from actual maintenance and operations to the early-stage design of new facilities or building components.

How can this logic system be integrated into a working organization? Figure 1.11 illustrates a master plan for the development of a comprehensive LCC system, assembled for a large government agency.

Phase of development	Concept	Design and development	Construction	Owning and operating
Problem definition and initial guidance	Develop manual method for LCC; notify field offices of LCC; test case applications	Review procedures and recommend LCC system; conduct seminars	Prototype LCC application	Field notification of intent to retrieve LCC data
Application trials and refinement	Review and revision of manual method; project application at headquarters, using manual LCC	Briefings and seminars and manual application on test cases	Select trial project	Develop data system; preliminary collection from various sources
Development of LCC model	Develop computerized initial cost retrieval system	Trial project application using computer LCC model	Review contractor applications	Review of LCC data
Field office installation and application	Formalize historical LCC application for computer retrieval	Project application at all centers	Recommendations for field implementation	Continuing update and refinement of LCC data

Facility cycle ⟶ (Time ↓)

Figure 1.11: **Master Plan for Development of a Comprehensive LCC System**

This plan was set up for an estimated 5–10-year program. The chart is organized horizontally in accordance with the facility cycle, going from the early concept, to design and development, to construction, to owning and operating. Vertically, the columns start from the present and descend to a future time.

As the chart indicates, the first step is to set objectives and issue guidance. Management must direct that LCC will be a part of the everyday decision-making process. The program should span all in-house activities and be built into all architectural and engineering contracts that have significant follow-on costs. In the early concept phase, notification should be given to all personnel, with procedures developed and defined in a convenient manual. In the design development phase, the recommended procedures should be reviewed and comments submitted, with seminars conducted for selected personnel. For the construction and owning and operating cycles, personnel should be notified about the intent to retrieve data. The first phase should take from 1–2 years. The next phase consists of applications, trials, and refinement.

In the planning and budgeting phase, parameter estimates are developed. During design development, manual applications on test cases should be conducted. The next phase consists of the development of an automated LCC system, with the last phase centered on installation and application. Because the number of variables that can be handled by manual procedures is limited, an electronic spreadsheet format is recommended.

To show the magnitude of the problem, an analysis of a typical initial-cost database yields upwards of 30,000 units of cost for a large construction project (based on RSMeans, one of the foremost cost data firms in the United States). How many items would the more complicated life cycle database, including replacement, maintenance and operation, and salvage costs contain? Certainly at least more than the initial unit cost items. The only feasible approach to handling this volume of data is to use an automated system. It is recommended that LCC be performed on a project basis with a concurrent effort devoted to developing a data system under which some field response can be realized. The ultimate objective will be to have an LCC system with a reliable database, updated by actual construction costs and supported by actual data from the organization's and facility's personnel. Initial data bank can be found in Part 2 of the book.

Summary

Until recently, only initial-cost dollars have been of prime concern in the building cycle. Because of recent trends regarding facility obsolescence, environmental sustainability, operational staff effectiveness (including re-engineering total quality management), and value engineering, the use of LCC techniques is being expanded. The process offers the owner the opportunity to impose resource limits and to establish designer benchmarks. The establishment of benchmarks such as energy budgets, maintenance, and other owning and operating targets is now possible.

As experience grows in these areas, more efficient performance from the facilities can be expected.

In performing LCC, the ultimate objective is optimum design decisions. The goal can be achieved through an organized LCC analysis that will optimize the total cost of ownership and be tempered by non-monetary criteria. Owners must require specific efforts. Designers must continually seek out innovative ideas and perform life cycle cost analyses of these alternatives to meet the challenge.

The best opportunity for saving life cycle dollars is in the earliest stages of the design process. The theory, for the most part, is well established. Owners must take the responsibility for setting realistic goals in the planning and budgeting phase and for providing assistance, as necessary, to designers, so that LCC does not become just another paperwork exercise. Owner personnel must become familiar with LCC procedures and be prepared to suggest areas for study to the designer. LCC should be performed during value engineering reviews to develop design alternatives at the earliest possible stages of facility design, so that any changes may be made with minimum effort and optimum effect.

Federal, state, and local governments, as well as industry organizations, such as General Motors and General Electric, are committed to procuring facilities and their operational and maintenance needs at the lowest total cost of ownership. This goal becomes realistically achievable through implementation of LCC during the planning and budget, schematic, and design development stages of a project. As the English philosopher and economist John Stuart Mill asked in 1857, "Towards what ultimate point is society tending by its industrial progress? When the progress ceases, in what condition are we to expect that it will aid designers to improve the use of our resources and, in the process, create a better environment?"

1. P. A. Stone, *Building Design Evaluation: Costs-in-Use,* 2nd Ed., E. & F. N. Spon, London, 1975, p. 177.
2. *Life Cycle Budgeting as an Aid to Decision Making,* Building Information Circular, Department of Health, Education, and Welfare, Office of Facilities Engineering and Property Management, Washington, 1972, p. 1 (draft).
3. Abraham Maslow, "Maslow's Hierarchy of Needs," film, Salenger Educational Media, Santa Monica, CA.

Chapter 2: Life Cycle Costing Fundamentals

I am disturbed when a consulting engineer tells me and my committee that he can make his life cycle cost figure come out any way he wants.
Senator Robert Morgan
Chairman, Senate Subcommittee on Buildings and Grounds

Suppose two design alternatives are being considered. Design A has design and construction (initial) costs of $2 million and no other costs. Design B has initial costs of $1 million, along with costs of $1 million to be expended over the life of the project. The alternative designs would seem to be equal in cost—$2 million for each. However, the lower initial cost of design B frees $1 million to be invested, and to earn interest, until it is needed. This has the effect of reducing the total cost of design B by the amount of interest earned. On the other hand, the general level of prices may rise over the life of the project, because of inflation. This would increase the total cost of design B, which includes periodic expenditures that are increased by inflation, but it would not affect the cost of design A. Thus, the total costs of the two designs are not really equal; they differ by an amount that depends on the earning power of money, the inflation rate, and even on what is meant by the *life* of the construction project. In addition, costs that are to be incurred at different times must be converted to equivalent costs at a common point in time before they can be summed to provide the total cost of ownership for the two alternatives and, thus, provide a basis for comparison. These concepts—*interest*, *inflation*, *unit* life, and *equivalent costs*—and the computations that follow from them form the foundation of life cycle cost analysis and are the subject of this chapter.

Life cycle cost analysis (LCCA) is defined as a cost-centered engineering economic analysis whose objective is to systematically determine the costs attributable to each of one or more alternative courses of action over a specified period of time. The key elements of such an analysis are those that affect the manner in which the analysis will be conducted and, therefore, the effectiveness of its results in a particular situation. A decision concerning each key element must be made before the LCCA can be performed. In particular, the definition of the LCCA suggests six questions that must be answered:

1. What analysis approach is to be used?
2. What is a realistic discount rate for use in the analysis?
3. How are the effects of inflation and increases in individual costs to be taken into account?
4. Over what specific period of time are the total costs of ownership to be determined?
5. When is that time period to begin?
6. What types of costs are to be included in the analysis, and what costs (if any) may be ignored?

Each of these questions concerns one key element. Possible answers are discussed in this chapter.

Time Value of Money

A sum of money may be invested to earn a return (interest) for its owner. For this reason, a sum of money in hand today is worth more than the same sum at a later date. The difference is the amount of money that the sum could earn in the interim. For example, if $100 is deposited in a savings account paying 7% annual interest compounded (paid) annually, then that $100 will earn $100 × 0.07 = $7 in one year, and the original sum will grow to $107. In this example, $100 in hand now is worth as much as $107 one year from now. Moreover, to have $100 a year from now, a person would have to deposit only $93.46 at the 7% interest rate, because $93.46 × 1.07 = $100. This ability of money to earn money and thus increase in amount over time is referred to as the *time value of money*. At an interest rate of 7% compounded annually, $100 grows to $107 in one year. Over the second year it grows to $107 × 1.07 = $114.49, and over the third year to $114.49 × 1.07 = $122.50. Figure 2.1 extends this example to an interest-earning (investment) period of 25 years.

In determining the total cost of ownership, the design professional is concerned with a number of sums that are invested at various times. *Cash flow diagrams* are of help in sorting out and keeping track of both outlays of money and money received. *Interest formulas* are simple mathematical equations that can be used to compute the amount to which a single investment or a series of equal investments will grow. *Interest tables* may be used for the same purpose and require a minimum of computation. All three are discussed.

Cash Flow Diagrams

Cash flow diagrams help in the visualization of cash flows, that is, expenditures (or outlays) and income (or receipts). A horizontal line is used as the time axis. The choice of time unit is arbitrary, but the scale is usually graduated in years beginning with 0 (now) at the left. Cash flows are represented by vertical arrows whose lengths are proportional to the cost magnitudes, and whose locations on the time-line indicate when they occur. These conventions are illustrated in Figure 2.2.

Whether a cash flow is shown as an income or an expenditure depends on the point of view of the person preparing the diagram. For instance, to an investor who places $100 in a bank and allows it to remain there, earning 7% interest for 10 years, the cash flow diagram would look like Figure 2.3. On the other hand, to the bank, the cash flow would look like Figure 2.4.

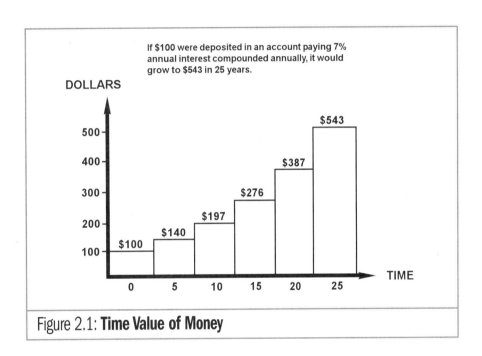

Figure 2.1: **Time Value of Money**

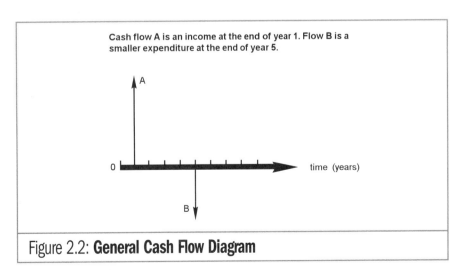

Figure 2.2: **General Cash Flow Diagram**

Costs to an owner may also be represented on a cash flow diagram. The owner of an asset, say an automobile, might prepare a cash flow diagram like the one in Figure 2.5. The long arrow pointing downward (at time 0) represents the cost of purchasing the automobile; the shorter uniform downward pointing arrows represent costs incurred from year to year, such as gasoline costs and annual maintenance and repair costs. Various one-time costs—brake replacement, for example—are indicated by the follow-on downward arrow. The upward arrow at the right represents the resale value of the car when the owner sells it or trades it in. Cash

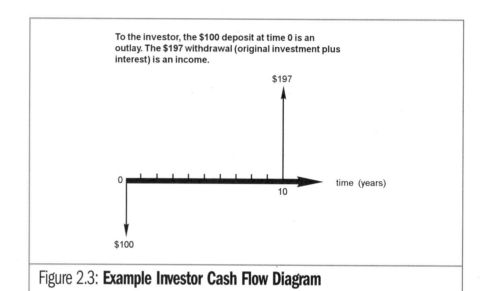

Figure 2.3: **Example Investor Cash Flow Diagram**

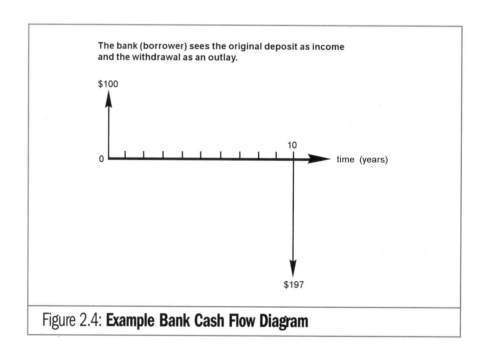

Figure 2.4: **Example Bank Cash Flow Diagram**

flow diagrams are an aid in applying the interest formulas that are discussed next.

Interest Formulas

To quantify the impact of the interest rate in relating dollars spent today to dollars spent in the future, six commonly-used interest tables are presented. Some of these formulas address situations involving a single present sum of money or the present worth of a single future sum of money, given the interest rate and the length of time of the cash flow. Others address situations dealing with constant annual payments, such as those involved in paying off a mortgage loan. The six interest formulas are

- Single compound amount (SCA)
- Single present worth (SPW)
- Periodic payment (PP)
- Present worth of annuity (PWA)
- Uniform sinking fund (USF)
- Uniform compound amount (UCA)

The symbols used in these formulas are:

i = interest rate per period
n = number of interest periods
P = present amount
F = future amount
A = uniform sum of money in each period

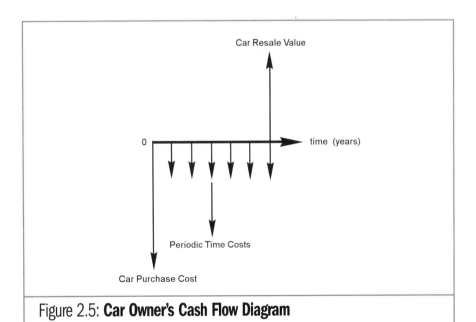

Figure 2.5: **Car Owner's Cash Flow Diagram**

Single Compound Amount

$$SCA = (1 + i)^n \qquad (2\text{-}1)$$

This factor may be used to determine the single future amount of a single present amount for an interest rate i for n periods. SCA represents a single future amount, P is present amount, and F is the future amount to be discounted.

$$F = P \times SCA \qquad (2\text{-}2)$$

An example using the above formulas is presented in Figure 2.6.

Single Present Worth

$$PW = \frac{1}{(1 + i)^n} \qquad (2\text{-}3)$$

This factor may be used to determine the present worth of a future amount discounted at interest rate i for n periods. PW represents single present worth, P is present amount, and F is the future amount to be discounted. Note that PW is the reciprocal of SCA.

$$P = F \times PW \qquad (2\text{-}4)$$

An example using the above formulas is presented in Figure 2.7.

Periodic Payment

$$PP = \frac{i(1 + i)^n}{(1 + i)^n - 1} \qquad (2\text{-}5)$$

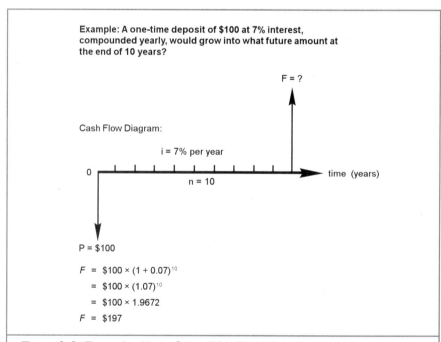

Figure 2.6: **Example: Use of the SCA Formula**

The PP factor may be used where a present amount at i % interest rate is returned in n equal periodic installments. PP represents uniform capital recovery, P represents present amount, and A, the uniform sum of money in each time period, is the unknown.

$$A = P \times PP \qquad (2\text{-}6)$$

An example using the above formulas is presented in Figure 2.8.

Present Worth of Annuity

$$PWA = \frac{(1+i)^n - 1}{i(1+i)^n} \qquad (2\text{-}7)$$

The PWA factor may be used where a present amount at i % interest is returned in n equal periodic installments. PWA represents uniform present worth, A represents the uniform sum of money in such time period, and P, the present worth of the installments, is the unknown. Note that PWA is the reciprocal of PP.

$$P = A \times PWA \qquad (2\text{-}8)$$

An example using the above formulas is presented in Figure 2.9.

Uniform Sinking Fund

$$USF = \frac{i}{(1+i)^n - 1} \qquad (2\text{-}9)$$

The USF factor may be used where n equal periodic installments, invested at i % interest rate, accumulates to a known future sum of money.

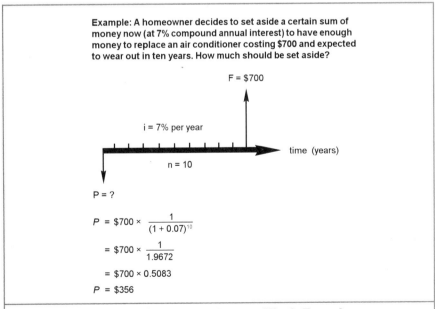

Figure 2.7: **Example of the Single Present Worth Formula**

Example: A store owner decides to purchase new entry doors for $1,100. However, rather than pay this amount at the present time, the owner decides to borrow the money and repay the loan over the next 10 years, making one payment at the end of each year. How much would this annual payment be if the interest rate is 7%, compounded annually?

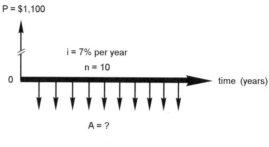

$$A = \$1,100 \times \frac{0.07(1+0.07)^{10}}{(1+0.07)^{10}-1}$$

$$= \$1,100 \times \frac{0.07(1.9672)}{(1.9672)-1} = \$1,000 \times \frac{0.1377}{0.9672}$$

$$= \$1,100 \times 0.14238$$

$$A = \$157$$

Figure 2.8: **Example Use of the Periodic Payment Formula**

Example: The store owner of the previous example anticipates an annual maintenance cost of $50 for the new entry doors. What amount should be set aside today so that there will be enough money to pay each year's maintenance cost over the next 10 years? The interest rate is 7% compounded annually.

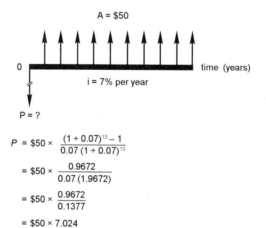

$$P = \$50 \times \frac{(1+0.07)^{10}-1}{0.07(1+0.07)^{10}}$$

$$= \$50 \times \frac{0.9672}{0.07(1.9672)}$$

$$= \$50 \times \frac{0.9672}{0.1377}$$

$$= \$50 \times 7.024$$

$$P = \$351$$

Figure 2.9: **Example Use of the Present Worth of Annuity Formula**

USF represents uniform sinking fund, F represents the known future sum of money, and A, the uniform sums of money in each time period, is the unknown. The payments are at the *end* of each period.

$$A = F \times \text{USF} \qquad (2\text{-}10)$$

The USF is very commonly used to save money on a periodic basis to pay for some future anticipated cost, such as college expenses. An example using the above formulas is presented in Figure 2.10.

Uniform Compound Amount

$$\text{UCA} = \frac{(1+i)^n\ 1}{i} \qquad (2\text{-}11)$$

The UCA factor may be used where n periodic installments, invested at $i\%$ interest rate, amount to a future sum of money that is to be determined. UCA represents uniform compound amount, A represents the uniform sum of money in each time period, and F, the future sum of money, is the unknown. Note that UCA is the reciprocal of USF.

$$F = A \times \text{UCA} \qquad (2\text{-}12)$$

An example using the above formulas is presented in Figure 2.11.

Interest Tables

The formulas themselves are easily derived and are available in most financial or engineering economic texts. Rather than apply a formula each

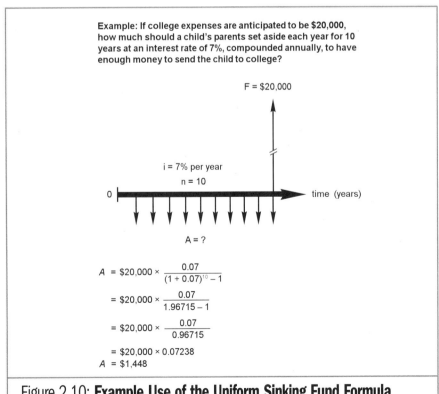

Figure 2.10: **Example Use of the Uniform Sinking Fund Formula**

time, however, it is much more convenient to use tables of factors which have been computed using the formulas. Appendix A, Tables 1 through 5 (6%, 7%, 8%, 10%, and 12%), presents six such factors, two dealing with single payments, and four with uniform annual payments. All six are interrelated, and they allow the movement of sums of money backward and forward in time. The factors that deal with uniform annual costs assume that these costs occur as single payments at the *end* of each period. There are other tables used for the beginning or middle of each period. The owner should be consulted to clarify which is to be used. Normally, end-of-year tables are used. The following is a summary of the procedures involved using these tables for each of the previous examples:

- *Single compound amount*, Eq. (2-1). Referring to Appendix A, Table 2 (7% compound interest factor) and given P to find F for an n of 10 years would yield a factor of 1.9672, the same as that found by solving the SCA formula.
- *Single present worth*, Eq. (2-3). Referring to Appendix A, Table 2 and given F to find P for an n of 10 years would yield the factor 0.5083, the same as that found by solving the PW formula.
- *Periodic payment*, Eq. (2-5). Referring to Appendix A, Table 2 and given

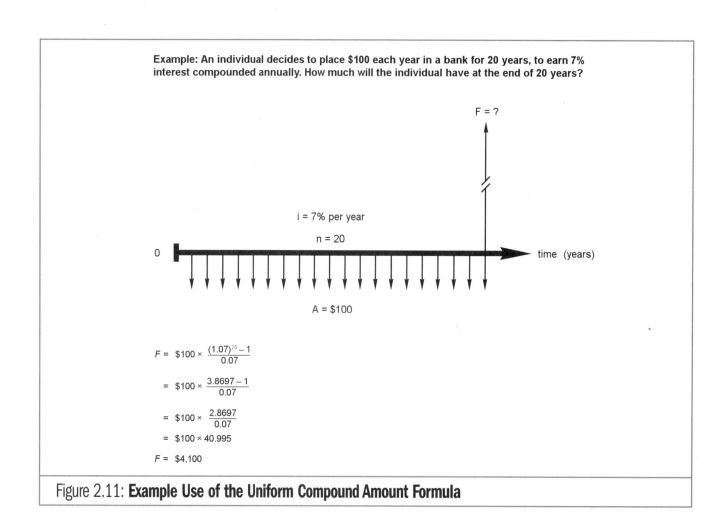

Figure 2.11: **Example Use of the Uniform Compound Amount Formula**

P to find *A* for an *n* of 10 years would yield the factor 0.14238, the same as that found by solving the PP formula.
- *Present worth annuity*, Eq. (2-7). Referring to Appendix A, Table 2 and given *A* to find *P* with an *n* of 10 years would yield the factor 7.024, the same as that found by solving the PWA formula.
- *Uniform sinking fund*, Eq. (2-9). Referring to Appendix A, Table 2 and given *F* to find *A* with an *n* of 10 years would yield the factor 0.07238, the same as that found by solving the USF formula.
- *Uniform compound amount*, Eq. (2-11). Referring to Appendix A, Table 2 and given *A* to find *F* with an *n* of 20 years would yield the factor 40.995, the same as that found by solving the UCA formula.

Basic Equivalent Approaches

Using the interest formulas or tables, it is possible to convert monies spent over various points in time to a common basis. Consider again the example of an investment of $100 in a savings account at a 7% interest rate. It was determined that this money would grow to $197 by the end of the tenth year. Just as $100 today is *equivalent* to $197 at year 10, it can be said that the $197 at year 10 is equivalent to $100 today (or at the present time), provided the interest rate is 7%. For purposes of converting monies at various points in time to a common basis, the concept of equivalence is very useful. Had the $197 been a cost at year 10, then it could be said that this cost is equivalent to $100 today. Life cycle costs include both costs today and costs in the future. Interest formulas (or tables) are used to make these expenditures "equivalent" with respect to time. The two methods most commonly used for converting present and future costs of an item, system, or facility to a common basis are the present worth and the annualized (equivalent uniform annual cost) methods. Both methods account for the time value of money, and therefore are interchangeable as measures of life cycle cost. Because future costs are "discounted" to a smaller value when converted to the present time, it is common practice to use the term "discount rate" in reference to the interest rate *i*, and the discount rate symbol *d* is sometimes used. Following are examples illustrating the use of both the present worth and the annualized methods.

Present Worth Method

The present worth method allows conversion of all present and future costs to a single point in time, usually at or around the time of the first expenditure. To illustrate the calculations involved, the following situation and cash flow are presented.

A lighting system is expected to cost $10,000 to install today; energy and lamp replacement costs are estimated to be $500 annually; a one-time replacement is anticipated to cost $3,000 at the tenth year; and the system is expected to have a salvage value of $2,000 after its life cycle of 20 years. What is the equivalent single payment for the lighting system over a 20-year stream of costs using a 10% discount rate? For purposes of this example, each uniform annual payment (energy and maintenance), and each one-time payment (replacement, salvage value), has been assumed to occur at the end of the year. This is commonly referred to as the *end-of-the-year convention*.

	Present worth
Initial cost	$10,000
Energy cost $P = A \times \text{PWA}$ (20 years, $i = 10\%$) $P = \$750 \times 8.514$ (Appendix A, Table 4)	6,386
Maintenance cost $P = A \times \text{PWA}$ (20 years, $i = 10\%$) $P = \$500 \times 8.514$ (Appendix A, Table 4)	4,257
Replacement cost $P = F \times \text{PW}$ (10 years, $i = 10\%$) $P = \$3,000 \times 0.3855$ (Appendix A, Table 4)	1,157
Salvage value $P = F \times \text{PW}$ (20 years, $i = 10\%$) $P = (\$2,000) \times 0.1486$ (Appendix A, Table 4)	(297)
Present worth of the flow of costs =	$21,503

The $21,503 can be viewed as equivalent to all the cash flows illustrated in Figure 2.12.

Annualized Method

The annualized method is also used to convert dollars expended over various points in time to an equivalent cost. Rather than being expressed as a one-time present worth cost, the annualized method converts all costs to an equivalent uniform annual cost. To illustrate the mathematics of this method, consider again the lighting example used in the present worth method. What is the equivalent single annual payment for the lighting system over a 20-year flow of costs, using a 10% discount rate?

	Annualized
Initial cost $A = P \times \text{PP}$ (20 years, $i = 10\%$) $A = \$10,000 \times 0.117$ (Appendix A, Table 4)	$1,170
Energy cost	750
Maintenance cost	500
Replacement Present worth of this cost is $1,157 ($3,000 × 0.3855 = $1,157) $A = P \times \text{PP}$ (20 years, $i = 10\%$) $A = \$1,157 \times 0.117$ (Appendix A, Table 4)	142
Salvage value Present worth of this cost is ($297) ($2,000 × 0.1486 = $297) $A = P \times \text{PP}$ (20 years, $i = 10\%$) $A = (\$297) \times 0.117$ (Appendix A, Table 4)	(35)
Equivalent uniform annual cost =	$2,527

This annualized cost of $2,527 may be mathematically converted to present worth by simply multiplying it by the PWA factor of 8.514 (20 years, $i = 10\%$).

$$\$2,527 \times 8.514 = \$21,515$$

This figure coincides with the $21,503 (minor difference due to rounding) calculated earlier and is equivalent to all the cash flows over the 10-year time frame illustrated in Figure 2.12.

Inflation

Another fundamental concept in the time value of money is inflation. Inflation may be considered as the general increase in price of the same goods and services over time (i.e., increase in cost without corresponding increase in value). Figure 2.13 graphically depicts rates of inflation the U.S. economy has experienced recently. Once the rate of inflation is known for a given item, the change in cost can be simply taken into account. Changes in cost may be expressed (or approximated) as a constant incremental rate or as a constant compound rate, but the latter is more accurate because inflation rates such as those in Figure 2.13 are estimated as compound rates. For an example of the difference, consider a $1,000 generator, the cost of which is anticipated to increase at an increment of $100 per year for the next 10 years, as illustrated in Figure 2.14 (Line A). The resulting

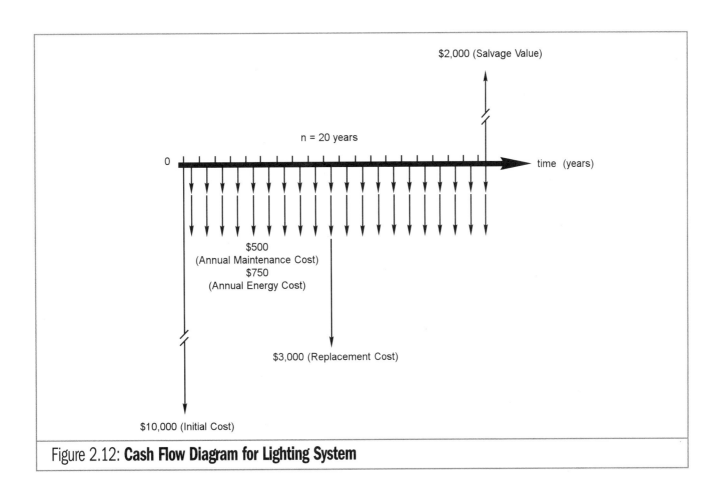

Figure 2.12: **Cash Flow Diagram for Lighting System**

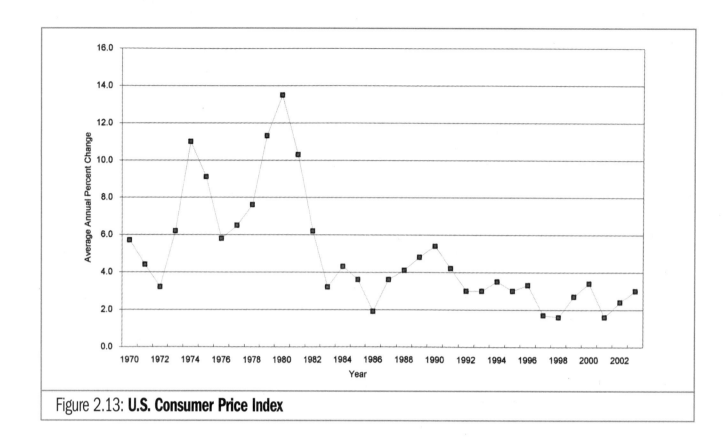

Figure 2.13: **U.S. Consumer Price Index**

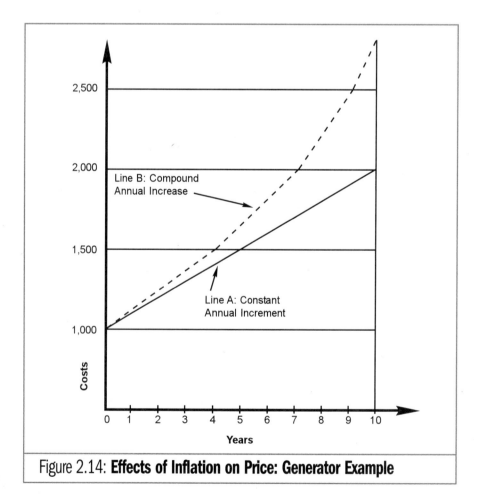

Figure 2.14: **Effects of Inflation on Price: Generator Example**

impact would be a cost for the generator of $2,000 by the tenth year ($1,000 + 10 × $100). On the other hand, if the cost were to increase at a rate of 10% compounded annually for 10 years, the resultant cost at year 10 would be calculated as follows:

$$\begin{aligned} F &= P \times \text{SCA for 10 years and } i = 10\% \\ &= \$1{,}000 \times 2.5937 \\ &= \$2{,}594 \end{aligned}$$

Line B of Figure 2.14 represents the yearly change in the anticipated price of the generator.

As another example of the impact of inflation and how it may be taken into account, consider a cooling tower with an expected replacement cost today of $7,000 which will be replaced 5 years in the future. If the rate of inflation is projected to be 7% for the first 2 years and 9% for the remaining 3 years, what cost should be anticipated for replacing the cooling tower in 5 years? The cooling tower cost (considering the effects of inflation) would be calculated as follows:

Cooling-tower cost = $7,000 (1.07) (1.07) (1.09) (1.09) (1.09) = $10,379

The compounded calculation yields the price that should be expected. The simplistic approach of adding each year's percentage of inflation to produce an aggregate five-year percentage of inflation, i.e., 41% (7% + 7% + 9% + 9% + 9%), understates the final result, as the following calculation shows:

$$\$7{,}000 \,(1.41) = \$9{,}870$$

In general, the higher the yearly escalation figures, or the longer the overall escalation period, the greater will be the distortion introduced by simply adding yearly inflation rates to produce an aggregate rate.

Life Cycle

Life of an Element

The life of a construction element is sometimes loosely defined, in parallel with human life expectancy, as some average number of years over which the element is expected to last. However, more precise definitions are available:

- The *technological life* of an item is the estimated number of years until technology causes the item to become obsolete.
- The *useful life* of an item is the estimated number of years during which it will perform its function according to some established performance standard.
- The *economic life* of an item is the estimated number of years until that item no longer represents the least expensive method of performing its function.

The economic life is the most important of the three from the viewpoint of cost minimization; however, the technological and useful lives of an item must obviously be considered when its economic life is estimated.

Life Cycle Cost

During its economic life, an item is subject to purchase, use, repair, maintenance, perhaps modification, and finally disposal. These processes comprise the *life cycle* of the item, and the costs of these processes make up the *life cycle cost,* or total *cost of ownership,* of the item. In attempting to predict the life cycle cost of a particular construction element, the design professional may be hampered by the fact that the economic life of the element can only be estimated and is not known for certain. To overcome this difficulty, the analyst may compare results using several reasonable estimates—say 15, 20, and 25 years for an economic life that seems to be about 20 years—to determine the effect of varying the economic life. The following table shows such a comparison for an air-conditioner compressor with initial cost of $2,000 and combined maintenance and energy costs of $700 per year, assuming a discount rate of 10%. Table 4 of Appendix A, for the end-of-period convention, was used to obtain the PW costs. The results show that in this case the estimated economic life does make a difference in the total cost of ownership. It should be noted that the difference is smaller for the longer economic lives, in this example and in general.

	Estimated Economic Life					
	15 years		*20 years*		*25 years*	
Life Cycle Cost	PW Factor	Cost	PW Factor	Cost	PW Factor	Cost
Initial cost = $2,000	1.0	$2,000	1.0	$2,000	1.0	$2,000
Energy cost/year = $600						
Maintenance cost/year = $100	7.606	$5,324	8.514	$5,960	9.077	$6,354
Total cost of ownership		$7,324		$7,960		$8,354

Discount Rate

The discount rate (or interest rate, as discussed earlier) is the time value of money. Often it is established as the nominal rate of increase in the value of money over time, or, more commonly, it is established as the actual rate of increase in the value of money, i.e., that rate over and above the general economy inflation rate. Much has been written on discount rates and the methods of determining them, but there is no universally accepted method or resulting rate used by various organizations. It is normally the prerogative of the owner or policy maker to select the discount rate. Both the private and the public sectors utilize a variety of discount rates for a variety of different reasons. Any of the following approaches are sometimes used in selecting the discount rate for the reasons discussed.

The discount rate may be established as the *cost of borrowing money* in the marketplace, usually the highest interest the organization expects to pay to borrow the money needed for a project. This is a relatively simple way to select the discount rate. However, for private industry, it does not take into account the risk of loss associated with the loan or the expected return from the investment itself. This method does, however, indicate the marketplace value of money over time.

The discount rate is also commonly established as the *minimum attractive rate of return* stipulated by the owner or policy-maker. This rate usually includes the basic cost of borrowing the money plus an increment that reflects the risk associated with the endeavor requiring the money (the investment). Because it is not easy to quantify risk as a percent increment, this selection criterion may be difficult to apply. However, it is a better indicator of the value of money to its user than the simple cost of borrowing money.

The discount rate is sometimes established as the rate of return that could be earned from some alternative investment opportunity which is foregone in favor of the project in question. This is sometimes referred to as the *opportunity rate of return*. For example, if it were possible to earn a 20% rate of return elsewhere, then the discount rate for the project in question would be set at 20%. This discount rate selection criterion is a realistic one, since it is based on the actual earning power of money. However, it is difficult to apply to public sector projects.

The use of an *after-inflation discount rate* in comparative analyses is based on the assumption that private industry will seek a certain set rate of return over and above the general inflation rate, no matter what the inflation rate may be. The after-inflation discount rate is equal to the average rate of return in the private sector, less the inflation rate. Because the effect of inflation is removed from the discount rate, there is no need to predict the inflation rates for future years. However, all costs in an LCCA—present and future—must be stated in terms of constant dollars (constant purchasing power with reference to a base year). One problem in the use of the after-inflation discount rate in an LCCA is the artificiality of the resulting total costs of ownership, which are in constant dollars rather than "real" (current) dollars. Nevertheless, all alternatives are treated in the same way, and as long as the LCCA is used only for the *comparison* of alternatives, the use of the after-inflation discount rate and constant dollars produces the same result as any other reasonable method of analysis. This method is favored by the authors.

Dealing with Inflation and Cost Growth

If the results of an LCCA are to be meaningful, the effects of inflation and cost growth on the total cost of ownership for the various design alternatives should be taken into account—especially when these effects have more of an impact on one alternative than on others. Inflation, as discussed earlier in this chapter, is a general increase in the prices of goods and services over time in the economy as a whole, without a corresponding increase in value. Cost growth, on the other hand, is an increase (or decrease) in the price of an individual item with or without a corresponding increase (or decrease) in value. An inflating economy may affect all design alternatives equally, but generally it has the greatest impact on alternatives for which the ratio of future costs to initial cost is largest. No such general statement can be made for cost growth. For example, growth of the cost of the labor required to replace light fixtures will probably not substantially affect one lighting alternative more than any other. However, growth of the cost of energy will have the most effect on the least energy-efficient alternative, and vice versa.

Constant Dollars and Current Dollars

The price of an item at any time is the number of dollars needed to purchase the item at that time. This price is stated in *current dollars*. As the price of the item rises, its cost in current dollars also rises. Thus, when a future cost is to be predicted in current dollars, the effects of inflation and cost growth must somehow be included in the prediction. If this is done well, then the predicted cost will accurately represent the outlay in dollars that will have to be made at that future time (the then current dollars). However, the prediction of inflation and cost-growth effects is usually quite difficult.

For purposes of economic analysis, future prices and costs may also be stated in terms of dollars of constant purchasing power, called *constant dollars*. Since, the purchasing power of the dollar changes from year to year, constant dollar costs must be referred to a particular *base year*; the purchasing power of the constant dollar is then the purchasing power that a dollar had in that base year. Future prices that are expressed in constant dollars may not be realistic, but their use eliminates the need to predict inflation rates, and they do provide accurate comparisons of future costs. Since the primary purpose of an LCCA is the comparison of alternatives, the constant dollars approach is most commonly used. When it is used, present (base year) prices are modified only for individual items whose cost growth is greater than or less than the general inflation rate. This growth in the costs of individual items over and above the general inflation rate is commonly referred to as *differential escalation*. Examples of the modification of nonrecurring and recurring costs to include inflation and cost growth in an LCCA follow.

Nonrecurring Costs and Inflation

For an LCCA that is performed using current dollars, future nonrecurring costs may be found by adjusting the present price with the appropriate SCA factor. This adjusted price is then used as the future cost in future dollars. As an illustration, consider a carpet example. Suppose the cost in current dollars of the one-time replacement 8 years from now will be subject to a 6% compound annual inflation rate and a 1% compounded annual cost growth. Both may be taken into account with the SCA factor $i = 7\%$, $n = 8$ years, which is 1.7182 (Appendix A, Table 2). Then the price in current dollars of the economy carpet costing $20 per yard will be

$20 \times 1.7182 = \$34.36$ per yard in 8 years. In an LCCA performed in constant dollars, only the differential escalation rate of 1% (cost growth) would be included in the computation, rather than the combined rate of 6% + 1% = 7%.

Recurring Cost and Inflation

If a recurring energy cost of $500 this year increases at a rate of 10% per year due to inflation and/or cost growth, it will be $500 \times 1.10 = \$550$ next year, $\$500 \times (1.10)^2 = \$605$ the following year, and so on. The following procedures may be used to take these increases into account in a life cycle cost analysis. First, each year's annual cost is adjusted to a single inflated cost using the appropriate SCA factor. Next, the equivalent present worth of each of these inflated costs is found, using the appropriate PW factor. Finally, these individual present worths are summed to obtain an equivalent single present cost. As an example, consider the effects of fuel cost increases for a hot water heater over a period of 5 years. It is expected that the present annual energy (fuel) cost of $45.00 will differentially escalate at an annual compounded rate of 7%. In addition, the discount rate of 10% must be taken into account. For year 1, the inflated value is $\$45.00 \times 1.070 = \$48.15$ (1.070 is the SCA factor for 1 year using a 7% inflation rate). The present worth of this inflated value is $\$48.15 \times 0.9091 = \$43.77$. The PW factor for 1 year at a 10% discount rate). For year 2, the inflated value is $\$45.00 \times 1.1449 = \$51.52$. The present worth of this inflated value is $51.52 \times 0.8264 = \$42.58$. If this procedure is followed for each of the 5 years and the individual present worth's are summed, the following results are obtained:

End of Year	Uninflated Energy Cost		7% SCA Factor		Inflated Energy Cost		10% PW Factor		Discounted Energy Cost
1	$45.00	×	1.0700	=	$48.15	×	0.9091	=	$43.77
2	45.00	×	1.1449	=	51.52	×	0.8264	=	42.58
3	45.00	×	1.2250	=	55.12	×	0.7513	=	41.42
4	45.00	×	1.3108	=	58.99	×	0.6830	=	40.29
5	45.00	×	1.4206	=	63.12	×	0.6209	=	39.19
Hot water heater 5-year energy cost (present worth equivalent)									$207.25

Because this procedure is cumbersome, the following formula may be used to account for both the discount rate and the escalation rate:

$$P = A \, \frac{v(v^n - 1)}{v - 1} \qquad (2\text{-}13)$$

where $v = \dfrac{1 + e}{1 + i}$

e = escalation rate (this is the differential escalation rate in the constant dollar approach)

The variables i, n, A, and P are defined earlier in this chapter. Note that, where $e = i$, the formula simplifies to

$$P = A \times n$$

If A is the amount of the first cash flow, Eq. (2-13) can be restated as follows:

$$P = A \frac{v^n - 1}{v - 1} \qquad (2\text{-}14)$$

Both formulas allow conversion of a series of increasing cash flows to an equivalent single value. Referring again to the hot water heater example, and given the following information, use Eq. (2-13) to calculate the hot water heater energy life cycle cost (PW):

e = 7% (energy differential annual rate of cost increase)
i = 10%
n = 5 years
A = $45.00 (annual energy cost one year prior to first payment in the series)

Solution:

$$v = \frac{1+e}{1+i}$$

$$= \frac{1+0.07}{1+0.10} = \frac{1.07}{1.10}$$

$$= 0.9727$$

The energy cost (PW) is

$$P = A \frac{v(v^n - 1)}{(v - 1)}$$

$$= \$45.00 \ \frac{0.9727\,[(0.9727)^5 - 1]}{0.9727 - 1}$$

$$= \$45.00 \ \frac{0.9727\,(0.8709 - 1)}{-0.02727}$$

$$= \$45.00 \ \frac{0.9727\,(-0.1291)}{-0.02727}$$

$$= \$45.00 \ \frac{(-0.1256)}{-0.02727}$$

$$= \$45.00 \ (4.6054)$$

$$= \$207$$

This process can be simplified by the use of specially prepared interest tables which include escalation. For example, referring to Appendix A, Table 9, given a discount rate of 10%, an escalation rate of 7%, and an n value of 5 years would yield the factor 4.605, which was just calculated using Eq. (2-13). Simply multiplying this factor by $45.00 yields the present worth of the energy annual cost including escalation, or $207 as previously calculated. Whether the constant dollars or current dollars approach is used, these formulas may be used to convert an annual series of increasing costs to an equivalent single amount. If the constant dollars approach is used, e becomes the differential escalation rate, and i becomes the discount rate after the effects of the general economy inflation have been taken into account. The following section discusses commonly-used methods for selecting the analysis period in a life cycle cost analysis.

Analysis Period

The analysis period is the number of years over which the total cost of ownership will be determined for the various design alternatives. Like other key elements, the analysis period should be established before the LCCA is begun and should be chosen with regard for the design situation at hand, the objectives and perspective of the owner, and established organizational policy. Following are discussions of a number of the more commonly-used criteria for establishing the analysis period.

- *Component life.* If the several alternatives being considered all have the same economic life, then that life (or a multiple of it) may be used as the analysis period. This criterion is a simple one that has the advantage of representing the life of the item or system under consideration.
- *Common multiple of component lives.* If the design alternatives have different economic lives, it may be possible to choose, as the analysis period, a common multiple of these lives. For example, the economic lives of two carpeting alternatives are 8 years and 12 years. A 24-year analysis period was selected as a common multiple of the two lives. The use of this criterion simplifies calculations involving unequal lives, and it eliminates residual values. However, not all combinations of economic lives have a common multiple that may be used as a realistic analysis period.
- *Facility life.* In some situations, the analysis period may be based on the technological or useful life of the facility as a whole. This criterion has the advantage of reflecting the "total facility" life and allowing the comparison of alternative life cycle costs over that life. Using facility life has the disadvantage of not always reflecting the life of the item being considered in the analysis or the mission life of the facility, as discussed next.
- *Investment or mission life.* The analysis period is sometimes established by limiting it to some investment or mission life for the facility. This is the expected number of years until the owner's investment objective is fulfilled, and it depends very much on that objective. For example, a speculative owner who plans to build an office facility for immediate sale may wish the analysis period to be relatively short. A corporate owner who plans to build and use an office facility would probably select a relatively long analysis period in keeping with the facility's long mission life. In the public sector, a government clinic may be viewed as having a relatively short mission life because of expected changes in the health care delivery process, and a relatively short analysis period might be used. This criterion for analysis period selection most truly reflects the

investment objective, provided all parties involved in the project agree on that objective.

- *Arbitrary life.* At times a somewhat arbitrary analysis period is selected even when it appears there is good reason to maintain a facility for an indefinite time. The analysis period might simply be established by *organizational policy* or as the limit of a *planning horizon.* For example, an organization may not be able to plan any further than 20 years in the future and may be willing to consider only alternatives that are cost-effective within the 20-year planning horizon. The federal government usually establishes the analysis period on the basis of organizational policy that can, at times, be somewhat arbitrary. This "criterion" provides a commonality among projects and among organizational units, but it does not take into account such important considerations as component life and facility life or mission. It should be noted that, no matter how the analysis period is selected, costs that are to be incurred far in the future (beyond, say, 25 years) become inconsequential both in size and in their effect on the LCCA analysis, whether the present worth or annualized analysis approach is used. This is illustrated in Figure 2.15, which shows the cumulative total ownership cost in present worth as the area under the curve. The authors recommend an analysis period of 25 to 40 years.

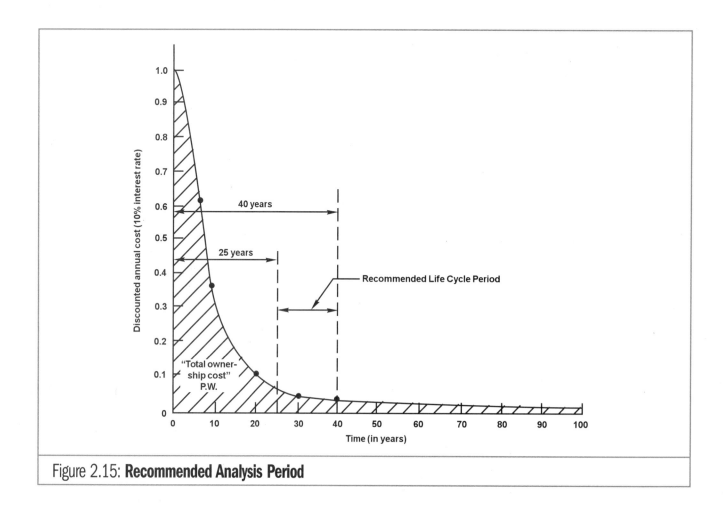

Figure 2.15: **Recommended Analysis Period**

Present Time

In an LCCA, the present time marks the beginning of the analysis period. It is also the time for which baseline costs are quoted. It sets the base year in a constant dollar analysis, and, in the present worth approach, it represents the time to which all life cycle costs are discounted for combining and comparison. Present time is usually chosen from among the points identified in Figure 2.16.

The most common choices for the present time are point B (during design), point E (halfway through construction), and point G (the beginning of occupancy). The selection of point B normally results in the most realistic baseline costs, which can be developed from contractor quotes or other sources of today's costs. These costs, in turn, provide the most accurate projections of future costs. Baseline costs for points subsequent to point B are increasingly unreliable, as are projections made from them. Nonetheless, point E is often selected as the present time, because budgets are often based on the midpoint of the construction phase. Then the costs expected to be paid to the contractor become the initial costs of the LCCA. Point A is the beginning of design; point C is the end of design; points D and F are the beginning and end of construction, respectively; and point G is the beginning of occupancy. Any of these could be chosen as the present time, but for all except point G, two steps would be required to discount recurring costs. This is because the recurring costs begin at point G, and the formulas for both the present worth method and the annualized method automatically assume this point to be the present time. If an earlier time, such as point B, is chosen, then the second step is the discounting of all costs from point G to the chosen present time by use of the PW formula.

Costs

The importance of costs in LCCA is obvious: they are the raw material of the analysis, and they form the basis for the comparison of alternatives. However, various costs are treated in different ways, and, depending on the situation, some need not be included in the LCCA. As is true for all key elements, improper treatment of costs in an LCCA can result in erroneous conclusions.

Types

All significant costs attributable to an alternative are normally considered for inclusion in an LCCA. This would include all the construction, construction-related, and procurement costs at the beginning of the

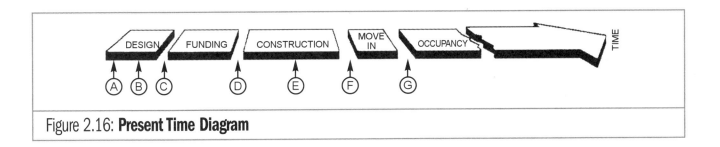

Figure 2.16: **Present Time Diagram**

analysis period; all the disposal, demolition, and other terminal costs or salvage values at the end of the analysis period; and all the various types of costs incurred between construction and the end of the analysis period. Whether or not any particular type of cost should be included depends primarily on two factors: (1) whether that type of cost is relevant for the particular facility under design and for the specific construction element under consideration and (2) whether the projected magnitude of that type of cost is significant in comparison to the other relevant costs for that LCCA. For convenience in discussion, these costs have been divided into the following categories:

- Initial project (investment)
- Energy
- Operation and maintenance
- Alteration and replacement
- Terminal costs or salvage value
- Associated (staffing, denial of use, etc.)

Initial project costs, energy costs, operation and maintenance costs, and replacement costs have been found to be both relevant and significant in most life cycle cost analyses. The relevancy and significance of other types of costs, such as alterations, design/redesign costs, terminal costs, downtime costs, and functional use costs, are usually established on a case-by-case basis. Each of these categories may include the following types of costs:

> *Initial project (investment) costs.* Costs associated with the initial design and construction of the facility. Construction costs include costs of labor, material, equipment, general conditions (job overhead), contractor's main office overhead, and profit. Other costs include those for design as well as special studies and tests, land, project administration, construction insurance, permits, fees, financing, etc., as required.
>
> *Energy costs.* Costs associated with the ongoing energy consumption of the facility. These include costs of electricity, oil, natural gas, coal, and other fuels necessary for the ongoing operation of the facility and its components.
>
> *Operation and maintenance costs.* All costs associated with operation, maintenance, repair, and custodial services for ongoing requirements at the facility. These include personnel costs, e.g., salary of a power plant operator, supplies and contract services (e.g., security, safety) necessary for ongoing operation, routine maintenance and repair, cleaning, grounds care, and trash removal.
>
> *Alteration and replacement costs.* Costs associated with planned additions, alterations, major reconfiguration, and other improvements to the facility to meet new functional requirements, and replacement costs required to restore the facility to its original performance. These include costs of facility components, redesign, demolition, relocation, and disposal, as well as costs of labor, materials, equipment, overhead, and profit.

Terminal costs or salvage values. Dollars associated with the demolition and/or disposal of the facility at the end of the analysis period. These include costs of demolition and disposal and salvage values of facility elements recovered as part of replacement, alteration, or improvement of facilities. Any trade-in value would be included in this category.

Associated costs. Other identifiable costs associated with an LCCA not previously mentioned. These include functional use costs, downtime costs, tax implications, etc. *Functional use costs* include the staff, materials, etc. required to perform the function of the organization using the facility or installation. *Downtime costs* are salaries that must be paid, lost sales of products which cannot be produced, and similar costs that are incurred during periods of major repair and replacement, when part or all of the facility cannot be used.

Monetary benefits, are normally considered as negative costs in the LCCA, include all benefits that can readily be quantified in terms of dollars, e.g., salvage values and other forms of income, cost reductions, and marketable byproducts (such as fly ash and sulfur products from an air pollution control system). The decision as to whether any particular type of monetary benefit should be included in an LCCA is usually based on the same criteria as those specified above for costs, primarily relevance and significance.

Common Costs

Since the purpose of an LCCA is to select the best alternative from among a number of alternatives (in essence a comparative analysis), it is not necessary to estimate all costs. Those types of costs which are common to all alternatives, with no differences in magnitude, can be excluded. The results will not be affected by this exclusion in terms for the relative economic ranking of the alternatives. Consider, for example, two alternative hot water systems that have the same impact on, say, the size of the maintenance staff. In this case it is not necessary to even consider maintenance staff costs (no matter how high or low they may be) in the life cycle cost analysis. Care is taken, however, to include differential costs. If the two hot water heater systems just noted place different requirements on the cost of the building's electrical system, these costs are included in the analysis in order to develop an accurate comparison.

Collateral Costs

As a part of an LCCA, it is at times necessary to include costs that are not directly attributable to the system being analyzed. For example, in the evaluation of various lighting systems, some alternatives may have differing ceiling and HVAC interface costs. Because these costs are also an important consideration in the selection of the most life cycle cost-effective lighting system, they are included in the analysis. The ceiling and HVAC costs in this example are commonly referred to as *collateral costs* to the lighting system. At times, these collateral costs are small enough to be excluded from the analysis, particularly since only the differential or incremental costs are important to the results of the LCCA. However, the assumption that such costs can be neglected in any given situation may not be valid, and probably should not be made without some evidence, e.g., preliminary calculations, to establish the validity of the assumption.

Sunk Costs

These are costs that have been incurred before an LCCA is begun. Only the costs that are expected to be incurred during the life cycle of the analysis (after the present time) are included in the cost estimates of alternatives. Therefore, sunk costs have no direct bearing on the results of the LCCA, since the asset or benefit provided is available regardless of which alternative is selected. Consequently, sunk costs are never included in a life cycle cost analysis.

Consider, for example, a study of two HVAC systems. Even though a large amount of design time may have been spent on one system, this design cost would not be included in the analysis. On the other hand, the redesign costs associated with the other alternative are normally included in the LCCA. Sunk costs are illustrated in Figure 2.17.

Continuing Costs

Costs are incurred in a facility throughout its life cycle. Ideally, these costs should be reflected in the life cycle analysis at the time in which they occur. This is rarely if ever done in practice, however. The standard procedure is to accumulate costs over some period of time, usually over the fiscal year or the calendar year for post-construction costs and over the construction period for construction-related costs, and to charge all of the costs incurred during that time as a single lump sum cost. These lump sum costs are

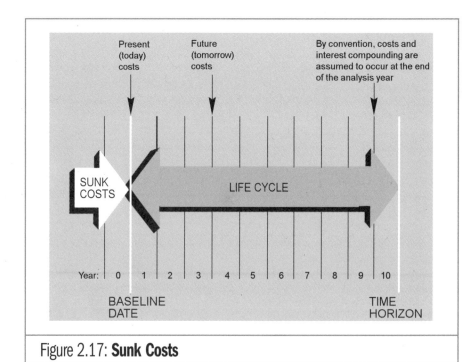

Figure 2.17: **Sunk Costs**

sometimes charged at the midpoint of the time period, but may also be charged at the end or the beginning of the time period. The additional accuracy provided by charging all the costs at the exact time that they are actually incurred is very small for the typical LCCA. Therefore, the additional computational effort required, particularly in view of the numerous uncertainties that characterize the typical LCCA is not warranted.

Summary

This chapter is intended as a review of the key fundamentals of life cycle costing. The time value of money and the concept of equivalence allow dollars spent over various points in time to be converted to a common basis. The present worth and the annualized methods are commonly used to convert dollars spent today and dollars spent in the future to an equivalent life cycle cost. Inflation is an important consideration in life cycle costing because of the effect it has on costs. The life cycle, or time frame, over which costs are projected also influences the value of the life cycle cost.

The Bibliography contains a list of books that provide a more comprehensive discussion of the fundamentals of life cycle costing. These books provide derivations of the interest (discount) formulas, complete explanations of the basic analysis approaches (present worth, annualized costs, etc.), and discussions of inflation and the life cycle.

Chapter 3 Economic Analysis

The architectural profession has been under severe pressure to find better overall answers to client demands for simultaneous handling of the intricate balance of time, quality and cost of projects. The ultimate goal is a facility that is delivered in the shortest possible time, at the lowest possible cost, with the highest possible quality and/or performance.
Philip J. Meathe, FAIA
Current Techniques in Architectural Practice
The American Institute of Architects, 1976

There is a lack of uniform acceptance and understanding of both economics terminology, and the role that certain aspects of economics should play in project design. As a logical follow-on to the review of LCC fundamentals in Chapter 2, it is important to discuss the terminology, types of analyses, benefits and costs, economic analysis approaches, and the relationship of economic analysis (EA) to value engineering (VE). Thus, owners, design professionals, and all others involved in financing, design, and construction of a project should be able to communicate more effectively about various aspects of economic studies as applied to facility planning, design, construction, operations, and maintenance.

Terminology

An *economic study* or *economic analysis* is an examination of the costs and benefits (both monetary and non-monetary) expected to result from a particular course of action or from alternative courses of action. An *engineering economic study* (or analysis) is an economic study dealing with facilities applications. A *life cycle cost analysis* is a specific type of engineering economic analysis—the type of most concern to facilities design professionals. These distinctions are examined in more depth in the following paragraphs.

Economic Study/Analysis

In the most *general* economic analysis, several courses of action are examined to determine and compare the costs that will be incurred and the benefits (both monetary and non-monetary) that will accrue from each over a designated period of time. The objective of the analysis is to find the course of action that provides the greatest benefit at the least cost.

In some situations, this objective may be attained with a less general, less exhaustive study. A *restricted* study may be performed if the benefits offered by the several alternatives might be so similar in nature and extent that they are not expected to affect the comparison, or they might be considered only secondary to cost. In this case, the benefits would be ignored in the analysis (at least initially), and a *cost-centered study* would be performed to determine which alternative would involve the least cost. These costs would include both the initial project cost and all subsequent costs. The benefits would be considered in a later final analysis only if the costs of two or more alternatives were found to be essentially equal.

Another example of a restricted study is one that focuses on a viable alternative. A *single-alternative study* would be performed to determine the costs and benefits that would result from the implementation of that alternative. Still another example of a restricted economic analysis is the *initial-cost study* (cost estimate), which might be performed when the initial cost is of prime importance, and all other costs and the benefits of the various alternatives are either similar or insignificant in comparison to initial costs. In all cases, the situation at hand and the objective of the analysis dictate the extent to which an economic analysis may be restricted.

Engineering Economic Study/Analysis

An engineering economic study or analysis deals with facilities planning, design, construction, and operation. As such, it may be applied in the most general form or in a restricted form as warranted by the problem at hand. Most engineering economic studies involve an evaluation of alternatives, although some deal with the economic impact of a single course of action. Engineering economic analyses are used throughout the facilities procurement process—from the very early planning stages (when investment-type decisions, such as build versus renovate or lease, are made), through design, to the final stages of construction, when actual on-site conditions or changed requirements may dictate the need for design changes.

Life Cycle Cost Analysis

A life cycle cost analysis (LCCA) is a cost-centered engineering economic analysis whose objective is to systematically determine the costs attributable to each of one or more alternative courses of action over a specified period of time. An LCCA may be restricted to initial costs only, but usually costs are analyzed over a designated period of time that is related to the life cycle of the construction element for which these costs are incurred. In most cases an LCCA is performed as the basis for a decision among investment alternatives (although doing nothing is sometimes a viable alternative in such a study). Further, an LCCA may be performed as an independent study or as part of a more comprehensive study, for example, when it is used in value engineering. The process of conducting a life cycle cost analysis is sometimes referred to as *life cycle costing*.

Office Economic Analysis Example

The following example, using a hypothetical office building, illustrates the significance of these economic techniques as they are applied to projects.

During the planning stage of the project, the total initial cost was estimated to be $48,486,000. The feasibility study indicated gross income from office rentals of $8,850,000 and operating costs of $3,110,000. The loan amount was determined by assuming that 75% of the project cost would be capitalized at 10% of net income (using constant dollars). Dividing the before-tax stabilized cash flow by the required equity investment yielded a return on investment of 18.1%. (See Figure 3.1.)

Several scenarios are presented as to what might happen during the course of designing the project. First, the project construction cost might rise 10% over that originally planned. This added increase in initial cost reduces the return on investment (ROI) to 12.1%. On the other hand, the designed facility operating costs for energy, maintenance, taxes, etc., might be 10% higher. If this were the case, the ROI would become 13.4%.

In the third situation, both the construction costs and the operating costs are assumed to be 10% higher than originally planned. The net result is an ROI of only 9.6%. In all three cases, the resulting ROI is less than the owner would have accepted during the planning phase, and represents a poor return on investment.

In the final situation, through systematic application of economic studies during project design, the return on investment has increased to a dramatic 30.2%. This result is achieved by monitoring project costs throughout design and applying both life cycle costing and value engineering techniques to reduce the construction cost by 5% and the operation cost by 5%. (This ROI includes the added cost of professional time for the life cycle costing team.) The 5% reductions in construction and operating costs are realistic expectations of cost improvement for any facility following the life cycle costing process described in this text, provided it is applied during early project design.

Types of Economic Analysis

Economic analyses can be categorized in several ways. Various professionals and technical organizations tend to use the classification systems that best suit their own needs. Two systems that are commonly used categorize economic analyses according to (1) their principal purpose and (2) the feasible alternatives. The systems apply, by implication, to LCCAs as well. Following is a discussion of the types of economic analysis. Figure 3.2 summarizes the types of economic analysis, their emphasis and objectives, and some sample applications.

	As Planned (Budgeted)	Construction Costs + 10%	Operation Costs + 10%	Construction & Operation Costs	Economic Studies Construction-5% Operation-5%
Total Construction Cost	34,757,000	38,233,000	34,757,000	38,233,000	33,020,000
Indirect Costs (includes Economics)	9,249,000	9,711,000	9,247,000	9,247,000	9,062,000
Land Cost	4,480,000	4,480,000	4,480,000	4,480,000	4,480,000
Total Project Cost	48,486,000	52,424,000	48,486,000	52,424,000	46,562,000
Less Mortgage Loan*	40,583,000	40,583,000	38,384,000	38,384,000	41,686,000
Equity Investment Required	7,903,000	11,841,000	10,102,000	14,040,000	4,876,000
Gross Income	8,850,000	8,850,000	8,850,000	8,850,000	8,850,000
Operating Costs	3,110,000	3,110,000	3,421,000	3,421,000	2,954,000
Net Income*	5,740,000	5,740,000	5,429,000	5,429,000	5,896,000
Less Mortgage Payment (Debt Service)	4,305,000	4,305,000	4,072,000	4,072,000	4,422,000
Before Tax Stabilized Cash Flow	1,435,000	1,435,000	1,357,000	1,357,000	1,474,000
(Return on Equity Investment)	18.1%	12.1%	13.4%	9.6%	30.2%

*Loan amount determined by 75% of capitalized (@ 10%) net income
0.75 × $5,740,000 × 9.427 (PWA) = $40,538,000
0.75 × $5,429,000 × 9.427 (PWA) = $38,384,000
0.75 × $5,896,000 × 9.427 (PWA) = $41,686,000

Figure 3.1: **Hypothetical Office Building Economic Analysis**

FEASIBLE ALTERNATIVES		PRINCIPAL PURPOSE Emphasis:	
	Objective:	**Primary** Save money or other economic benefit	**Secondary** Satisfy business or service requirement
Investment (Feasibility Phase)	Determine (1) whether an investment is justified, and if so, (2) the most economical strategic course of action	Replace existing high cost facilities Install energy saving devices	Do nothing to start new venture Lease, renovate or build new facilities
Design (Design Phase)	Seeks the most economical design solution which satisifes the "required function"	(not applicable)	Single story layout or multistory building Type of mechanical HVAC system

Figure 3.2: **Types of Economic Analysis and Sample Applications**

Primary versus Secondary Economic Analysis

The distinction between primary and secondary analysis is based on the principal purpose of the analysis.

A *primary economic analysis* is one that is undertaken principally to save money or to achieve some other economic benefit. A cost-centered economic analysis is an obvious example of a primary economic analysis, but a primary analysis may also include non-monetary benefits if they are of major concern. Typical situations in which a primary economic analysis would be used are:

- *Replacement of an existing facility or construction element with one that has lower operating and maintenance costs.* Here the same function is expected to be performed at lower cost. An example is replacing nursing towers in a hospital with newly-designed ones in order to achieve a more staff-efficient layout.
- *The installation of energy-saving devices, such as additional insulation or heat exchangers, in existing facilities to reduce heating and cooling costs.* Here a function is being modified or upgraded to reduce costs.

Note that, in both these situations, the objective of the analysis is to reduce an existing cost.

A *secondary economic analysis* is one that is performed principally to determine the most effective way of satisfying a new or newly-identified business or service requirement. Satisfaction of this functional requirement (e.g., AIDS research, manufacture of a new type of vehicle) is of principal concern, but the criterion for effectiveness is economic. A secondary economic analysis may be very general in nature, or it may be restricted in some manner; it would not, however, be restricted to a single-alternative study except in unusual circumstances.

The decision as to whether to build, renovate, or lease space for a particular function typically requires a secondary economic analysis. Decisions as to framing system type, lighting, heating and cooling systems, and space layout and partitioning in a new building would also be based on secondary economic analyses. In each case a requirement is to be satisfied, and the secondary analysis is used to determine the most economical method of doing so. Except for the build-renovate-lease analysis situation, most secondary analysis is performed during the schematic and design development phases. Primary analysis, on the other hand, is performed during the project feasibility stage.

Investment versus Design Economic Analysis

The distinction between investment and design economic analysis is based on whether one particular course of action—to do nothing—is a feasible alternative.

An *investment economic analysis* is undertaken to determine whether an outlay of funds (an investment) is warranted and/or which of several strategic alternative courses of action (including doing nothing) is most economical in meeting certain objectives. Such analyses would be used to determine whether to install energy saving devices or replace existing high-cost facilities (also the subjects of primary economic analysis); whether to lease, renovate, or build new (the subject of secondary analysis); and whether the installation of a solar space or water heating system would be economically justified (a subject for primary economic analysis). Investment analyses are most commonly performed early in the project justification stage, before design begins. This could be part of a project feasibility study. An investment economic analysis is usually very general, but the subject of the analysis may allow some restriction.

A *design economic analysis* is an economic analysis (not a design analysis) performed to examine alternative courses of action for providing a required function, and to highlight the most economical alternative. Because the function that gave rise to the analysis must be performed, doing nothing is *not* an acceptable alternative. Design economic analyses are undertaken during the schematic and design development phases of a project, and they are always secondary economic analyses by definition. A design economic analysis might be used to determine whether a single-story or multistory building would be more economical and to decide on the heating and lighting systems, partitioning, entrance locations, etc. Depending on the item under consideration, this type of analysis may include certain non-monetary benefits but it is often restricted to cost-centered considerations of alternatives.

Benefits and Costs

As stated earlier in this chapter, the objective of an economic study is to seek the alternative course of action that provides the greatest benefits for the least cost. Benefits are normally considered to include those functions, advantages, or improvements in a condition that are derived from the installation or utilization of construction elements. Benefits are of two types: those that *can* be expressed in terms of dollars (monetary) and those that *cannot* be expressed in terms of dollars (non-monetary). Examples of non-monetary benefits are:

- Aesthetics, image
- Expansion potential
- Spatial relationships
- Flexibility, versatility
- Safety
- Reduction of environmental pollution
- Conformity with political considerations

The benefits and costs involved in engineering economic analyses are viewed and treated differently by the private and public sectors. A discussion of these differences is useful in explaining how and why benefits and costs are treated differently in the analysis of project designs.

Private Sector Treatment

A private owner-investor generally considers benefits as a full and equal partner with costs in analysis of potential investment opportunities. The private investor may be interested solely in the financial return of the investment, via improved earning power or increased profit. In this case, only the monetary costs and benefits are considered. On the other hand, the investor may be willing to pay (invest) additional money (over and above the investment required for the minimum desired benefits) to derive additional, non-monetary benefits. For example, a company may spend more than the necessary minimum for site and building features (such as landscaping, site size and location, and architectural style) to enhance its public image or to attract workers—both non-monetary benefits. The decision rests with the investor, who is free to choose a more costly alternative that provides marginal non-monetary benefits.

Public Sector Treatment

The public sector approach to benefits and costs stems from a recognition of the need to minimize taxes and their impact on private spending and investment. This has led to a philosophy and regulations that limit public sector investment spending to the amount necessary to meet minimum functional and technical requirements. For investment analysis purposes, alternatives are considered equivalent in providing benefits if they meet these minimum requirements. Benefits exceeding the minimum requirements are usually not sought, and they are not considered if they lead to additional costs. For design analysis purposes, the only alternatives that may be included in an LCCA are those that at least meet the minimum functional and technical requirements. Alternatives are considered to provide equal benefits if they satisfy these requirements. Monetary benefits are treated as negative costs. Non-monetary benefits beyond the minimum requirements are not normally considered unless competing alternatives are otherwise essentially equal. Thus for nearly all government designs, non-monetary benefits beyond the minimum functional requirements established for the project hold little interest for the designer or project manager. Lowest life cycle cost is the principal criterion for selecting one design alternative over others. The exception to these general rules occurs when a project contains specific requirements regarding one or more non-monetary benefits.

Economic Analysis Approaches

The analysis approach is the manner in which the fundamental concepts described in Chapter 2 are incorporated into the economic analysis. Several approaches are available, and each includes an associated evaluation criterion. The economic analysis approaches most commonly used in design economic analysis (present worth and annualized LCCA) are discussed in detail in Chapter 4. Other approaches best suited to investment economic analyses (payback period, return on investment, savings to investment ratio) are discussed in general terms in the following sections. These approaches may also be useful in the design phase, e.g., in performing a special investment analysis in response to a directive emanating from an owner or, in the case of public sector projects, a congressional committee or executive department.

Payback Period

The payback period is the time, usually in years, required for the expected savings due to an investment accumulating (paying back) the invested amount. In the payback period approach, this time is used as a measure of the effectiveness of investment alternatives. The payback period may be calculated as a *simple payback period* or *a discounted payback period*.

When the simple payback period is used, the time value of money is ignored in return for computational ease. The simple payback period is calculated as:

$$\text{Simple payback period} = \frac{\text{initial cost}}{\text{annual savings}}$$

For example, suppose a new nursing tower is being considered for a hospital in order to reduce staffing costs. The tower is estimated to cost $20,000,000 to construct, and it is expected to reduce nurse staffing costs by $5,000,000 per year. The simple payback period is then:

$$\text{Simple payback period} = \frac{\$20,000,000}{\$5,000,000} = 4 \text{ years}$$

The *discounted payback period* is obtained by first converting each year's savings to an equivalent present worth at the time of the initial expenditure, and then accumulating these equivalent present worth values. The discounted payback period is the time required for the accumulated equivalent present worth values to equal the initial expenditure. Because savings are discounted to their present worth, this approach takes account of the time value of money. For the nursing tower example, the cumulative equivalent present worth of the annual savings would be computed as shown in the following table.

Cumulative discounted savings process (discount rate = 10 %)

Year	Present worth savings, $	Cumulative PW savings, $
1	$5,000,000/1.10 = 4,545,000$	4,545,000
2	$5,000,000/(1.10)^2 = 4,132,000$	8,677,000
3	$5,000,000/(1.10)^3 = 3,757,000$	12,834,000
4	$5,000,000/(1.10)^4 = 3,415,000$	15,849,000
5	$5,000,000/(1.10)^5 = 3,105,000$	18,954,000
6	$5,000,000/(1.10)^6 = 2,822,000$	21,776,000

Interpolating between the fifth and the sixth year, the discounted payback period would be roughly 5.3 years, since

$$\frac{\$5,000,000}{(1.10)^{5.3}} = \$1,045,000$$

which yields a cumulative PW savings of $19,999,000.

An alternative approach to the above calculations is to use economic tables for discounted payback such as those in Appendix A (Tables 11 to 15). Using Table 14 for a discount rate of 10% and an escalation rate of 0% for annual staffing escalation, enter the row with a ratio of initial cost to first year savings (simple payback) of 4. This gives a discounted payback of 5.36 as calculated above. On the other hand, if nursing salary increases are estimated to rise 6% differentially per year, that period would be reduced to 4.42 years. This method produces a more realistic picture of investments made, especially when operating costs are expected to increase with time.

Return on Investment

This approach, also referred to as the *internal rate of return,* is used most often to evaluate private sector investment alternatives, but it is sometimes also used by the public sector. The expected annual savings due to an investment (the return on that investment) are expressed as a discounted percentage of the investment. (This discounted percentage may be viewed as analogous to an interest rate.) The present worth annuity (PWA) factor is used to discount the equal periodic savings. For example, suppose an investment of $1,000,000 in a new heating, ventilating, and air-conditioning system is expected to result in an annual energy cost reduction of $150,000 over a period of 10 years. To compute the rate of return on the investment, the PWA factor is first found:

$$\text{PWA} = \frac{P}{A} = \frac{1,000,000}{150,000} = 6.667$$

The PWA factor of 6.667 is somewhere between the PWA factor for 8% and 10 years (6.710 in Appendix A, Table 3) and the PWA factor for 10% and 10 years (6.145 in Table 4). Interpolating between 8% and 10% gives a rate of return of approximately 8.4%. More extensive tables would allow more accurate determination of the discounted rate of return.

Savings to Investment Ratio

In this approach, the savings to investment ratio (SIR) is used as a measure of investment effectiveness. The SIR is calculated by dividing the present worth of the annual cost savings by the initial cost. If the SIR is higher than 1, the investment can be considered cost-effective; the higher the ratio, the greater the dollar savings per dollar spent. By comparing the SIRs of various alternatives, the most effective course of action can be readily determined. The SIR is similar in nature to the ratio of benefits to costs used to compare various alternatives for Army Corps of Engineers civil works projects. As an example, suppose the installation of shelters on loading docks is proposed to reduce heat loss at a northern facility. The estimated cost of one shelter alternative is $15,000 and the estimated annual savings is $4,200 for a period of 8 years. The discount rate is 10%. For 8 years and 10%, the PWA factor (from Appendix A, Table 4) is 5.335. Then the SIR is

$$\begin{aligned} \text{SIR} &= \frac{\text{annual savings} \times \text{PWA}}{\text{investment cost}} \\ &= \frac{\$4,200 \times 5.335}{\$15,000} \\ &= 1.494 \end{aligned}$$

The SIR in excess of 1 indicates that the investment is economically viable. However, the alternative with the highest SIR should be implemented.

Value Engineering

The methods of value engineering provide an important tool for improving the decisions of managers and design professionals. If that tool is to be used effectively, the relationship between economic analysis (especially LCCA) and VE analysis must be understood. Some professionals perceive the two to be mutually exclusive, in fact, opposing each other, so that the results of one automatically negate the results of the other. Others perceive the two analysis methods as duplicative. Neither of these perceptions is correct.

Comparison of LCCA and VE

Both LCCA and VE are analysis techniques whose goal is to reduce the cost of ownership of a facility.

LCCA focuses on the costs of feasible design alternatives—alternatives that meet the minimum functional and technical requirements of the project design—with the objective of identifying the least-cost alternative. Little attempt is made to challenge the necessity for a particular function, or to change any previously identified functional requirement. VE analysis, on the other hand, focuses on the functions themselves, identifying the essential functions, and eliminating or modifying nonessential functions that represent unnecessary costs. Moreover, in too many cases VE analysis is most often limited to savings in initial cost, whereas LCCA affects savings in the total cost of ownership.

Combined Use of LCCA and VE

Because the emphasis of LCCA is on costs, while the emphasis of VE is on functions, the two methods of analysis are different and distinct. However, because functions give rise to costs, the two methods complement each other. For example, LCCA is used to identify high-cost, but marginal functions—candidates for elimination or modification through VE analysis—in projects whose budgets are in danger of being exceeded. Conversely, VE is very useful when a particular alternative, selected through LCCA, has a high (over-budget) initial cost, but low subsequent costs. Then, if a VE analysis can produce a sufficient reduction in the initial costs of other design features, the selected alternative can be implemented to minimize the total cost of ownership of the project as a whole.

Conclusion

It is important to be aware of the terminology, types of analyses, benefits and costs, economic analysis approaches, and the relationship of economic analysis to value engineering. This knowledge allows you to communicate more effectively as you work in facility planning, design, construction, operations, and maintenance.

Chapter 4: Life Cycle Cost Analysis

The object of the world of ideas is not the portrayal of reality—this would be an utterly impossible task—but rather to provide us with an instrument for finding our way about in this world more easily.

Hans Valhinger
1876

Life cycle cost analysis (LCCA), as defined in Chapter 3, is a cost-centered engineering economic analysis whose objective is to systematically determine the costs attributable to each of one or more alternative courses of action over a specified period of time. The choice of analysis approach fixes the manner in which the concepts presented in Chapter 2—primarily the time value of money—will enter the analysis.

The approach chosen determines the yardstick that will be used to compare the total costs of ownership of the various design alternatives. The two most commonly used LCCA approaches are the *present worth approach* and the *annualized approach*. (The relevant computations were introduced in Chapter 2.) Both approaches result in the conversion of dollars spent today and dollars spent in the future to equivalent costs suitable for comparison of design alternatives.

This chapter illustrates the process of LCCA in assessing design alternatives using both the present worth and annualized approaches on a variety of project design examples. To assist analysts in this LCCA process, both manual calculation worksheets and microcomputer spreadsheets are illustrated.

Present Worth Method Calculations

To illustrate the process of LCCA, two examples will outline the similarities and differences between the present worth and the annualized methods of analysis. The present worth method allows all costs to be converted to equivalent costs at one point in time.

Rooftop Air-Handling System

Consider a choice between two air-handling units. A 10% discount rate, a 24-year life cycle, and a differential energy escalation rate of 2% per year are assumed.

Other relevant data are:

Type of cost	Alternative 1	Alternative 2
Initial cost	$15,000	$10,000
Energy (annual)	1,800	2,200
Maintenance (annual)	500	800
Useful life	12 years	8 years

The solution begins by converting all annual or recurring costs to the present time. Using the present worth annuity (PWA) factor (Appendix A, Table 4), the recurring costs of maintenance would be:

Alternative 1:

$$\text{Maintenance (present worth)} = \$500 \times 8.985 = \$4,492$$

Alternative 2:

$$\text{Maintenance (present worth)} = \$800 \times 8.985 = \$7,188$$

According to Appendix A, Table 9, the present worth of the energy costs for each alternative would be:

Alternative 1:

$$\text{Energy (escal.} = 2\%) = \$1800 \times 10.668 = \$19,202$$

Alternative 2:

$$\text{Energy (escal.} = 2\%) = \$2200 \times 10.668 = \$23,470$$

Replacement, or nonrecurring, costs are considered next. When one or more alternatives have a shorter or longer life than the life cycle specified, an adjustment for the unequal life is necessary. If the life of an alternative is shorter than the project's life cycle, the item continues to be replaced until the life cycle is reached. On the other hand, if the item life is longer than the specified life cycle, then a terminal or salvage value for the item is recognized at the end of the life cycle. This treatment (using the present worth factors in Appendix A, Table 4) is illustrated as follows.

Alternative 2:

$$\text{Replacement}(n = 8) = \$10,000 \times 0.4665 = \$4,665$$

Alternative 1:

$$\text{Replacement}(n = 12) = \$15,000 \times 0.3186 = \$4,779$$

Alternative 2:

$$\text{Replacement}(n = 16) = \$10,000 \times 0.2176 = \$2,176$$

The salvage value for both systems equals zero, since they both complete replacement cycles at the end of the 24-year life cycle. A summary of present worth life cycle costs follows.

Types of costs	Alternative 1 (PW)	Alternative 2 (PW)
Initial cost	$15,000	$10,000
Maintenance (recurring) cost	4,492	7,188
Energy (recurring) cost	19,202	23,470
Replacement (nonrecurring), year 8	0	4,665
Replacement (nonrecurring), year 12	4,779	0
Replacement (nonrecurring), year 16	0	2,176
Salvage, year 24	0	0
Total present-worth life cycle costs	$43,473	$47,499

The first alternative should be selected on the basis of this life cycle cost analysis.

Annualized Method Calculations

The annualized method expresses all life cycle costs as annual expenditures. Recurring costs are expressed as annual costs and require no time adjustment, while initial costs require equivalent-cost conversions.

Dental Clinic Feasibility Study

Let's consider a situation in which prospective owners are preparing a feasibility study to determine whether they should lease or build a dental clinic needed for a life cycle of 15 years. Were they to build, the financing rate of interest would be 12%. Leasing would cost $27.00/S.F./year over the next 15 years, and would include utilities, maintenance, and replacement costs. The following data summarize the costs for the build option:

	Build Option
Initial construction cost (including land)	$120.00/S.F.
Energy cost (per year)	$2.50/S.F.
Maintenance cost (per year)	$4.00/S.F.
Carpet replacement (year 8)	$3.00S.F.
Resale/salvage (year 15)	$65.00/S.F.

The LCC analysis would begin by converting initial costs to an annualized cost using the periodic payment schedule (Appendix A, Table 5). Initial costs for the build option would be converted to an annualized figure by the following calculations:

$$\text{Initial cost (annualized)} = \$120.00 \times 0.1468 = \$17.62/\text{S.F.}$$

No conversion is necessary for energy and maintenance costs since they are already expressed as annualized figures. Replacement costs would then be converted—first to present-worth values (using the information in Appendix A, Table 5), and then expressed as annualized costs (using the periodic payment schedules as in the above initial cost calculations). The replacement cost calculation would proceed as follows:

$$\text{Carpet replacement, year 8 (PW)} = \$3.00/\text{S.F.} \times 0.4039 = \$1.21/\text{S.F.}$$
$$\text{Carpet replacement (annualized)} = \$1.21/\text{S.F.} \times 0.1468 = \$0.18/\text{S.F.}$$

The resale/salvage value (treated as a negative cost) would also be converted to an annualized cost:

Resale/salvage, year 15 (PW) = ($65.00/S.F.) × 0.1827 = ($11.88/S.F.)
Resale/salvage (annualized) = ($11.88/S.F.) × 0.1468 = ($1.74/S.F.)

The parentheses indicate values that are income, not costs.

Following is a summary of the life cycle costs expressed as annualized figures:

	Build Option (annualized LCC)
Initial construction cost	$17.62/S.F.
Energy cost	$2.50/S.F.
Maintenance cost	$4.00/S.F.
Carpet replacement (year 8)	$0.18/S.F.
Resale/salvage (year 15)	($1.74/S.F.)
Total life cycle costs	$22.56/S.F.

From these figures, the owners determine it would be more economical to build ($22.56/S.F./year) rather than lease ($27.00/S.F./year).

LCCA Calculation Worksheets

Formats for PW and Annualized Methods

Two LCCA worksheets have been developed to assist in the above calculations and serve as a reminder of the step-by-step process. These present worth and annualized formats are also useful in organizing all the data and economic assumptions. The authors have used these formats for over 15 years in preparing life cycle cost analyses.

PW Worksheet—Rooftop Air-Handling System

To simplify the present worth example, a uniform worksheet (using the present worth method) is presented in Figure 4.1. Each cost element for the rooftop air-handling example discussed earlier in this chapter is entered on the present worth format worksheet. This worksheet is divided into three major categories. The upper third is devoted to initial costs. In this example, the purchase cost is listed for each air-handler alternative. Note that the estimated cost and the present worth cost are the same for the initial cost category. The total initial cost for each alternative is then determined and recorded in the appropriate place on the form.

The middle third of the worksheet is used for recording replacement and salvage values. In this example, replacement costs occur at years 8 and 16 for the economy air handler, and year 12 for the performance air handler. These values are listed under the Estimated-Cost column of the worksheet. To calculate the present worth of replacement costs, the present worth (single amount) factor must be obtained for years 8, 12, and 16 from Appendix A, Table 4, which is the 10% discount rate of this example. Once these factors are obtained, the estimated cost of repair is simply multiplied by the appropriate factor to arrive at the present worth equivalent cost. For example, $10,000 × 0.4665 = 4,665 is the present worth for the year 8 replacement cost for the economy air handler. The total replacement and salvage present worth cost is then determined and recorded in the appropriate place on the worksheet.

The final third of the worksheet records annual costs such as maintenance and energy costs. In this example, these costs are listed for each alternative. To convert these estimated annual costs to an equivalent present worth cost, the present worth annuity factor must be obtained from Appendix A, Table 4. For electrical energy costs, the fuel is anticipated to escalate differentially by 2% per year. Therefore, from Appendix A, Table 9, under the 2 column heading for a 10% discount rate and a 24-year life cycle, the factor is 10.668. Multiplying this factor by the estimated annual energy cost for each alternative yields the equivalent present worth cost. For example, the performance air handler cost is $1,800 × 10.668 = $19,202. Other annual costs are calculated similarly. The total annual present worth cost is then determined and recorded in the appropriate place on the worksheet.

Life-Cycle Costing Estimate
General Purpose Work Sheet

Study Title: Rooftop Air Handler System
Discount Rate: 10% Date:
Economic Life: 24 years

			Alternative 1 Describe: performance air handler		Alternative 2 Describe: economy air handler		Alternative 3 Describe:		Alternative 4 Describe:	
			Estimated Costs	Present Worth	Estimated Costs	Present Worth	Estimated Costs	Present Worth	Estimated Costs	Present Worth
Initial/Collateral Costs										
A. Air Handler System			15,000	15,000	10,000	10,000				
B.										
C.										
D.										
E.										
F.										
G.										
Total Initial/Collateral Costs				15,000		10,000				
Replacement/Salvage (Single Expenditure)	Year	PW Factor								
A. replace economy AH	8	0.4665			10,000	4,665				
B. replace performance AH	12	0.3186	15,000	4,779						
C. replace economy AH	16	0.2176			10,000	2,176				
D.										
E.										
F.										
G.										
H.										
Salvage air handlers	24	0.1015	0	0	0	0				
Total Replacement/Salvage Costs				4,779		6,841				
Annual Costs	Diff. Escal. Rate	PWA W/Escal.								
A. maintenance	0%	8.985	500	4,492	800	7,188				
B. energy	2%	10.668	1,800	19,202	2,200	23,470				
C.										
D.										
E.										
F.										
G.										
Total Annual Costs				23,694		30,658				
LCC Total Present-Worth Life-Cycle Costs				43,473		47,499				
Life-Cycle Present-Worth Dollar Savings				4,026		—				

PW – Present Worth PWA – Present Worth Of Annuity

Life-Cycle Cost Analysis
Using Present-Worth Costs

Figure 4.1: **PW Worksheet—Rooftop Air-Handling System**

The total present worth life cycle cost for each air handler is established by totaling the present worth of the initial, replacement/salvage, and annual operating costs. As a result of life cycle costing, one can see that although the performance air handler was the most expensive initially, it costs much less to operate. These calculations show that, over its life, the performance air handler will save $4,026 over the economy air handler.

PW Worksheet—Data Processing Center Layout

To demonstrate the versatility of the present worth worksheet, a second example is presented. This LCCA involves three building layouts for a computer data center as illustrated in Figure 4.2. Figure 4.3 presents the life cycle cost analysis for these three layout alternatives. Note that the level of detail contained on the PW worksheet depends on the subject being analyzed. In this layout LCCA, only major building systems are listed. In the LCCA of the rooftop air handling system, further estimating detail is recorded on the worksheet.

Annualized Worksheet for the Dental Clinic Feasibility Study

The annualized worksheet (Figure 4.4) has been completed for the dental clinic feasibility study discussed earlier in this chapter. The form shown is broken down into *input data* (requiring cost conversion) and *output* (annualized costs). The input data contain initial project costs and replacement costs, as well as salvage values. The replacement costs and salvage value occurs in the future and therefore must be converted to present worth using the economic tables in Appendix A. The output portion of the worksheet permits the final economic conversions necessary to calculate the equivalent annual cost of the initial project costs and the replacement costs. This portion of the worksheet also permits entering annual costs, such as for energy and maintenance. All annualized costs are then totaled to arrive at the total life cycle cost.

Using the same information from the example of the dental clinic feasibility study discussed earlier in this chapter results in the completed annualized worksheet shown as Figure 4.4. The input data section shows the initial cost of the build option of $120.00/S.F. and $0/S.F. for the lease option. The carpet replacement at year 8 is shown next. Note the PW factor of 0.4039 is also entered on the worksheet, as is the present worth amount calculated as before:

$$\$3.00 \times 0.4039 = \$1.21/\text{S.F.}$$

The building salvage value of $65.00/S.F. is entered next. This figure is converted to present worth using the 15-year PW factor of 0.1827 as follows:

$$(\$65.00) \times 0.1827 = (\$11.88/\text{S.F.})$$

Note that parentheses are used to show $11.88/S.F. is income (negative), not a cost.

The output portion of the worksheet converts these present worth costs to an equivalent annual cost. The initial project cost is first calculated by multiplying the $120.00/S.F. by the periodic payment (PP) factor for 15 years as follows:

$$\$120.00 \times 0.1468 = \$17.62/\text{S.F.}$$

The replacement annual cost is next calculated on the worksheet by multiplying the $1.21/S.F. by the PP factor of 0.1468, which results in an annualized cost of $0.18/S.F. for the carpet replacement in the build

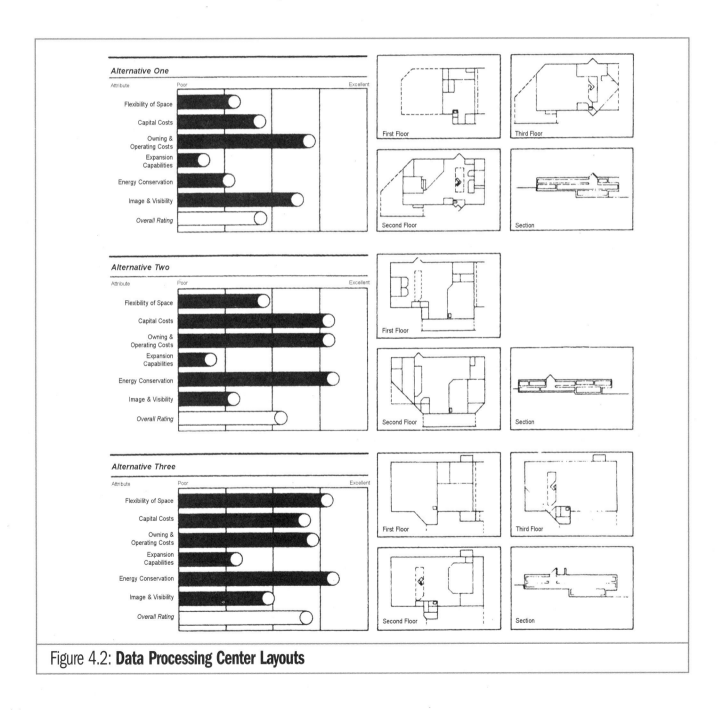

Figure 4.2: **Data Processing Center Layouts**

option. The salvage value is calculated similarly as follows:

$$(\$11.88) \times 0.1468 = (\$1.74/\text{S.F.})$$

The annual costs are then entered on the form. These include the maintenance and energy annual costs for the build option and the leasing annual cost for the lease option.

The final step consists of adding all annualized costs together for each alternative. In this example, the lease option total annualized life cycle cost is $27.00/S.F. The total for the build option is $22.56/S.F. According to this analysis, the build option appears the best choice.

LCCA Spreadsheet

Format for PW and Annualized Methods

Life cycle cost analysis calculations are performed easily on electronic spreadsheets, which can be made to appear very similar to the worksheets just presented to achieve the same benefits. In addition, spreadsheets

Life-Cycle Cost Analysis
Using Present-Worth Costs

Life-Cycle Costing Estimate
General Purpose Work Sheet

Study Title: Building Layout
Discount Rate: 10% Date:
Economic Life: 25 years (constant dollars)

			Alternative 1 Describe: sketch		Alternative 2 Describe: sketch		Alternative 3 Describe: sketch		Alternative 4 Describe:	
			Estimated Costs	Present Worth	Estimated Costs	Present Worth	Estimated Costs	Present Worth	Estimated Costs	Present Worth
Initial/Collateral Costs										
A. Structural			10.74/sf	$472,600	9.30/sf	$409,200	10.74/sf	$472,600		
B. Architectural			16.10/sf	708,400	15.20/sf	668,800	15.26/sf	671,400		
C. Mechanical			12.40/sf	545,600	12.02/sf	528,900	12.40/sf	545,600		
D. Electrical			6.50/sf	286,000	6.50/sf	286,000	6.50/sf	286,000		
E. Equipment			0.25/sf	11,000	0.25/sf	11,000	0.25/sf	11,000		
F. General Conditions			6.82/sf	300,000	6.42/sf	282,500	6.70/sf	292,800		
G. Site			9.00/sf	396,000	8.80/sf	387,200	9.00/sf	396,000		
Total Initial/Collateral Costs				2,719,600		2,573,600		2,677,400		
Replacement/Salvage (Single Expenditure)	**Year**	**PW Factor**								
A. Carpeting/Interiors	8	0.466	177,100	82,500	167,200	77,900	167,900	78,200		
B. Lighting System	10	0.385	50,600	19,500	50,600	19,500	50,600	19,500		
C. HVAC System	12	0.319	193,000	61,600	184,800	58,900	193,000	61,600		
D. Carpeting/Interiors	16	0.218	177,100	38,600	167,200	36,400	167,900	36,600		
E. Lighting	20	0.149	50,600	7,500	50,600	7,500	50,600	7,500		
F. Roofing	20	0.149	58,500	8,700	63,100	9,400	58,500	8,700		
G.										
H.										
Salvage (Resale Value)	25	0.092		NIC		NIC		NIC		
Total Replacement/Salvage Costs				218,400		209,600		212,100		
Annual Costs	**Diff. Escal. Rate**	**PWA W/Escal.**								
A. Energy-Elect. (computer, etc.)[1]	1%	9.894	71,500	707,400	69,000	682,700	68,000	672,800		
B. Energy-Nat. Gas (heat)	2%	10.819	15,000	162,300	13,000	140,600	13,500	146,100		
C. Maint.-Custodial	0%	9.077	35,000	317,700	30,000	272,300	33,000	299,500		
D. Maint.-Architectural	0%	9.077	11,000	99,800	10,500	95,300	11,000	99,800		
E. Maint.-Mechanical	0%	9.077	15,400	139,800	14,500	131,600	15,000	136,200		
F. Maint.-Electrical	0%	9.077	8,800	79,900	8,800	79,900	8,800	79,900		
G. Maint.-Site	0%	9.077	4,400	39,900	4,200	38,100	4,400	39,900		
Total Annual Costs				1,546,800		1,440,500		1,474,200		
LCC: Total Present-Worth Life-Cycle Costs				4,484,800		4,223,700		4,363,700		
LCC: Life-Cycle Present-Worth Dollar Savings				—		261,100		121,100		

PW – Present Worth PWA – Present Worth Of Annuity (1) Note: Staffing, Insurance, Taxes, etc. are the same for all schemes.

Figure 4.3: **PW Worksheet–Data Processing Center Layout**

permit the design professional to quickly perform a variety of "what if?" variations to the numbers in order to see the effects. Documentation and reporting is very easily accomplished as well. Graphic presentation of the results can also be incorporated in the spreadsheet to visually summarize

Life Cycle Cost Analysis
Using Annualized Costs

Item: Dental Clinic Feasibility Study   Date: _____

		Original	Alt. No. 1	Alt. No. 2
Input Data — Collateral & Instant Contract Costs	**Initial Costs**		Lease	Build
	Base Cost $/GSF			120.00
	Interface Costs			
	a. _____			
	b. _____			
	c. _____			
	Other Initial Costs			
	a. _____			
	b. _____			
	c. _____			
	Total Initial Cost Impact (IC)		0	120.00
	Initial Cost Savings			
Input Data — Salvage & Replacement Costs	Single Expenditures @ 12% Interest			
	Present Worth Carpet			
	1. Year 8 _____ Amount			3.00
	PW = Amount x (PW Factor 0.4039) =			1.21
	2. Year _____ Amount			
	Amount x (PW Factor _____) =			
	3. Year _____ Amount			
	Amount x (PW Factor _____) =			
	4. Year _____ Amount			
	Amount x (PW Factor _____) =			
	5. Year 15 _____ Amount			(65.00)
	Amount x (PW Factor 0.1827) =			(11.88)
	Salvage Amount x (PW Factor _____) =			
Output — Life Cycle Costs (Annualized)	**Annual Owning & Operating Costs**			
	1. Capital IC x (PP 0.1468) =		0	17.62
	Recovery 15 Years @ 12 %			
	Replacement Cost: PP x PW			
	a. Year 8 $1.21 x 0.1468		0	0.18
	b. Year _____			
	c. Year _____			
	d. Year _____			
	e. Year _____			
	Salvage: 15 ($11.88) x 0.1468		0	(1.74)
	2. Annual Costs			
	a. Maintenance			4.00
	b. Operations (Energy)			2.50
	c. Leasing Cost		27.00	0
	d. _____			
	e. _____			
	3. Total Annual Costs		27.00	22.56
	Annual Difference (AD)			
	4. Present Worth of Annual Difference			
	(PWA Factor _____) x AD			

PP · Periodic Payment to pay off loan of $1.
PWA · Present Worth of Annuity (What $1 payable periodically is worth today).
PW · Present Worth (What $1 due in future is worth today).

☐ Future Costs
☐ Present Costs

Figure 4.4: Annualized Worksheet—Dental Clinic Feasibility Study

the results. Add-on software packages permit probabilistic risk analysis to be performed quite easily on the spreadsheet.

PW and Annualized Spreadsheet—Nursing Tower

The spreadsheet shown as Figure 4.5 combines both the PW worksheet and the annualized worksheet shown earlier. It is organized in the three primary sections:

Section 1   Initial Project Cost

Section 2   Replacement Cost/Salvage Value

Section 3   Annual Costs

The bottom portion of the spreadsheet contains the total life cycle costs expressed in present worth and as an annualized cost. The discounted payback period is also calculated for Alternatives 2, 3, and 4 compared to Alternative 1.

LIFE CYCLE COST ANALYSIS (LCCA)
Project/Location: City Hospital, Detroit, Michigan

Subject: Nursing Towers Study
Description: Investment Analysis
Project Life Cycle = 25 Years
Discount Rate = 10.00%
Present Time = Date of Occupancy

				Alternative 1 Remain in Existing Nursing Towers GSF = 200,000		Alternative 2 Renovate Existing Nursing Towers GSF = 200,000		Alternative 3 Build New Nursing Towers GSF = 180,000		Alternative 4	
INITIAL COSTS	Quantity UM		Unit Price	Est.	PW	Est.	PW	Est.	PW	Est.	PW
Construction Costs											
A. Existing Nursing Towers	200,000 GSF		$0.00	0	0		0		0		0
B. Renovate Nursing Towers	200,000 GSF		$50.00		0	10,000,000	10,000,000		0		0
C. Build New Nursing Towers	180,000 GSF		$120.00		0		0	21,600,000	21,600,000		0
D. Demolish Existing Towers	200,000 GSF		$5.00		0		0	1,000,000	1,000,000		0
E.					0		0		0		0
F.					0		0		0		0
G.					0		0		0		0
Total Initial Cost					0		10,000,000		22,600,000		0
Initial Cost PW Savings (Compared to Alt. 1)							(10,000,000)		(22,600,000)		0
REPLACEMENT COST/ SALVAGE VALUE											
Description		Year	PW Factor								
A. Air Handling Units		5	0.6209	2,000,000	1,241,842		0		0		0
B. Re-roof Nursing Towers		10	0.3855	250,000	96,385	250,000	96,385		0		0
C.			1.0000		0		0		0		0
D.			1.0000		0		0		0		0
E.			1.0000		0		0		0		0
F. Salvage Value(New @ 50%)		25	0.0923	0	0	0	0	(10,800,000)	(996,796)		0
Total Replacement/Salvage Costs					1,338,227		96,385		(996,796)		0
ANNUAL COSTS											
Description		Escl. %	PWA								
A. Energy Annual Cost		0.000%	9.077	850,000	7,715,484	750,000	6,807,780	650,000	5,900,076		0
B. Maintenance & Repair Cost		0.000%	9.077	1,250,000	11,346,300	1,100,000	9,984,744	800,000	7,261,632		0
C. Nurse Staffing Annual Cost		5.000%	14.437	7,776,000	112,258,324	6,480,000	93,548,603	5,180,000	74,781,136		0
D.		0.000%	9.077		0		0		0		0
E.		0.000%	9.077		0		0		0		0
F.		0.000%	9.077		0		0		0		0
Total Annual Costs (Present Worth)					131,320,108		110,341,127		87,942,844		0
Total Life Cycle Costs (Present Worth)					132,658,335		120,437,512		109,546,048		0
Life Cycle Savings (Compared to Alt. 1)							12,220,823		23,112,286		0
Discounted Payback (Compared to Alt. 1)			PP Factor				4.08 Years		4.49 Years		0.00 Years
Total Life Cycle Costs (Annualized)			0.1102	14,614,713	Per Year	13,268,369	Per Year	12,068,477	Per Year	0	Per Year

Figure 4.5: **PW and Annualized Spreadsheet—Nursing Towers**

A nursing tower example is presented to help explain the spreadsheet application. A Midwestern hospital association is interested in upgrading its nursing tower facilities because of technological obsolescence, rising energy and maintenance costs, and increasing staffing costs. There are three alternatives to be considered: remain (do nothing), renovate the existing facilities, or build new nursing towers. Following are the estimated costs of these three alternatives:

Alternative 1—Remain in existing nursing towers	Cost, $
Replace air-handling units within 5 years	2,000,000
Re-roof nursing towers—10 years	250,000
Nursing tower value assumed in 25 years (salvage)	0%
Annual energy consumption	850,000
Maintenance/repair costs per year	1,250,000
Nursing staff annual costs	7,776,000

Alternative 2—Renovate nursing towers *[200,000 gross square feet (GSF)]*	Cost, $
Renovation construction cost	50/GSF
Re-roof nursing towers in 10 years	250,000
Nursing tower value assumed in 25 years (salvage)	0%
Estimated annual energy costs	750,000
Maintenance/repair costs per year	1,100,000
Nursing staff annual costs	6,480,000

Alternative 3—Build new nursing towers *(180,000 GSF)*	Cost, $
New construction cost	120/GSF
Demolish existing nursing towers	5/GSF
Nursing tower value assumed in 25 years (salvage)	50%
Estimated annual energy costs	650,000
Maintenance/repair costs per year	800,000
Nursing staff annual costs	5,180,000

The hospital administrator has requested the architect/engineer to perform a life cycle cost analysis of these three situations using the present worth method of analysis. The economic criteria for the analysis are as follows:

Project life cycle	25 years
Discount rate	10% compounded annually
Inflation approach	Constant dollars
Present time	Date of occupancy

Differential escalation rates are:

Energy	0% per year
Maintenance/repairs	0% per year
Staffing	5% per year

What is the life cycle cost of each of these alternatives? What should the hospital administrator do?

The spreadsheet in Figure 4.5 presents the results. From this analysis, it can be seen that Alternative 3—Build New Nursing Towers—will result in the lowest life cycle cost, although it has the highest initial cost investment. The reduced staffing annual costs, as compared to the present situation, are the primary reason for this savings. The discounted payback period is 4.49 years. Figure 4.6 is a graphic presentation of the life cycle cost analysis shown in Figure 4.5, generated using the same data from the computer spreadsheet.

More Complex LCC Analyses

The authors almost exclusively use the Present Worth method. It is easier to use various escalation factors and simpler formats. Figure 4.7 is the latest version of a worksheet for handling more complex LCC analyses. The spreadsheet allows user input for Replacement/Salvage Costs either at one time in the future, or in cyclical times. For example, an LCCA might be performed for a new car purchase. Under Replacement/Salvage, one might estimate that tires will be rotated every 1.5 years, and replaced every 3 years. The life cycle is 9 years. In lieu of entering 1.5 years 5 times for rotations, and entering 3 years twice for replacement, one needs only to enter 1.5 years for rotation and 3 years for replacement in the cycle column under Occurrence in the Replacement/Salvage Costs section. The spreadsheet will perform the necessary calculations. For spreadsheet and examples, see the book's Web site: **www.rsmeans.com/supplement/67341.asp**

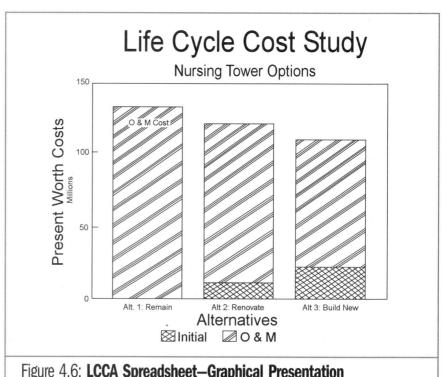

Figure 4.6: **LCCA Spreadsheet—Graphical Presentation**

Conclusion

While the present worth method allows all costs to be converted to equivalent costs at one point in time, the annualized method expresses all life cycle costs as annual expenditures. The approach you choose will determine the yardstick used to compare the total costs of ownership of the various design alternatives.

Life Cycle Costing - General Purpose Worksheet				*Original Design*		*Option 1*	
Study Title:							
Discount Rate: Date: 18 mar.03				Estimated Costs	Present Worth	Estimated Costs	Present Worth
Life Cycle (Yrs.)							
INITIAL / COLLATERAL COSTS	Initial/Collateral Costs						
	A. Stucco						
	B.						
	C.						
	D.						
	E.						
	F.						
	G.						
	H.						
	I.						
	J.						
	Total Initial/Collateral Costs						
	Difference						
REPLACEMENT / SALVAGE COSTS	Replacement/Salvage (Single Expenditures)	Occurance Year -or- Cycle	Inflation/ Escal. Rate	PW Factor			
	A.						
	B.						
	C.						
	D.						
	E.						
	F.						
	G.						
	H.						
	I.						
	J.						
	Total Replacement/Salvage Costs						
ANNUAL COSTS	Annual Costs		Inflation/ Escal. Rate	PW Factor			
	A.						
	B.						
	C.						
	D.						
	E.						
	F.						
	G.						
	H.						
	I.						
	J.						
	Total Annual Costs						
	Sub-Total Replacement/Salvage + Annual Costs (Present Worth)						
	Difference						
LIFE CYCLE COSTS	Total Life Cycle Costs (Present Worth)						
	Life Cycle Cost PW Difference						
	Payback - Simple Discounted (Added Cost / Annualized Savings)						N/A
	Payback - Fully Discounted (Added Cost+Interest / Annualized Savings)						N/A
	Total Life Cycle Costs - Annualized				Per Year:		Per Year:

Figure 4.7: **Life Cycle Costing Worksheet**

Chapter 5: Estimating Life Cycle Costs

Of the input data required for LCCA, specific project information, initial costs, and specific site information are usually available. It is unusual, however, for detailed facility components' dates to be available—especially data regarding useful life, maintenance and operations. Although such input is needed to calculate roughly 25% of the total costs, few designers have access to comprehensive information in a format that facilitates LCCA.

Alphonse J. Dell'Isola
Value Engineering: Practical Applications, 1997

Estimating life cycle costs requires data and, at times, educated guesswork. *Analyzing* life cycle costs places a greater emphasis on procedures and technical skills and is intended to require less guesswork. Project decisions are made throughout the design process in an expanding fashion. On the other hand, accurate costs must be generated from a detailed cost base. This concept is illustrated in Figure 5.1. The level of decision-making and the degree of information available determine which costing method is most appropriate. Systems costs, based on historical project data and predefined subsystems of usual types of construction, provide the link between detailed costs and early-stage estimates.

The clustering of cost items into systems becomes even more significant when estimates are prepared at the budget and early design stages. Systems costing allocates funds to the various functional elements of a facility and allows the designer to make early cost comparisons among alternatives. When maintenance and operation data are provided, a total life cycle cost analysis is possible. A clearly defined accounting framework is required to adequately track the life cycle and costs from project inception through construction and occupancy.

This chapter outlines techniques that simplify economic assessment of design alternatives. The discussion includes estimating procedures for initial project costs; energy, maintenance, repair, and custodial costs; alteration and replacement costs; and associated costs (staffing, denial of use, etc.). Figure 5.2 illustrates the range of these costs for a variety of facility types.

Estimating Framework

A standard accounting system is one of the fundamental components of any cost control and estimating approach. This framework serves several purposes:

1. It allows cost data from different stages in project development to be related uniformly and consistently.
2. It provides a frame of reference within which life cycle cost data may be collected.
3. It provides a checklist for project estimating and for referencing specifications.
4. It facilitates communication among all members of a project team.
5. It allows designers, owners, and value engineers to quickly identify and differentiate high-cost and low-value study areas.

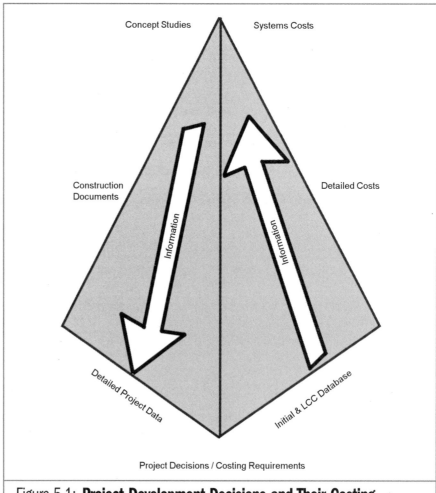

Figure 5.1: **Project Development Decisions and Their Costing Requirements**

The need for a consistent framework becomes even more important when the system is fully automated.

For many designers, the standard estimating framework is the 16-division Construction Specifications Institute (CSI) MasterFormat. This format is now widely used as a data communication medium in the building industry. Specifications, product data, and project management systems are commonly structured around it. Because it resembles the form by which building projects are procured (that is, the familiar subtrade or contract package), the MasterFormat classification system is also used as a cost control and estimating framework. While this format may be suitable for use at the final design stages, and during construction, it does not respond adequately for use during project budgeting and early design. Because the CSI system is heavily product- and materials-oriented, it does not relate well to decisions made at the early design stages, such as whether to build two or three stories, or whether to use a masonry or metal panel exterior wall.

For some years, managers involved with design-phase construction cost control have realized the inadequacy of the CSI format for their purpose and have moved to defining building systems based on an elemental

Facility Types—Cost Per Building Gross Square Foot

	Corp. Office $/ GSF		Financial $/ GSF		Medical $/ GSF		University $/ GSF		Research $/ GSF		Industrial $/ GSF	
INITIAL COSTS:	Low	High	Low	High	Low	High	Low	High	Low	High	Low	High
Initial Project Cost	**165.48**	**301.92**	**191.61**	**336.76**	**314.19**	**631.51**	**193.04**	**348.37**	**140.00**	**645.86**	**140.00**	**259.65**
Construction Cost (incl. Site)	123.95	169.62	143.53	189.19	208.76	287.05	143.53	195.72	169.62	293.57	91.33	130.48
Design Fees	6.20	13.57	7.18	15.14	16.70	28.71	8.61	15.66	11.87	29.36	3.20	6.52
Construction Administration	2.48	6.78	2.87	7.57	4.18	11.48	2.87	7.83	3.39	11.74	1.37	3.91
Site	6.20	25.44	7.18	28.38	10.44	43.06	7.18	29.36	8.48	44.04	2.74	13.05
Reservation Costs:												
Const. Contingency	5.58	10.18	6.46	11.35	9.39	17.22	6.46	11.74	7.63	17.61	4.11	7.83
Furnishings/ Equip.	12.40	33.92	14.35	37.84	41.75	143.53	14.35	39.14	33.92	146.79	9.13	65.24
Interium Financing	7.44	25.44	8.61	28.38	12.53	43.06	8.61	29.36	10.18	44.04	5.48	19.57
Other	1.24	16.96	1.44	18.92	10.44	57.41	1.44	19.57	8.48	58.71	0.91	13.05
ANNUAL COSTS:	$/ GSF/Year		$/ GSF/Year		$/ GSF/Year		$/ GSF/Year		$/ GSF/Year		$/ GSF/Year	
Energy/ Fuel Costs	**1.93**	**3.59**	**2.04**	**3.35**	**2.69**	**4.27**	**2.00**	**3.29**	**2.24**	**3.56**	**2.28**	**6.52**
Maintenance, Repair and Custodial	**2.92**	**6.83**	**2.54**	**5.12**	**3.46**	**7.21**	**2.17**	**4.54**	**3.13**	**6.90**	**2.41**	**5.41**
Cleaning (Custodial)	1.15	2.24	1.04	1.93	1.39	3.35	0.91	1.70	1.30	3.12	0.78	1.83
Repairs & Maintenance	1.40	3.46	1.25	2.48	1.57	2.84	1.04	2.22	1.46	2.65	1.37	2.61
Roads & Grounds Maintenance	0.38	1.13	0.24	0.71	0.50	1.02	0.21	0.63	0.38	1.13	0.26	0.98
Alterations and Replacements	**3.72**	**8.48**	**4.31**	**9.46**	**6.26**	**25.83**	**4.31**	**9.79**	**5.09**	**26.42**	**2.74**	**11.74**
Alterations	1.24	3.39	1.44	3.78	2.09	14.35	1.44	3.91	1.70	14.68	0.91	6.52
Replacements	2.48	5.09	2.87	5.68	4.18	11.48	2.87	5.87	3.39	11.74	1.83	5.22
Associated Costs	**117.25**	**211.83**	**120.30**	**217.64**	**166.67**	**391.55**	**44.58**	**99.85**	**149.70**	**492.09**	**56.34**	**240.86**
Administrative (Bldg Mgt)	0.57	1.36	0.48	1.20	0.63	1.28	0.39	0.91	0.59	1.44	0.52	1.17
Interest (Debt Service)	15.11	39.39	17.50	43.94	28.70	82.40	17.63	45.45	12.79	84.27	12.79	33.88
Staffing (Functional Use)	97.86	163.10	97.86	163.10	130.48	293.57	26.10	52.19	130.48	391.43	39.14	195.72
Denial-of-Use Costs	(Lost Income)		(Lost Income)		(Lost Income)		(Lost Income)		(Lost Income)		(Lost Income)	
Other Costs:												
Security	0.09	0.28	0.30	0.89	0.25	0.51	0.05	0.16	0.30	0.89	0.26	0.91
Real Estate Taxes	3.23	6.64	3.75	7.41	5.45	11.24	N/A	N/A	4.43	11.49	2.38	5.11
Water & Sewer	0.21	0.40	0.23	0.37	0.90	1.42	0.22	0.37	0.90	1.42	1.12	3.56
Fire Insurance	0.16	0.66	0.19	0.74	0.27	1.12	0.19	0.77	0.22	1.15	0.12	0.51

Figure 5.2: **Life Cycle Cost Range for a Variety of Facility Types**

format. This approach involves the separation of a building into its component or functional *parts*, *elements*, or *subsystems* (terminology varies widely), in an attempt to relate these more closely to the language of design. As a consequence, several forms have proliferated, including UNIFORMAT, the most-used elemental system, and those developed by *Progressive Architecture* and *Engineering News-Record* magazines, the State University Construction Fund (New York), and the U.S. military

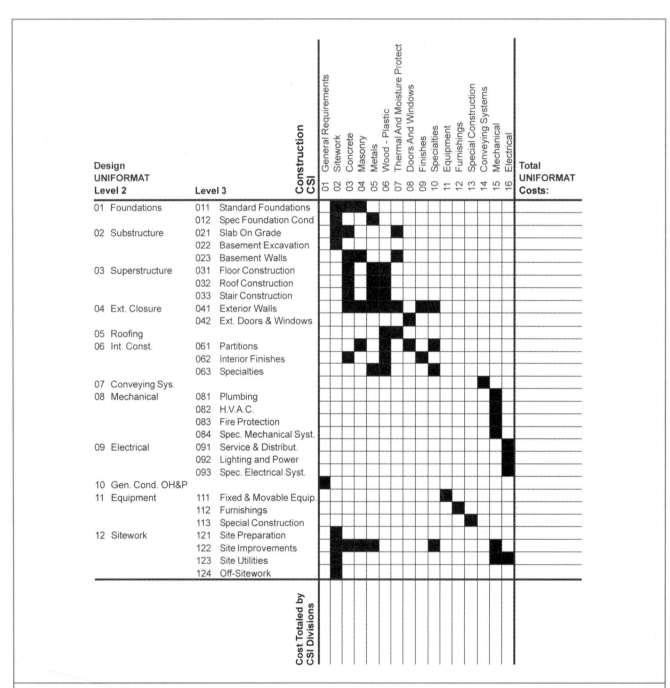

Figure 5.3: **Relationship Between UNIFORMAT and the Construction Specifications Institute**

(Corps of Engineers and Navy Facilities Engineering Command). In addition, many consultants have developed their own formats.

Responding to the need for a more design-stage-friendly classification system, the U.S. General Services Administration (GSA) in 1973 issued a draft outline of the Uniform Building Component Format (UNIFORMAT). Concurrently, the American Institute of Architects (AIA) began development of the MASTERCOST system, intended to be a national building cost databank.

Because the two formats were similar, GSA and AIA merged the UNIFORMAT and MASTERCOST classification systems, and it is now known as the GSA UNIFORMAT (also referred to as UNIFORMAT I.). The various levels of detail and their relationship with CSI divisions are presented in Figure 5.3. Figure 5.4b provides a further breakdown of the GSA UNIFORMAT categories. In October 1999, the National Academy of Standards and Technology (NIST) of the U.S. Department of Commerce issued an updated version, called "UNIFORMAT II Elemental Classification for Building Specification, Cost Estimating, and Cost Analysis NISTIR 6389." Figure 5.4a shows a breakdown of its categories. The use of either format will assist early project estimating and control.

Initial Costs

Initial project costs are the owner's costs associated with the initial development of a facility and project support (fees, land, furnishings, etc.), as well as construction costs. These costs, itemized in Figure 5.5, are often referred to as *project costs*. Cost items associated with the initial development of the facility may include design, consulting, legal, and other professional fees; construction costs, including all furnishings; equipment; land costs; and construction-phase financing. As debt incurred to finance the initial costs is amortized, the amortization payments represent the financing costs considered as an annual charge.

Decisions based on the initial investment cost of a facility are familiar to all designers. The percentages of such costs are illustrated for a typical office

	UNIFORMAT II	
	A	Substructure
	B	Shell
	C	Interiors
	D	Services
	E	Equipment & Furnishings
	F	Special Construction & Demolition
	G	Building Sitework

Figure 5.4a: **UNIFORMAT II Categories**

LEVEL 2	LEVEL 3	LEVEL 4
01 Foundations	011 Standard Foundations	0111 Wall Foundations
		0112 Col. Foundations & Pile Caps
	012 Spec. Foundation Cond.	0121 Pile Foundations
		0122 Caissons
		0123 Underpinning
		0124 Dewatering
		0125 Raft Foundations
		0126 Other Spec Foundation Cond.
02 Substructure	021 Slab on Grade	0211 Standard Slab on Grade
		0212 Structural Slab on Grade
		0213 Inclined Slab on Grade
		0214 Trenches, Pits & Bases
		0215 Foundation Drainage
	022 Basement Excavation	0221 Excavation for Basements
		0222 Structure Fill & Compact
		0223 Shoring
	023 Basement Walls	0231 Basement Wall Construction
		0232 Moisture Protection
		0233 Basement Wall Insulation
03 Superstructure	031 Floor Construction	0311 Susp. Basement Floor Construction
		0312 Upper Floor Construction
		0313 Balcony Construction
		0314 Ramps
		0315 Special Floor Construction
	032 Roof Construction	0321 Flat Roof Construction
		0322 Pitched Roof Construction
		0323 Canopies
		0324 Special Roof Systems
	033 Stair Construction	0331 Stair Structure
04 Ext. Closure	041 Exterior Walls	0411 Exterior Wall Construction
		0412 Exterior Louvers and Screens
		0413 Sun Control Devices (Ext)
		0414 Balcony Walls and Handrails
		0415 Exterior Soffits
	042 Ext. Doors & Windows	0421 Windows
		0422 Curtains Walls
		0423 Exterior Doors
		0424 Storefronts
05 Roofing		0501 Roof Coverings
		0502 Traf. Topng & Paving Membr
		0503 Roof Insulation & Fill
		0504 Flashing & Trim
		0505 Roof Openings
06 Int. Construction	061 Partitions	0611 Fixed Partitions
		0612 Demountable Partitions
		0613 Retractable Partitions
		0614 Compartments & Cubicles
		0615 Int. Balustrades & Screens
		0616 Interior Doors & Frames
		0617 Interior Storefronts
	062 Interior Finishes	0621 Wall Finishes
		0622 Floor
		0623 Ceiling Finishes
	063 Specialties	0631 General Specialties
		0632 Built-in Fittings
07 Conveying Systems		0701 Elevators
		0702 Moving Stair & Walks
		0703 Dumbwaiters
		0704 Pneumatic Tube Systems
		0705 Other Conveying Systems
		0706 General Construction Items
08 Mechanical	081 Plumbing	0811 Domestic Water Supply System
		0812 Sanitary Waste & Vent System
		0813 Rainwater Drainage System
		0814 Plumbing Fixture

Figure 5.4b: **UNIFORMAT Level 4, Cost Accounting System** (continued)

LEVEL 2	LEVEL 3	LEVEL 4
	082 HVAC	0821 Energy Supply
		0822 Heat Generating System
		0823 Cooling Generating System
		0824 Distribution Systems
		0825 Terminal and Package Units
		0826 Controls and Instrumentation
		0827 Systems Testing & Balancing
	083 Fire Protection	0831 Water Supply (Fire Protect)
		0832 Sprinklers
		0833 Standpipe Systems
		0834 Fire Extinguishers
	084 Spec. Mechanical Syst.	0841 Special Plumbing Systems
		0842 Spec. Fire Protection System
		0843 Misc. Spec. System and Devices
		0844 Gen. Const. Items (Mech.)
09 Electrical	091 Service & Distribution	0911 High Tension Service & Dist.
		0912 Low Tension Service & Dist.
	092 Lighting and Power	0921 Branch Wiring
		0922 Lighting Equipment
	093 Spec. Electrical System	0931 Communications & Alarm System
		0932 Grounding Systems
		0933 Emergency Light & Power
		0934 Electric Heating
		0935 Floor Raceway System
		0936 Other Spec. System & Devices
		0937 General Construction Items
10 Gen Cond OH&P		1001 Mobilization & Initial Expenses
		1002 Site Overheads
		1003 Demobilization
		1004 Main Off. Expense & Profit
11 Equipment	111 Fixed & Movable Equip.	1111 Built-in Maintenance Equipment
		1112 Checkroom Equipment
		1113 Food Service Equipment
		1114 Vending Equipment
		1115 Waste Handling Equipment
		1116 Loading Dock Equipment
		1117 Parking Equipment
		1118 Detention Equipment
		1119 Postal Equipment
		1120 Other Specialized Equipment
	112 Furnishings	1121 Artwork
		1122 Window Treatment
		1123 Seating
		1124 Furniture
		1125 Rugs, Mats & Furn. Accces
	113 Special Construction	1131 Vaults
		1132 Interior Swimming Pools
		1133 Modular Prefab Assemblies
		1134 Special Purpose Rooms
		1135 Other Special Construction
12 Site Work	121 Site Preparation	1211 Clearing
		1212 Demolition
		1213 Site Earthwork
	122 Site Utilities	1221 Parking Lots
		1222 Roads, Walks, Terraces
		1223 Site Development
		1224 Landscaping
	123 Site Utilities	1231 Water Supply & Dist. Systems
		1232 Drainage & Sewage Systems
		1233 Heating & Cooling Dist System
		1234 Elec. Dist. & Lighting System
		1235 Snow Melting Systems
		1236 Service Tunnels
	124 Off-Site Work	1241 Railroad Work
		1242 Marine Work
		1243 Tunneling
		1244 Other Off-Site Work

Figure 5.4b: **UNIFORMAT Level 4, Cost Accounting System**

building in Figure 5.6. The following brief examples illustrate the relationship of first cost to future follow-on costs:

- The selection of cooler, but more expensive light fixtures that will reduce the size and energy requirement of the proposed heating, ventilating, and air-conditioning (HVAC) system.
- The purchase of additional land to permit a single-story facility, thus eliminating all first and future follow-on costs associated with a multi-story building, such as elevators, stairs, and supported floors.

Budget Phase

Budget and conceptual estimates may be prepared using single-unit costs based on broad-accommodation parameters (for example, cost per student or per hospital bed) and cost per square foot of gross floor area or cubic foot of building volume. See Figure 5.7 for the AIA standard method of measurement for gross square feet. Efficiency ratios and conversion factors for transforming net program space to gross square feet (GSF), together with historical costs per square foot, are also useful in preparing estimates.

This information can be expanded to define costs per square foot of functional activities programmed for each space—for example, cost per square foot of wet laboratories, versus space used for offices and secretarial spaces. These costs are also tempered by the basic criteria and design parameters of the building systems and components to be selected. Costs

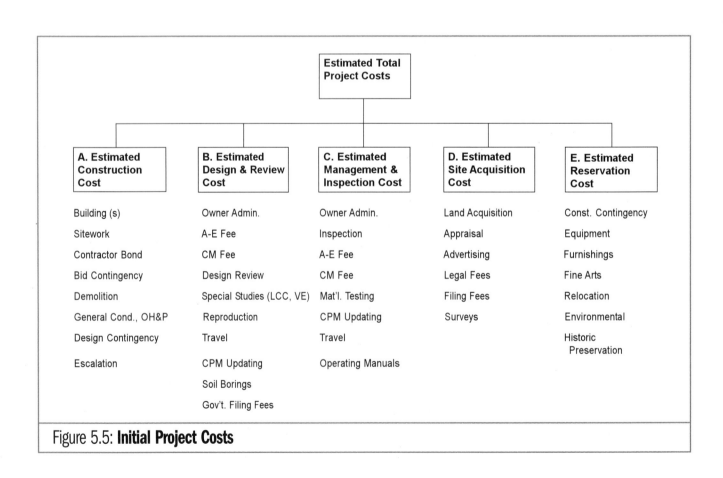

Figure 5.5: **Initial Project Costs**

are then organized into a systems format, such as GSA UNIFORMAT or UNIFORMAT II. A second method of estimating the budget is called *parameter estimating*. The data can be derived from contractor-reported actual costs.

Parameter costs can be useful in developing feasibility studies, calculating preliminary budgets, aiding design decisions, assisting in value engineering, and determining and checking bids. Given an approximate building

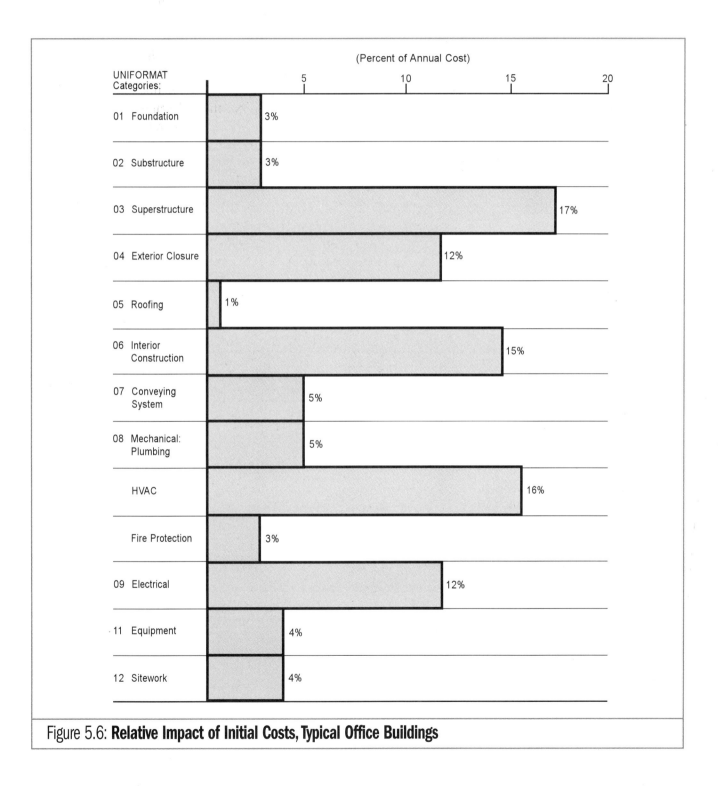

Figure 5.6: **Relative Impact of Initial Costs, Typical Office Buildings**

configuration, designers use building parameters to prepare a preliminary estimate for a building. In addition, costs of alternatives may be derived with the aid of parameters.

For example, to estimate by the parameter method, the total cost of a specific work item, such as foundation construction, is divided by a physical parameter (measure), such as the basement area. The resulting unit cost is used to estimate the price of a similar work item in another building. Ideally, the other building should be in the same geographical area, though this is not required. Figure 5.8 shows the budget cost estimate in GSA UNIFORMAT of a research laboratory prepared from historical parametric cost data. Figure 5.9 is the costs presented as a cost model. To show the differences between the two UNIFORMAT systems, Figure 5.8

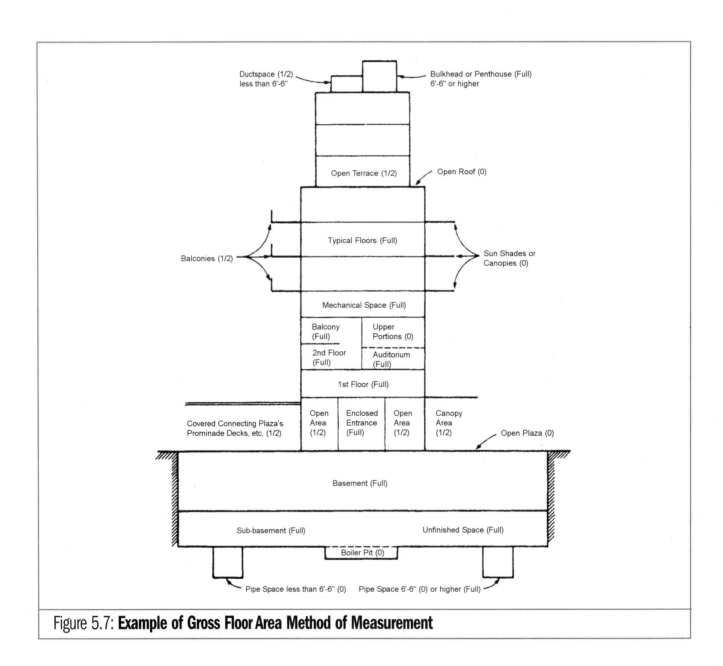

Figure 5.7: **Example of Gross Floor Area Method of Measurement**

illustrates a recent cost model of a hospital in UNIFORMAT II. Also see the cost model on this book's Web site: **www.rsmeans.com/supplement/ 67341.asp**

Design Phase

In the early design phase, it is normal to estimate costs based on the components for each building subsystem. For example, the cost comparison between steel and concrete frames (in terms of the project's scope and characteristics) can be determined from previous experience on

CONSTRUCTION SYSTEMS COST SUMMARY

Project Name: **University Teaching Laboratory**
Project Location: **Virginia**

Bldg.Constr.Type: Steel Frame Structure   GSF: 74,775   VOL(CUFT): 1,001,985
Bldg. Type 732   Bldg. Cost Index - Current: 1437.00   Orig: 1437.00   City Cost Index: 0.853

SYSTEM	SUBSYSTEM	QUANTITY	UNIT	UNIT COST	SYS.COST	COST/GSF
01 Foundation	011 Standard Foundations	19,305	FPA	5.39	$103,991	$1.39
	012 Special Foundations	0	FPA	0.00	$0	$0.00
02 Substructure	021 Slab on Grade	9,092	SQFT	5.33	$48,492	$0.65
	022 Basement Excavation	57,600	BCF	0.17	$9,553	$0.13
	023 Basement Walls	3,770	BWA	25.66	$96,752	$1.29
03 Superstructure	031 Floor Construction	70,381	UFA	17.03	$1,198,577	$16.03
	032 Roof Construction	17,847	SF	8.17	$145,783	$1.95
	033 Stair construction	16	FLT	5,769.09	$92,305	$1.23
04 Ext. Closure	041 Exterior Walls	39,226	XWA	30.68	$1,203,319	$16.09
	042 Ext Doors & Windows	5,878	XDA	41.68	$244,976	$3.28
05 Roofing	050 Roofing	15,311	SF	10.43	$159,757	$2.14
06 Interior Const	061 Partitions	77,087	PSF	4.52	$348,121	$4.66
	062 Interior Finishes	286,965	TFA	2.48	$711,179	$9.51
	063 Specialties	74,775	SF	2.01	$150,452	$2.01
07 Conveying Sys	071 Elevators No. 2	9	LO	15,683.37	$141,150	$1.89
	072 Other	0	LS	0.00	$0	$0.00
08 Mechanical	081 Plumbing	260	FXT	3,084.58	$801,990	$10.73
	082 HVAC	340	TON	12,294.25	$4,180,045	$55.90
	083 Fire Protection	74,775	AP	1.93	$144,358	$1.93
	084 Spec Mech Systems	74,775	SF	0.00	$0	$0.00
09 Electrical	091 Service & Distribution	979	KW	629.14	$615,929	$8.24
	092 Lighting & Power	74,775	SF	5.48	$409,550	$5.48
	093 Spec Elec Systems	74,775	SF	4.08	$305,113	$4.08
10 Gen. Cond., OH&P	100 General Conditions	15.00%	PCT	120,047	$1,800,700	$24.08
11 Equipment	111 Fixed Equip.(Casework	1	LS	883,727	$883,727	$11.82
	112 Furnishings	1	LS	0.00	$0	$0.00
	113 Special Construction	1	LS	9,623.89	$9,624	$0.13
12 Sitework	121 Site Prep/Demolition		ACR	0.00		$0.00
	122 Site Improvements		ACR	0.00		$0.00
	123 Site Utilities		ACR	0.00		$0.00
	124 Off Site		LS	0.00		$0.00
Sub Total					$13,805,443	$184.63
Design Contingency			PCT	0.00		$0.00
Sub Total					$13,805,443	$184.63
Escalation	Const. Midpt. 9 Months		PCT	0.00		$0.00
Total Building Estimate:		**1,001,985**	**CUFT**	**13.78**	**$13,805,443**	**$184.63**

Figure 5.8: **Budget Cost Estimate for University Teaching Laboratory**

similar structures, regardless of programmatic requirements. This is true for most of the major subsystems, such as HVAC, electrical distribution, lighting, and plumbing.

As the design progresses, more detailed elemental subsystem cost information is required. This level of cost information allows selection of components and the system specifications, which can later be used for the preparation of more detailed estimates. More detailed cost information allows the designer to control costs by suggesting changes in the project, should the estimates indicate costs in excess of the budget.

As part of the construction documents, it is normal to use composite unit rates for construction components, assemblies, and systems (known as assemblies estimates). Detailed unit rates are often required at this point for pre-bid estimates, final cost checks, and verification of contractor's bid break-downs as a part of contract negotiations.

Sources of Information

Unit prices for various building elements are contained in a number of estimating publications. Costs are reasonably detailed, and the books are

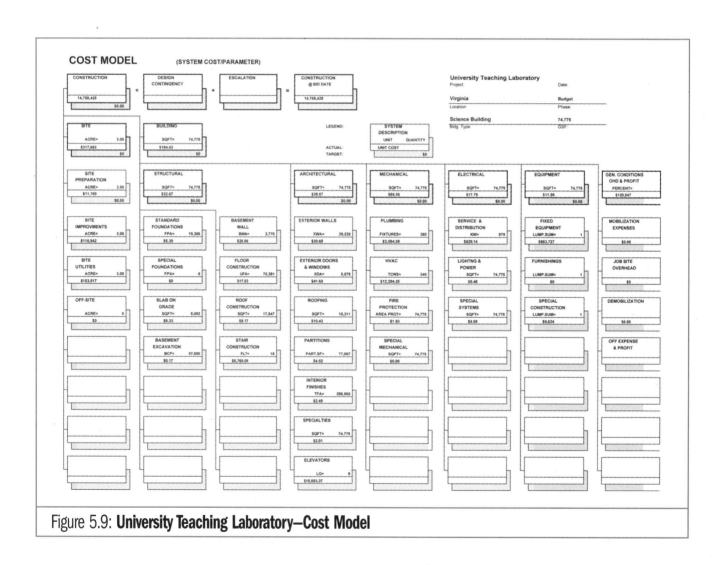

Figure 5.9: **University Teaching Laboratory—Cost Model**

often marketed to owners, construction contractors, and design-builders, as well as to the design professions.

Among the best known is *Building Construction Cost Data*, published by RSMeans of Kingston, Massachusetts. For mechanical and electrical costs, RSMeans publishes *Mechanical Cost Data* and *Electrical Cost Data* in print and electronic (CostWorks) formats. The RSMeans data is assembled in accordance with both the MasterFormat and the UNIFORMAT II categories and includes location modifiers to adjust national average costs to specific locations.

Such volumes can be helpful in preparing estimates once design documentation is developed to the point that an accurate quantity takeoff can be made.

Value appraisal manuals are published mainly for the appraisal profession to assist in the preparation of replacement cost estimates for existing structures. They also achieve a steady sale within the construction industry, as a tool for developing preliminary cost estimates from minimum details. These publications supply regional cost modifiers and some issue residential building valuation manuals.

One is the *Marshall Valuation Service*, published by Marshall and Swift Publication Company of Los Angeles. Another is, *Boeckh Building Valuation Manual* by the American Appraisal Company of Milwaukee, with a bimonthly *Building Cost Modifier*.

The quarterly magazine, *Design Cost Data* is quite useful for preparing budget cost estimates for projects. It presents a variety of case studies with associated costs broken down according to CSI format. The electronic version can be manipulated to adjust for alternative locations and cost escalation.

RSMeans data specifically geared toward early stage estimates includes Means *Square Foot Costs* and *Assemblies Cost Data*. These are very useful guides for making design assumptions and pricing the results.

Another source is *Engineering News-Record*, which for many years has published a parameter cost series, with trade cost analyses of several projects in quarterly cost issues.

Several cost consultants offer computerized estimating systems, some of which are specifically designed to assist architects and owners with early cost estimates. These systems usually contain extensive cost databases, usually marketed as a package with cost estimating services. One of the oldest, by Management Computer Controls (MC^2), of Memphis, Tenn., is a system based on estimating parameters that are essentially the same as the building elements listed in the GSA UNIFORMAT elemental breakdown. RSMeans provides automated estimating in the form of *CostWorks* for 20 specialized databases. All cost estimating, whether manual or automated, relies on historical data and experience in one form or another, the collection and retrieval of which is a costly, but essential process.

Initial Costs—Summary

Most cost estimating of construction projects to date has been trade-oriented. Certainly, the ability to modify designs and to substitute materials and specifications is a key tool in overall cost control. The UNIFORMAT elemental framework provides a more meaningful approach at the early stages.

Designers are called on continually to offer greater cost control services and information, and the degree of liability assumed in so doing cannot be underestimated. The use of elemental estimating techniques can expand capabilities to address problems in this area and provide information for earlier decision-making. The availability of a broad construction cost database and improved control procedures for capital costs and other life cycle costs will have tremendous impact in the coming years.

Energy Costs

Energy consumption currently represents approximately $1.75 to $3.50 per square foot per year of annual life cycle costs of a typical office facility. Energy costs for buildings are normally broken down into the categories of heating, cooling, ventilation, lighting, domestic hot water, and other miscellaneous equipment loads. This section addresses various estimating methods, energy-modeling techniques, and energy savings opportunities.

Figure 5.10 illustrates the relative cost impact of energy on the various GSA UNIFORMAT system categories of a typical office facility. For example, this chart shows that the building's exterior closure influences about 9.5% of the total energy consumed in a facility because of the heat gain and loss through the skin system. Lighting is the most significant influence, contributing as much as 48% from direct lighting energy and the indirect influence of additional heat in the space, which must then be cooled.

Resources to meet human comfort needs include the building itself, the energy used, and the skills of facility managers to operate the building. A shortage in one area may be made up by an abundance in another. Thus, a lack of energy can be compensated for by a more skillful operation or an improved building. Efficient energy use should not and need not be achieved at the expense of comfort.

Efficient energy use incorporates some or all of the following approaches:

- Collecting the natural energy (heat, light, nocturnal cooling, air movement) that flows through a building, and storing it for use when needed (such as storing in the floor mass the abundant solar heat that comes through a south-facing window).
- Conserving collected energy (as by double-glazing a south-facing window).
- Using waste heat from building processes, building equipment, and power-generating equipment.
- Using sustainable sources that function independently of the building design (such as active solar heating and cooling, solar energy from photovoltaics, wind, and geothermal and tidal energies).

- Efficiently converting nonrenewable energy sources into forms required by the building (such as carefully selected furnaces and chillers).

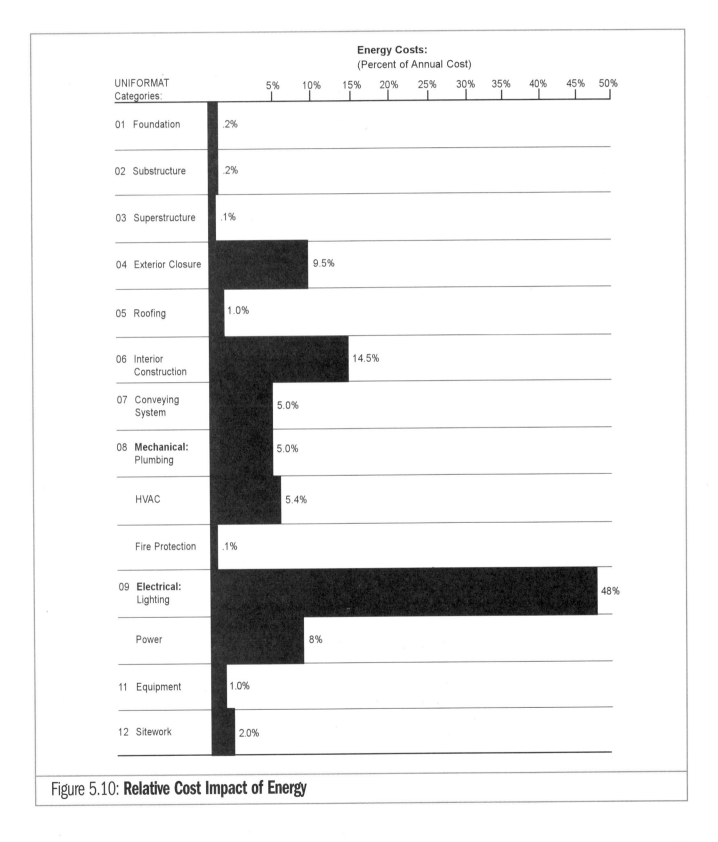

Figure 5.10: **Relative Cost Impact of Energy**

Energy Estimating Methods

The design mechanical engineer must estimate the energy consumption of many types of equipment. Some mechanical equipment always operates at the same energy consumption rate.[1] The energy consumption is found by multiplying the consumption rate by the hours of operation. Other mechanical equipment operates at variable rates. At least four energy estimating methods are available:

- Equivalent full-load hours
- Degree days
- Hour by hour
- Outside temperature bins

The *equivalent full-load hour method* calls for judgment, project information, and data from previous projects to estimate the equivalent number of full-load hours of operation per month for each item of equipment. These hours are multiplied by the hourly full-load rate of energy consumption to yield the projected energy usage. The estimate is improved by using average-load efficiency instead of full-load efficiency of the variable-load equipment items. Still, accuracy of the results depends mainly on the user's judgment. This method is best for estimating consumption of items such as elevators and office equipment. In HVAC energy estimating, it is useful for making quick, early-stage estimates.

The *degree-day method* is also common in early concept design work, chiefly for heating season estimates. It establishes a base temperature, usually 65°F, below which the building begins to require heating. Heating requirements are calculated for a 24-hour design-load heating period. The results are divided by the difference between the 65°F base temperature and the outside heating design temperature to obtain the Btu requirements per degree day. This value is multiplied by the number of monthly or annual degree days from weather data. These loads are then divided by the furnace efficiency to estimate the energy consumed to meet the load. This is an empirical method based on statistical samples of large numbers of buildings, with no assurance of accuracy when applied to a specific project.

Most HVAC systems are designed to handle extreme heating and cooling weather conditions which occur less than 5% of the time. The remaining 95% of the time, they operate less efficiently at part loads, with the average load at about 30%. A system that is highly efficient at full load is not necessarily efficient at reduced load. A thorough HVAC energy estimate should be *dynamic,* so it can evaluate system performance over a full range of likely operating conditions. Dynamic analysis methods include the *hour-by-hour* and *bin methods.*

The hour-by-hour method computes the instantaneous building load, residual stored loads, and resulting HVAC system performance separately for each of the year's 8,760 hours. It then adds them to obtain monthly and yearly consumption. Maximum demand (rate of energy use) is found by identifying the hour with the highest use. Building heat storage and temperature swing can also be simulated. Because of the amount of data, handled, hour-by-hour analyses are always done by computer.

A modified hour-by-hour analysis may be performed for one or more typical days of each month. The results for each day are multiplied by the number of assumed typical days to obtain the monthly energy consumption. This averaging, however, reduces the accuracy of the estimate by eliminating extremes of weather conditions and solar and internal loads.

The *outside temperature bin method* (a temperature bin is a range of outside temperatures, usually in 5°F increments) is based on the principle that the load on an HVAC system is related to outside temperature. Like the typical-day per month, hour-by-hour method, the bin method uses an averaging process and is therefore less correct than the true full-year, hour-by-hour method. It does take into account internal loads, energy use created by patterns of operation and occupancy, and varying solar loads. The HVAC energy consumption at a few outside temperatures is calculated, and the consumption at other temperatures and conditions is extrapolated. The bin method includes four steps:

- Plotting the energy profile (a graph of energy consumption vs. outside temperature).
- Finding the number of hours in each outside temperature bin.
- Determining energy consumption for each temperature bin.
- Totaling energy consumption for bins.

A separate peak demand estimate is required if utility demand charges are to be estimated.

Software programs are available from the federal government and commercial sources using the typical-day-per-month method, the hour-by-hour method; and the bin method. The programs vary widely in their ability to provide accurate results for a particular situation. A program may provide an adequate solution for one situation and an inadequate solution for a different type of building or HVAC system. The inherent accuracy of a computer program depends on:

- Method of analysis, with the true hour-by-hour method having the greatest advantage.
- Accuracy of load estimating. Many of the load-estimating methods incorporated into computer programs were developed to conservatively predict maximum design load conditions, and are inaccurate in estimating part-load conditions.
- Accuracy of system response and equipment performance. Many programs are weak and inflexible in this area. There are many types of variable-air-volume systems available, yet a number of computer programs are unable to distinguish one from another.

Several computer software programs are available that simulate a large number of weather conditions and perform a significant number of error-free calculations. However, the logic of each program is frequently hidden, or described in such a manner as to make it difficult for the user to understand, resulting in possible misapplication of the program.

In the case of energy analysis, *accuracy* is a relative term that is virtually impossible to measure. There are always differences in use and occupancy, hours of building systems operations, control-point settings, building and system maintenance conditions, and actual and estimated weather conditions. All of these factors both create and explain differences between actual and estimated energy consumption. Annual energy estimates for the same building, using the popular computer programs, will vary as much as 30%.

The design mechanical engineer should be able to prepare analyses using both manual and computer methods. Simple preliminary analyses can usually be done more quickly and with less expense using a manual method. A manual analysis may be the only accurate alternative for unusual situations or for systems that have not been adequately programmed for computer use. The computer is the best choice for presenting a final thorough analysis, especially if many alternatives are involved.

Energy Performance Standards

To better target energy performance standards for new buildings, the U.S. Department of Housing and Urban Development assigned the former AIA Research Corporation the task of establishing a baseline of current practice. This baseline has been used for the evaluation of alternative systems being considered in new buildings. It was established by estimating the energy consumption of residential and nonresidential buildings from building design data. These Building Energy Performance Standards (BEPS) were once being considered by the Department of Energy as a criterion to be followed throughout the United States.

The energy performance of buildings is a function of their physical characteristics, intended use, and climatic conditions. In order to calculate the annual energy consumption of the designed buildings, it was necessary for the research group to classify the variations in building type and climate across the country, define data requirements, identify a building sample, collect data on building designs, and employ consistent methods to calculate the energy performance from the several buildings.

Because of differences in design and construction practices and data availability, buildings were divided into two major groups for data collection and data analysis. Nonresidential building types included commercial and high-rise multifamily residential construction. Residential building types included single-family homes, low-rise multifamily housing, and mobile homes.

The development of an adequate climatic classification system for data collection depended on isolating the climatic variables affecting energy consumption. Yet it was not possible to judge the variables' effect on energy use without building energy performance data for comparison.

An *a priori* climatic classification system for stratification of the building sample was constructed by the study group. This ensured that the locations chosen for data collection encompassed the range of climatic variation across the country. The climatic stratification was based on combinations of heating degree days and cooling degree hours.

A revised system of heating and cooling degree-day regions (Figure 5.11) for data analysis and the tabulation of survey findings was developed by substituting more current cooling degree days for the cooling degree hours used for sample stratification, as follows:

1. <2,000 CDD; >7,000 HDD
2. <2,000 CDD; 5,500–7,000 HDD
3. <2,000 CDD; 4,000–5,500 HDD
4. <2,000 CDD; 2,000–4,000 HDD
5. <2,000 CDD; 0–2,000 HDD
6. >2,000 CDD; 0–2,000 HDD
7. >2,000 CDD; 2,000–4,000 HDD

CDD: cooling degree day
HDD: heating degree day

The intent of the building survey was to collect data on the current design practices in order to calculate the energy performance of buildings.

Use of design data from the sample buildings was appropriate for the study, since the performance standards would be used at the design, rather than the operational stage. The design data accurately represent the

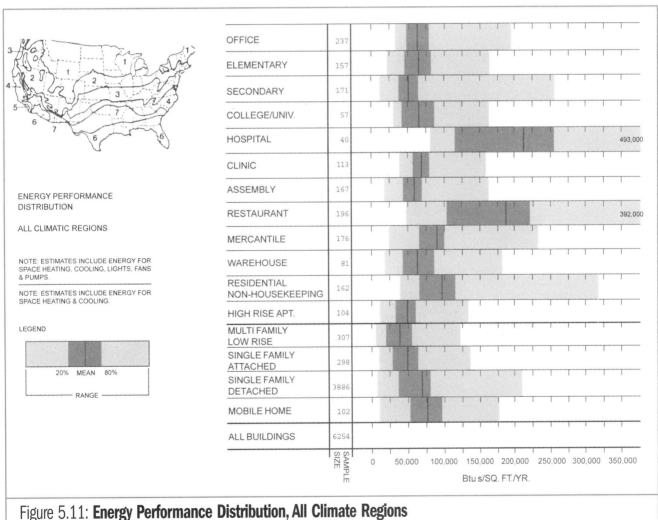

Figure 5.11: **Energy Performance Distribution, All Climate Regions**

variables within the control of the designers of actual buildings that, as designs, have satisfied their clients' functional and economic requirements.

The primary consideration in selecting a data analysis tool was that it produce accurate and reasonably detailed estimates of energy performance from the sample buildings, using data that were readily available and practical to collect from building designers. A modified version of the AXCESS energy analysis computer program was selected as the tool to estimate the energy consumption of the sample buildings. This program met both the technical evaluation criteria and the resource constraints of the project. By assuming standard values for selected variables for each building type and using simplified building modeling techniques, it was possible to limit requirements for the survey to 100–125 data points per building.

Figure 5.11 is a graphic presentation of the national summary statistics for nonresidential and residential performance estimates. The graph summarizes estimates of the designed energy performance of actual buildings throughout the continental United States on which construction took place in a two-year period. They are the result of processing data from 1,661 nonresidential and 4,593 residential data collection forms, representing 1,661 nonresidential and high-rise apartment buildings and 344,529 low-rise residential units.

The estimates for nonresidential and high-rise apartment buildings include energy for space heating and cooling, lights, pumps, and fans. They do not include energy required for hot water, vertical transportation, or equipment such as computers, copy machines, and ovens; or process energy required for manufacturing and production. The estimates for low-rise residential buildings include energy for space heating and cooling, but not for hot water. All of the estimates represent energy use within building boundaries.

Standard operating and occupancy profiles were assumed for each building type in order to provide a consistent basis for comparison. (For example: office buildings were assumed to be occupied nine hours per day and their heating and cooling equipment operated ten hours per day, five days a week.)

Since the sample buildings were designed after the 1973 oil embargo, it is reasonable to expect that energy conservation was a more important criterion in their design than in the generation of buildings immediately preceding them. For instance, a preliminary analysis of the building characteristics of one building type, office buildings, has indicated higher thermal resistance values and lower installed lighting values than for pre-1974 buildings.

Figure 5.12 contains a series of bar charts indicating the relative percentage of each of the four energy end-use components measured by the phase I analysis: lighting, heating, cooling, and fans. These charts help design mechanical engineers to aportion energy budgets for a given region of the country according to energy needs.

Energy Data for Analysis

The specific data and quantities for an energy estimate will vary from project to project. However, the information required may be organized into three types: *building*, *environmental*, and *mode of operation*. The building data include items such as:

- Floor areas, locations, and uses for air-conditioned and heated spaces and non-conditioned spaces

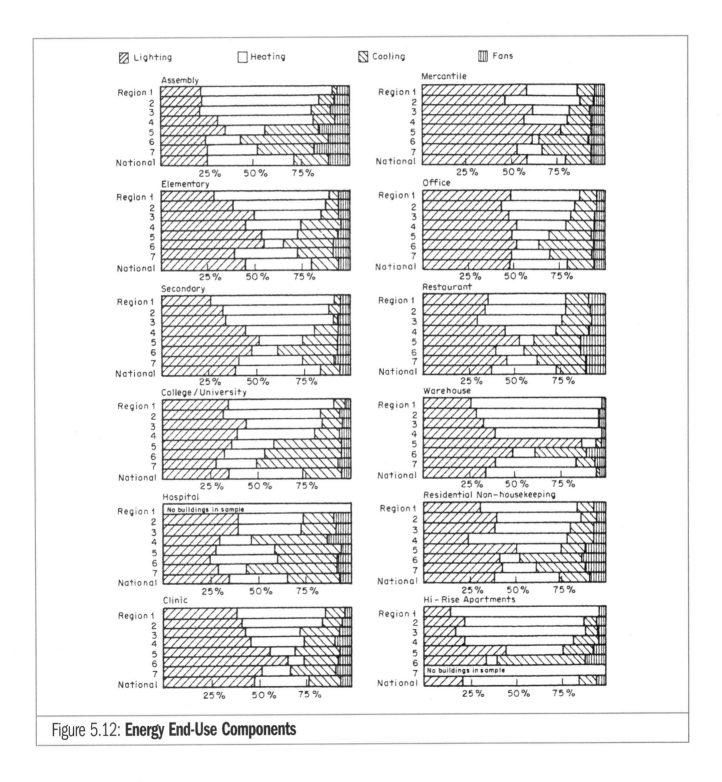

Figure 5.12: **Energy End-Use Components**

- Glass areas, including skylights, including their orientation and shading (note type of glass and operation of windows)
- Wall and roof surface areas, thermal resistance, and weight (in the case of massive buildings)
- Building orientation, configuration, and solar access
- Drawings, equipment schedules, and specifications for mechanical system design and distribution layout, including control diagrams for operation modes
- Electrical drawings and reflected ceiling plans for lighting layout, as well as fixture type and quantity

Environmental or weather data for the project may be obtained from the local office of the National Weather Service, the Environmental Data Service of the National Oceanic and Atmospheric Administration (NOAA), and the U.S. Department of Commerce. Monthly heating and cooling degree-days may also help in a comparison with similar buildings in other areas. Utility costs should be obtained from the utility company to make energy source comparisons. Both energy peak demand and billing demand costs should be determined.

The designer must know how and when each piece of equipment is to be operated. Operational data should be gathered from the client to establish:

- Building operating hours (weekend, weekday)
- System and equipment control schedules
- Control set points, such as discharge air temperatures, chilled water temperatures, and freeze protection temperatures
- Rooms, such as computer facilities, requiring special temperature control
- Quantities of exhaust, outside ventilation, and air movement

Information used in the analysis must be accurate, and any assumptions regarding use and operation must be documented. For instance, if HVAC equipment is assumed to operate 18 hours a day, 7 days a week, but really operates 12 hours a day, 5 days a week, then calculations will be very inaccurate. If cooling is assumed to operate at an indoor temperature of 78°F, but really operates at 72°F, results again will be misleading. If there are operable internal shading devices, are they to be taken as open or closed, or 50% closed?

If the building's energy use and costs during a specific year's operation are to be compared with actual records, it is important to specify the years, since weather, operations, use characteristics, and energy costs differ with various base years. All data must be related to the base year selected.

Energy Modeling

Since energy consumption represents a significant portion of total facility life cycle costs, a tool is required to assist design professionals in energy analysis. Depending on the progress of a facility design, the energy modeling concept can be used to develop a basic energy budget or to indicate areas where potential energy savings might exist. Unlike the cost model, the energy model uses energy units (EUs) instead of dollars. These energy units can be translated into dollars, depending on the specific source of energy/fuel chosen.

The energy model can be produced from a detailed energy analysis, as well as from basic performance assumptions. The level of detail should be consistent with the amount of information available and the purposes intended. Since, in most cases, the results of the model will be used to make comparisons, the accuracy of calculations need only be sufficient to assure accuracy of comparison.

Figure 5.13 presents an example of an energy model produced for a utility company service center and used in a recent study. After projected consumption was examined, each area was studied in regard to potential energy savings. The target value (upper figure in each block) represents an optimum consumption that should meet requirements. A comparison between the target value and estimated value indicates where energy-saving potential exists. In this case, a potential savings of some 50% exists in the heating systems for shops and warehouses.

There are several basic formats for developing an energy model. However, it must be understood that, unlike the cost model, no general-purpose energy model has yet been developed that covers all cases for all facilities.

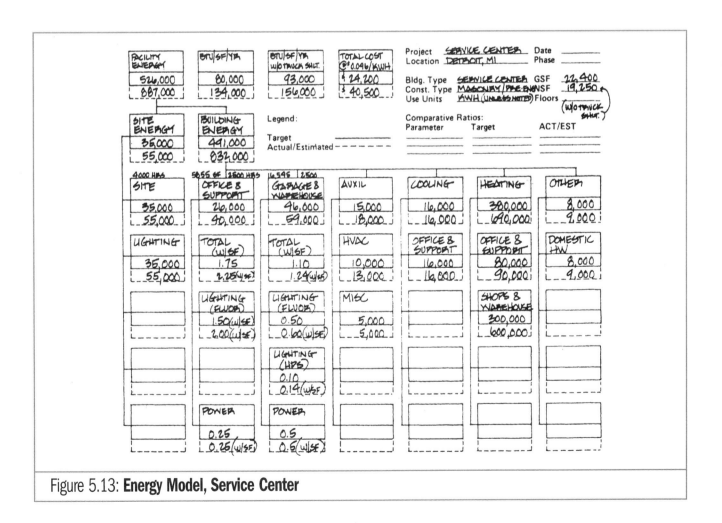

Figure 5.13: **Energy Model, Service Center**

The following set of procedures is intended to provide a guideline and checklist for use in the preparation of an energy model. Generally, energy consumption for a facility will fall into the following major categories.

Heating, cooling, and ventilation: ASHRAE has a number of texts to aid the user. These include the *ASHRAE Handbook of Fundamentals, 2001, Standard 90.1, Energy Standards for Buildings, IP Edition*, and the *ASHRAE Energy Standard 90.1 User's Manual*. The *Systems Handbook* provides excellent guides for the analysis of heating and cooling energy consumption. Appendix B, Tables B-I to B-5, contains information excerpted from RSMeans annual cost publications and other reference material useful for estimating heating and cooling energy. It should be noted that other, more exacting methods, such as the bin method or computerized techniques, are available. Use of such techniques is recommended when a more detailed analysis is required. There are also software programs available for this analysis. These include DOE-2, Energy 10, BLAST, Spark, Hot2000, and component simulation software, including RESFEN, DINDOW, and THERM (available from Lawrence Berkeley National Laboratory). Software for solar water heating includes FRESA and RETscreen. Many of these are available at no charge from the federal government.

Lighting: Appendix B, Tables B-6 to B-8, provides basic data for estimating lighting levels and energy required for those levels for various types of fixtures. Table B-6 can be used for gross estimates.

Power: Power use varies significantly from facility to facility. A useful guideline is as low as 0.25 watts per square foot for office structures to as high as 2 watts or more per square foot for specialty facilities (labs, etc.). Computer facilities, for example, may run as high as 10–50 watts per square foot for computer equipment areas.

Domestic hot water, equipment, and miscellaneous: Appendix B, Table B-9, presents basic guidelines for estimating hot water, equipment, and other loads. It should be noted that in many facilities domestic hot-water energy use is the highest single item. Pump energy can be calculated by using the formulas in Table B-10.

General: The accuracy of the above estimate guidelines depends on a sound assessment of building occupancy schedules and use. Appendix B, Tables B-11 and B-12, provides heating values for various fuels and a list of energy conversion factors. Metric conversion factors are contained in Table B-13. Refer to the bibliography (under "Estimating") for other sources of information helpful in determining energy consumption for various facilities.

Once the energy estimate has been developed, and the cost estimate prepared, one last step must be completed. The energy consumed should be reorganized into the UNIFORMAT system categories. For assistance in this process, refer back to Figure 5.12. The intent of this adjustment is to allocate energy loss to those areas controllable by the various disciplines. For example, heat loss and heat gain through the exterior closure and roof are most directly influenced by architectural decisions. Therefore, this portion of the energy consumed directly as a result of heat loss and gain is apportioned from the heating and cooling energy blocks. The HVAC category should include only the energy lost because of that equipment's inefficiency. It may also include those secondary pieces of equipment, such as cooling towers and pumps, used to produce cooling and heating.

Opportunities for Savings

The designer should review calculated loads from the analysis, as well as the relationships among and between loads and the system response. Final results of energy performance calculations in and of themselves will not help without a broad understanding of the building's energy performance. It is essential that the designer note the key variables that determined the building's energy use during the energy analysis.

While opportunities for improvement may come from any source, the biggest ones will emerge from methodically reviewing the analysis. This is done by means of a series of questions, which will vary from one project to another. First, the owner's concerns should be arranged in order of priority. Then, the factors that influence these concerns should be identified in order of their importance.

The designer must ask questions such as:

1. What is the owner's primary concern?
2. What determines the building's power or energy costs?
3. What modifications can be made to reduce consumption of the most significant item?
4. To which factors is the building's energy consumption sensitive?

Energy Costs—Summary

This section was prepared to provide the design engineers with a basic look at the concepts and concerns that enter into an energy-estimating effort. The procedures remain essentially the same, whether it is a new or a retrofit life cycle cost study.

Maintenance, Repair, and Custodial Costs

Significance and Opportunities for Savings

Of the categories of life cycle costs, energy has received the most attention over the past years. From the standpoint of owner cost impact, however, maintenance is usually the more significant cost item in an LCC study. The maintenance category alone averages (in a typical office building) between approximately $2.75 and $6.00 per gross square foot per year. Ironically, this is the cost factor that has received the least research and documentation for designers' use. Appendix C presents a summary of maintenance data for typical kinds of construction.

What then is maintenance? Maintenance has been defined to include the costs of regular custodial care and repair, annual maintenance contracts, and salaries of facility staff performing maintenance tasks. Replacement items of minor value, or having a life of less than five years, are included as a part of maintenance. Thus, tasks such as replacing light bulbs and repainting are normally included under the maintenance category.

Special care should be taken that system comparisons are based on comparable levels of maintenance. Estimates and data must refer to a uniform, optimum maintenance level. Much of the available historic data is not usable because one owner may be maintaining a facility in mint condition, while another may permit deferred maintenance.

Some of the decisions with major maintenance implications as follows:

1. The choice of exterior and interior finishes, for example, the selection of high-quality surfaces that do not require painting or other recurring maintenance.
2. Selection of light fixtures, floor covering materials, and other interior elements with minimum routine repair and replacement requirements.
3. The decision to plan for and implement a predictive maintenance program, which can have implications for selection of the facility's equipment.
4. The decision to perform most or all maintenance with full-time staff, or to contract for the services.

Many of the individual elements that make up a facility have a shorter life cycle than that planned for the entire facility. As a result, replacement costs can become a major consideration in the evaluation of alternative items of equipment and other limited-life building elements.

For better understanding of a breakdown of maintenance costs with respect to design systems, Figure 5.14 provides an approximate percentage of annual costs for each of the UNIFORMAT Level 2 categories discussed earlier. It can be seen that interior construction and mechanical systems have the most significant impact on maintenance costs. Designers must be especially careful when making decisions in these areas to consider the influence of maintenance on life cycle costs.

The opportunities for creative design for maintainability are enormous. Edwin Feldman, in his book *Building Design for Maintainability*, points out that for most buildings, maintenance costs will equal the original cost of construction (in undiscounted dollars) in as little as two or three decades. He also points out that, typically, architects, engineers, management, and interior designers "just don't seem to care." This apparent lack of interest is perhaps because there has not been enough exposure to the economic importance of the subject; or perhaps too many people do not recognize the magnitude of the opportunity to minimize maintenance costs.

The greatest possible return on the maintenance dollar can be achieved when these costs are included in LCC analyses during the design stages of a facility. It is recommended that maintenance and facility management personnel, as well as the owner, work with the architect, engineers, and interior designer in selecting materials and surfaces. Some architectural firms and contractors have begun to obtain recommendations on maintainability and to advise their clients on proper procedures.

Maintainability is an inherent characteristic of system design and installation. It is concerned with ease, economy, and safety in the performance of maintenance actions (both scheduled and unscheduled maintenance). Feldman defines maintainability as "a characteristic of design and installation which is expressed as the probability that an item will be retained or restored to a specified condition within a given period of time when maintenance is performed in accordance with prescribed procedures and resources."

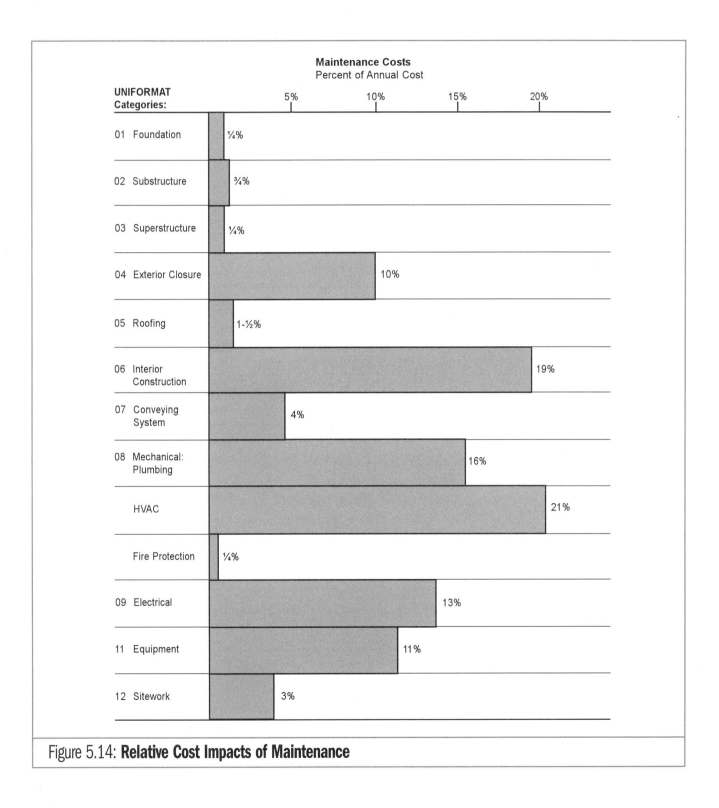

Figure 5.14: **Relative Cost Impacts of Maintenance**

Estimating Methods and Maintenance Data

The authors first published comprehensive data on life cycle costs for buildings in the United States in their book, *Life Cycle Cost Data,* published by McGraw-Hill in 1982.

A second source of life cycle cost data is from a computer program called "Building Maintenance, Repair and Replacement Database (BMDB) for Life Cycle Cost Analysis," available through the American Society for Testing and Materials (ASTM). This database provides designers with data related to maintaining a building in continuing use at its initial state of performance. It includes routine maintenance and repairs, as well as occasional high-cost repairs and replacements. The data are "constructed," rather than historical, in that the tasks to be performed are based on expert opinion; the hours required to do the tasks are based (for the most part) on time-motion studies; and the wage and equipment rates are specified by the user. Costs are expressed in dollars as of the date of the study. Data are presented in several levels of aggregation. The highest level of aggregation gives maintenance, repair, and replacement costs in present worth dollars per unit (such as per square foot) summed over the study period. The lowest level of aggregation gives a year-by-year listing of labor hours, equipment hours, and material costs per unit.

Data from the program, though developed by the U.S. Army, can be used by any organization to estimate maintenance costs for life cycle cost analysis during design and maintenance planning. To the extent that the maintenance standards of the user differ from those of the U.S. Department of Defense agencies for the type of building/system/component in question, the estimates will, of course, be over- or understated.

The database is constructed from typical systems and subsystems of a building, such as roofs and roof coverings. Alternative subsystem components include items such as built-up roofing, shingles, and tile, all roof covering components. The next item identified is maintenance tasks required to keep the building functional. Labor hours, equipment hours, and material cost required to perform each maintenance task were estimated, as well as when each task would be performed. This information comprises the core of the database. By assuming wage and equipment rates, and aggregating across tasks, it is possible to estimate labor, equipment, and materials costs per unit for a component over its life.

The type of data used by the Army to generate the computer database is consistent with Engineered Performance Standards (EPS) adopted by all Department of Defense agencies for determining required labor resources. When EPS does not cover a task, other sources are used to estimate the labor resources. Standard references, such as the Army's Corps of Engineers unit price manuals, are applied to determine the Washington, D.C. area material costs.

MAINTENANCE AND REPAIR DATA BASE FOR LIFE-CYCLE ANALYSIS COMPUTER INPUT-RESOURCES AND UNIT COSTS

PRESENT WORTH OF ALL 25 YEAR
MAINT. AND REPAIR COSTS (d=10%)
($PER UNIT MEASURE)

LOCATION: Detroit, Michigan
STUDY STARTS 0 YEARS BEFORE BENEFICIAL OCCUPANCY

COMPONENT DESCRIPTION	UM	ANNUAL MAINTENANCE AND REPAIR			UNIT COST $	REPLACEMENT AND HIGH COST TASKS				UNIT COST $
		LABOR Hours	MATERIAL $	EQUIPMENT Hours		YRS	LABOR Hours	MATERIAL $	EQUIPMENT Hours	
ARCHITECTURE										
ROOFING										
ROOF COVERING										
BUILTUP ROOFING	SF	0.00487	0.03814	0.00244	0.17	28	0.04938	0.84907	0.02469	2.21
PLACE NEW MEMBRANE OVER EXISTING - BUILTUP						14	0.02414	0.84269	0.01207	1.51
MOD. BIT./THERMOPLASTIC	SF	0.00245	0.03876	0.00123	0.11	20	0.5659	1.03421	0.02829	2.59
THERMOSETTING	SF	0.00173	0.02652	0.00086	0.07	20	0.03683	0.84269	0.01841	1.86
SLATE	SF	0.00253	0.01757	0.00126	0.09	70	0.06885	7.27776	0.03442	9.17
CEMENT ASBESTOS	SF	0.00246	0.04099	0.00123	0.11	70	0.05437	0.90569	0.02718	2.4
TILE	SF	0.00212	0.03533	0.00106	0.09	70	0.10169	3.70272	0.05084	6.5
ROLL ROOFING	SF	0.00757	0.01875	0.00378	0.23	10	0.04141	0.90295	0.02070	2.04
SHINGLES	SF	0.00262	0.0287	0.00131	0.1	40	0.04118	0.89734	0.02059	2.03
REPLACE NEW OVER EXISTING - SHINGLED ROOF						20	0.02996	0.52349	0.01498	1.35
METAL	SF	0.00199	0.01862	0.00099	0.07	30	0.36265	2.61744	0.18132	12.59
FIBERGLASS RIGID STP. ROOF	SF	0.00228	0.07548	0.00114	0.14	20	0.04543	7.24584	0.02272	8.5
CONCRETE, SEALED PANEL ROOF	SF	0.00596	0.01978	0.00298	0.18	60	0.06123	28.99804	0.03061	30.68
CONCRETE, SEALED PANEL RF4	SF	0.00552	0.01416	0.00276	0.17	300	0.04342	28.99804	0.02171	30.19
CONCRETE, SEALED POURED	SF	0.0138	0.10609	0.0069	0.49	500	3.81056	21.72028	1.90528	126.55
FIBERGLASS, RIGIF ROOF	SF	0.00468	0.07548	0.00234	0.2	20	0.04133	7.24584	0.02066	8.38

See NOTES on the last page of this table for Explanation of Column Headings

Figure 5.15: **Roofing Maintenance and Repair Data Example from ASTM Computer LCC Database**

Material costs in the task database are given in July 1985 dollars in the Washington, D.C. area. Material costs are adjusted for site location by applying a geographic location adjustment factor. Material costs are updated from 1985 to the date of the study by applying a time adjustment factor or construction cost index factor from July 1985 to the new point in time, the date of study.

Task frequencies are the most subjective portion of the database and are determined by applying professional experience, trade publication data, and data in manufacturers' literature. Heating, ventilating, and air conditioning maintenance varies by the climate zones.

The database has been reviewed by 13 different maintenance organizations and has been determined to accurately represent the resources required to perform the tasks. This database serves as the foundation for the cost estimates produced by the computer program.

The type of data generated by this computer program is illustrated in Figure 5.15 for roofing maintenance and repair. This printout lists a variety of roofing systems that might be considered by the design professional. Columns to the right of the component description are as follows:

Item	Description
1. U.M.	Unit of measure (square feet, in this example)

Annual Maintenance and Repair

2. Labor	Annual labor (expressed in hours per UM)
3. Material	Annual material (expressed in dollars per UM)
4. Equipment	Annual maintenance equipment (expressed in equivalent labor hours per UM)
5. Unit cost	Annual maintenance and repair unit cost (expressed in dollars per UM)

Replacement and High-Cost Tasks

6. Yrs	Years (first line of each roofing system is replacement life, second line is the year of high-cost-task occurrence)
7. Labor	
(1st line)	Replacement labor (first line of each roofing system is the replacement labor expressed in hours per UM)
(2nd line)	High-cost-task labor (second line of each roofing system is the high-cost-task labor expressed in hours per UM)
8. Material	
(1st line)	Replacement material (first line of each roofing system is the replacement material expressed in hours per UM)
(2nd line)	High-cost-task material (second line of each roofing system is the high-cost-task labor expressed in hours per UM)
9. Equipment	
(1st line)	Replacement equipment (first line of each roofing system is the maintenance equipment necessary to replace the system expressed in equivalent labor hours per UM)
(2nd line)	High-cost-task equipment (second line of each roofing

	system is the maintenance equipment necessary to complete the high-cost-task expressed in equivalent labor hours per UM)
10. Unit cost	
(1st line)	Replacement unit cost (first line of each roofing system is the replacement cost expressed in hours per UM)
(2nd line)	High-cost-task unit cost (second line of each roofing system is the unit cost of the high-cost expressed in hours per UM)

Life Cycle Cost Data in Part 2 of This Book

Part 2 of this book provides life cycle data on maintenance (labor, material, and equipment costs), operational (energy) demands, and replacement (parts' life in years) needs for selected building elements. Figure 5.16 presents some of the many sources utilized in the preparation of this data. Figure 5.17 illustrates the format used to organize this information. First, the UNIFORMAT category is given, and the item is described. Next, the item is further defined in terms of the units in which it is measured. There follows a description of the maintenance task associated with the item, including (in many cases) labor units for the amount of estimated time to perform the task and the frequency with which it is assumed the task will have to be repeated. Annual maintenance costs are listed next, broken down into separate columns for labor, material, and equipment. The energy demand for each life cycle data item is provided if applicable, and is expressed in the appropriate energy units (EU)—i.e., kilowatts (kW), Btu's per gallon (Btu/gal), and so forth. Finally, the remaining two columns provide useful data concerning replacement, in terms of each item's replacement life in years and the percentage amount of the item that must be replaced. Following is a discussion of this life cycle information and the assumptions made in its collection.

Of the life cycle data being addressed, the maintenance category, with overall cost averaging between approximately $2.75 and $6.00 per gross square foot per year, is the most significant. It is for this reason that maintenance has been broken down into several columns. It should be kept in mind that all task descriptions, as well as their labor units, must be referred back to their unit of measure. Also, depending on the nature of the building item, there may be more than one maintenance task described. It is the authors' intention to present as much known information as possible concerning the maintenance of a particular item, in order to assist the designer in estimating the specific costs of any project. The maintenance tasks listed may not necessarily apply to the item in a given situation.

Annual maintenance costs, expressed in dollars, are presented next to the maintenance description. Labor hours for each maintenance trade are translated into dollars per year per unit of measure, using 2003 figures based, in this example, on the city of Chicago. (The base wages, fringe benefits, and total hourly wages for maintenance in each of the trades are presented in Figure 5.18.) To this total wage rate, the reader may want to

	Type of LCC Data				
Sources of Data	Structural	Architectural	Mechanical	Electrical	Equipment, Site, and Other
1. Navy Public Works, San Francisco Bay, Oakland, California		•	•	•	
2. Operations Group, U.S. Postal Service, Washington, DC		•	•		
3. Public Buildings Service, General Services Administration, Washington, DC		•	•	•	
4. Civil Engineering Research Laboratory, Champaign, Illinois		•	•	•	
5. Mountain States Telephone & Telegraph Company, Denver, Colorado	•	•	•	•	•
6. Building Properties Department, Manufacturer's Bank, Detroit, Michigan	•	•	•	•	•
7. Director of Facilities, National Aeronautics and Space Administration, Washington, DC				•	
8. Building Owners and Managers Association International, Washington, DC		•	•	•	•
9. *ASHRAE Journal*, American Society of Heating, Refrigerating and Air-Conditioning Engineers, Inc. New York, New York			•		
10. *TRACE, A Handbook of Estimated, Installed, and maintenance Costs for Heating and Air Conditioning Systems*, Trane Air Conditioning Company, Commercial Air Conditioning Division, La Crosse, Wisconsin			•		
11. "Operation and Maintenance of Real Property," in *Maintenance Handbook*, Series MS-1, U.S. Postal Service, Washington, DC	•	•	•	•	
12. Report on Reliability Survey of Industrial Plants, Parts I, II, III," from Industrial and Commercial Power Systems Conference, Atlanta, Georgia, may 14-16, 1973, IEEE, New York, New York				•	
13. "Report on Reliability Survey of Industrial Plants, Part IV: Additional Detailed Tabulations of Some Data Previously Reported," from Industrial and Commercial Power Systems Conference, Atlanta, Georgia, May 14-16, 1973, IEEE, New York, New York				•	
14. California Recreation Department, San Jose, California					•
15. Albert Wahnon (ed.), Handbook of Contract Floor Covering, Bart Publications Inc., an affiliate of Hearst Business Communications, Inc., New York, New York		•			
16. Fred A. Wilson Company, Lathrup Village, Michigan (for roof and floor repair systems and materials)		•			
17. "Housekeeping, Postal Facilities," in *Maintenance Handbook*, Series MS-47, U.S. Postal Service, Washington, D.C.					•
18. Continental Manufacturing Company, St. Louis, Missouri (for sanitary maintenance equipment and supplies)		•	•	•	•
19. Robert Korte, "Rooftop Multizone Unit Maintenance Costs," in Bob Korte (ed.). *Heating/Piping/Air Conditioning* (journal), Rheingold Publishing Company, Chicago, Illinois			•		
20. Leonard A. McMahon (ed.), *Dodge Guide to Public Works and Heavy Construction Costs* (an annual), McGraw-Hill Information Systems Company, New York, New York					•
21. *Heating System Maintenance and Repair Requirements*, a report on the U.S. Army Administration Building written for the U.S. corps of Engineers by the Bendix Field Engineering Corporation, Columbia, Maryland			•		
22. *Ceramic Tower,* unpublished NASA Report, Washington, D.C.			•		

Figure 5.16: Sources of LCC Data *(continued)*

	Type of LCC Data				
Sources of Data	Structural	Architectural	Mechanical	Electrical	Equipment, Site, and Other
23. A/C Pipe Producers Association, Arlington, Virginia (also known as the Association of Asbestos Cement Pipe Producers)			•		•
24. Partial list of manufacturers:					
(a) Trane Company, La Crosse, Wisconsin			•		
(b) Endure-a-Lifetime Productrs, Inc., Miami, Florida (makers of Endura-a-Dors)		•			
(c) Crane Company, New York, New York			•		
(d) Johns-Manville Corporation, Denver, Colorado		•			
(e) Baltimore Aircoil Co., Inc., Baltimore, Maryland			•		
(f) Westinghouse Electric Corporation, Pittsburgh, Pennsylvania				•	
(g) United States Gypsum Co., Chicago, Illinois			•		
25. Ed Robinson "Rules of Thumb"			•		

Figure 5.16: **Sources of LCC Data**

Item Description	Unit of Measure	Maintenance Description	Maintenance Annual Cost, $			Energy Demand (EU)	Replacement Life, Years	% Replaced
			Labor	Material	Equipment			
Used to document UNIFORMAT category and describe specific facility items analyzed.	Given for each task.	Used to describe specific maintenance tasks and corresponding labor performance standard(s).	1	2	3	4	5	6
			1. Used to convert labor hours into annual costs. 2. Used to convert material requirements into annual costs. 3. Used to convert maintenance equipment into annual costs. 4. Used to record energy consumption requirements for the facility items. 5. Used to document replacement life of significant components of facility items. 6. Used to estimate percent of facility item cost replaced at the year specified (see Replacement Life).					

Figure 5.17: **Sample Life Cycle Cost Documentation Sheet, with Explanation of Various Categories**

add the cost of supervision and overhead (normally an additional 20–40%).

Figure 5.19 presents proper definitions of various labor trades with their corresponding duties. These responsibilities vary somewhat from location to location. Trade unions specify the responsibility for the various building maintenance tasks, and should be consulted if there are uncertainties.

Maintenance material costs are listed adjacent to the *labor costs* column and reflect the costs for minor replacement parts such as filters, linkages, contacts, seals, light fixture tubes, etc., as well as expendable supplies, storage devices, waste receptacles, and maintenance log books.

Maintenance equipment costs, although minor when compared to other maintenance costs, have been included as a separate category. Equipment required to perform maintenance tasks may be organized into three general categories:

• tools and devices,

Labor Trade	Wage Rate ($/Hr)	Supervisor & Overhead (30% Assumed)	Total Wage Rate ($/Hr)
Janitorial			
Base Wage	13.57		
Fringe Benefits	2.22		
Total	15.79	4.736327	20.5241
Engineers			
Base Wage	26.75		
Fringe Benefits	4.31		
Total	31.05	9.316081	40.3697
Elevator Operator			
Base Wage	13.96		
Fringe Benefits	2.22		
Total	16.18	4.853756	21.0329
Security			
Base Wage	12.59		
Fringe Benefits	2.22		
Total	14.81	4.442753	19.2519
Electricians (Maintenance)			
Base Wage	24.86		
Fringe Benefits	3.88		
Total	28.73	8.619332	37.3504
Electricians (Relamper)			
Base Wage	19.58		
Fringe Benefits	3.85		
Total	23.43	7.030118	30.4638
Window Washer			
Base Wage	14.39		
Fringe Benefits	1.76		
Total	16.15	4.845928	20.999
Fireman			
Base Wage	20.94		
Fringe Benefits	3.26		
Total	24.20	7.261063	31.4646
Painter			
Base Wage	28.31		
Fringe Benefits	6.13		
Total	34.45	10.3338	44.7798

Figure 5.18: **2003 Wages for Maintenance Labor Trades**

- testing equipment, and
- miscellaneous equipment.

The *tools* and *devices category* includes items such as hand tools, power tools, vacuum sweepers, floor scrubbers, and buffers. The *testing equipment* category would include items such as electrical meters, pressure gauges, velocity meters, and refrigeration testing apparatus. Finally, the *miscellaneous equipment* category includes items such as safety devices (hats, glasses, and shoes), scaffolding, and window-washing apparatus. Costs for equipment items are amortized over the equipment's useful life. Descriptions of both the material and the equipment items for a particular

Janitorial	• Vacuuming – horizontal and vertical, nonindustrial-type • Spot interior glass washing • Dusting • Sweeping and dry mopping sporadic • Trash removal – desk-type wastebaskets – empty and light containers – filled bags to designated spot on floor • Cleaning of ashtrays • Spot cleaning of walls sporadic • Damp wiping • Mirror and brightwork • Policing of corridors and washrooms • Regular towel and supply service from storage areas – hand carried or small lightweight carts • Wet mopping (sporadic) • Night washroom sanitation (sporadic) • Manual scrubbing (sporadic) • Clean up after construction, painting, and repair jobs • Furniture washing and polishing • Metal washing and polishing • Exterior policing including hosing and sweeping of sidewalks • Carpet and furniture shampooing – manual • Washroom sanitation • Cleaning light fixtures • Cleaning venetian blinds – washing drapes • Operating industrial-type vacuum cleaners and wet pickup machines • Wet mopping • Floor-type interior and exterior power machines and related wet mopping • Marble maintenance, exclusive of washing • Metal refinishing • Snow removal • Washing and polishing of vertical surfaces, baseboards, and ceilings • Incinerators, balers and compactors • Moving furniture and setting up for special events • Removal of old carpet • Loading and unloading of trucks, dock work, and moving supplies to and from storage areas • General handiwork
Janitorial (Special)	• High level work – 12 feet from floor level and over • Furniture crating and uncrating • Removal of tile affixed to floor • Moving and storing of construction equipment and material • Exterior metal refinishing – after on hour in one day – from first hour of work • Loading and unloading of trucks and dock labor – after two hours in one day – from first hour of work • Moving furniture – after three hours in one day – from first hour of work
Operating Engineers (Stationery Engineers)	Operating or assisting in operating: • Heating and ventilation equipment • Engines • Turbines

Figure 5.19: **Building Owning and Operating Labor Trades** *(continued)*

	• Motors
	• Combustion engines
	• Generators
	• Pumps
	• Air-compressors
	• Ice and refrigerating machines
	• Air-conditioning units
	• Fans
	• Siphons
	• Automatic and power-oiling pumps and engines
	• Operating and maintaining all instrumentation and appurtenances
	• Steam boilers
	• Handling, preparing and delivering fuel from storage
	• Operating repairs of all plants, machinery, and engines
	• Power plant equipment
	• Power driven engines connected with:
	— water treatment plants
	— garbage, sewage disposal plants
	— breweries and distilleries
	— office, municipal buildings
	— canneries and dairies
	— theatres
	— schools, hospitals
	— hotels and motels
	• Pumping and boosting stations
	• Operation of valves, gates, locks
Elevator Operators	• Operating passenger elevators
	• Operating freight elevators
Elevator Starters	• Starting and monitoring passenger elevators
	• Starting and monitoring freight elevators
Security	• Primary function – building security
	• Guarding lobbies
	• Making clock rounds, etc.
	• Excludes highly compensated loss prevention, detection, and investigative employees and security personnel employed by building tenants
Maintenance Electricians	Maintain, Repair and Replace:
	• Electrical wiring
	• Electrical appliances
	• Electrical equipment of any nature
	• Light and power wiring fixtures
	• Electrical related:
	— sprinkler systems
	— temperature controls
	— refrigeration
	— escalators
	— conveyors
	— private telephone systems
	— air-conditioning equipment
	— tube systems
	— radio and public address systems
	— elevators
	— motor generators
	• General electrical maintenance work
	• Excludes major construction or repair work
Fixture Cleaners and Relampers	• Clean and maintain fluorescent fixtures
	• Clean and maintain incandescent fixtures
	• Lamping and relamping of fluorescent fixtures
	• Lamping and relamping of incandescent fixtures
Window Washers	• Clean and maintain exterior windows on both sides
Firemen and Oilers	• Assist the engineer on any high-pressure steam generating plant
	• Assist the engineer on any high-pressure hot water heating plant
	• May include service to low-pressure plants

Figure 5.19: **Building Owning and Operating Labor Trades**

life cycle data item are provided when the costs are a significant part of the total maintenance cost.

Those life cycle data items that consume energy have a number listed for energy demand, expressed in energy units appropriate to them. For example, a commercial, gas-fired, hot-water generator (70% efficient and 500-gallon-per-hour recovery rate) will have an energy demand of 1,070 Btus per gallon (assuming 50°F incoming water temperature). If no energy demand exists, a *not applicable* (N/A) notation is listed. The energy demand number is used to calculate the energy consumption that the item requires for a specific project. From the number of hours an item will operate and the fuel costs it will generate, an estimate of annual energy costs can be made.

Replacement costs for a facility may be estimated by using the information contained in the remaining two columns of the life cycle data sheets. The replacement life is expressed in years and is based in part on the historical physical life, the technological life, and the economic life of the given item. Because this number is so difficult to predict for any item, no attempt has been made to establish confidence limits. Engineering judgment should be applied to the numbers presented to adjust for owner operating and maintenance standards that may vary from those listed in the maintenance description. The location and climatic environment of the facility and its relationship to other facilities will also affect replacement life and should be taken into account.

The last column, *percent replaced*, refers to the proportion of the item being replaced expressed as a percentage of its initial costs. This allows the designer to estimate the cost of replacement for a particular item simply by multiplying the percentage listed by the item's initial cost. In most cases, 100% of any item must be replaced after the number of years listed in the *replacement life* column. If, however, only a single component of an item will require replacement, a percentage of the overall item cost, representing the component cost, is listed.

Maintenance Costs—Summary

Maintenance costs are significant in any LCC analysis and should be included in all alternatives. As more research is conducted in this area, more confidence can be placed in the comparisons. In the interim, however, the data in this book can be used as a baseline and a format for further recording of information.

Alteration and Replacement Costs

Alteration costs are those involved with changing the function of the space. For example, when a tenant leaves an office, the owner must have the space redone to suit the functional requirements of the new tenant. Replacement costs are those expenses incurred by the owner to maintain the original function of the facility or space. Current alteration costs in a typical office building can range from $1.25 to $3.00 per gross square foot of space per year. Replacement costs range from $2.00 to $5.00 per gross square foot per year.

Alteration Costs

Alteration costs may include the expenditures for anticipated modernization or the changing of a facility to provide a function not originally intended. Estimates of health facility alterations have ranged as high as 10% of the usable space per year. The typical figure is considerably lower, but even annual changes of only 1% can be a significant cost consideration. The relative costs of alterations for the various GSA UNIFORMAT cost categories are approximately 72% for interior construction, 12% for HVAC, and 16% for electrical.

Even those facilities that do not anticipate high-cost alteration programs should consider whether the decision to minimize alterations will lead to inefficiency in space flexibility for the future. Thus, some evaluation of the probability of, need for, and extent of the future alterations should enter into a study. In a recent study of health facility costs, the General Accounting Office noted the following general guidelines for hospitals relative to interstitial space and its effect on alteration costs:

- *Surgery.* Disruption is totally intolerable, and, although few major remodeling projects were recorded, interstitial space would be justified.
- *Radiology.* Disruption is tolerated, and most changes are confined to small areas; therefore, interstitial space may not be needed.
- *Laboratory.* Frequency of change is not high, but work usually involves plumbing and other mechanical systems. Laboratory activities are relatively tolerant to disruption. When large-scale remodeling does occur, it is sufficiently extensive that interstitial space might cut costs and save time.
- *Emergency.* The infrequency of change, which is mostly small-area remodeling, as well as tolerance for disruption, does not justify using interstitial space.

Typical decisions that affect alteration costs include the following:

1. Whether to plan for movable partitions, interstitial space, utilities in the exterior walls, knockout panels, and other design options that facilitate alterations.
2. Whether to build spare capacity into the mechanical or electrical systems to accommodate additional demands, or at least size main and select secondary feeder for future needs.

The establishment of an *alterability factor* for various building components helps to single out systems for trade-offs. This factor might have a scale from 0 to 1, depending on the cost of altering the system. For example, for interior partitions, this scale might be developed as follows:

Interior partition	*Alterability factor*
Movable	0.1
Demountable	0.3
Gypsum board and metal stud	0.6
Masonry	1.0

Since a movable partition would cost very little to alter, it has been assigned an alterability factor of 0.1; and because a masonry partition would require total demolition and reconstruction at a new location, it has been assigned a factor of 1.0.

A given space or functional area may be assigned an alteration cycle. For example, an office rental space is known to change tenants every six years. Determining the annual cost of alteration thus becomes a matter of dividing the cost to alter by the alteration cycle.

Replacement Costs

Building replacement costs include the costs of replacing the many equipment or other facility elements having an estimated life cycle shorter than that planned for the entire facility. These costs can become a major consideration in the evaluation of alternative items of equipment and other limited-life building elements. Common limited-life elements found in facilities, as well as sample estimates of typical useful lives, are included in Part 2.

Typical decisions based on replacement costs include:

1. Whether to specify short-life (but lower capital investment) building elements, such as rooftop HVAC units, or shorter-life roofing materials.
2. Whether to specify short-life elements in areas that may undergo significant future alterations.

To help understand the breakdown of replacement costs with respect to design systems, Figure 5.20 provides an approximate percentage of annual costs for each of the UNIFORMAT categories for a typical office building. Replacement costs are most significant for interior construction and HVAC systems, each approximating 20% of the total replacement costs.

As discussed previously in this chapter, under "Maintenance," the performance of materials and systems can be specified, predicted, and measured. The life cycle data contained in Appendix C provides the replacement cycle, or mean time between replacements (MTBR), for various materials and subsystems. The MTBR can be determined by either scheduled or unscheduled actions, and usually generates space or repair part requirements.

Figure 5.21 presents various curves related to failure rates of components. The top curve illustrates a hypothetical distribution of failures. The normal distribution implies that 50% of similar items will fail at the mean life of the system. The lower curve illustrates that the failure rate has a lot to do with the running frequency of the system. As data are collected for various components, a more reliable database may be developed, taking into account differing modes of operation—high, normal, and low. The climatic conditions and functional applications will also vary the MTBR figures.

Part 2 contains replacement lives and percentages of replacement for a variety of construction systems. This information is contained in the last two columns of the data. The replacement life is expressed in years. This figure has been based in part on the historical physical life, the technological life, and the economic life of the given item. These figures are difficult to predict, and engineering judgment must be applied to adjust for owner operating and maintenance standards. The location of the facility, its relationship to other facilities, and the microclimate will also affect the replacement life and must be taken into account.

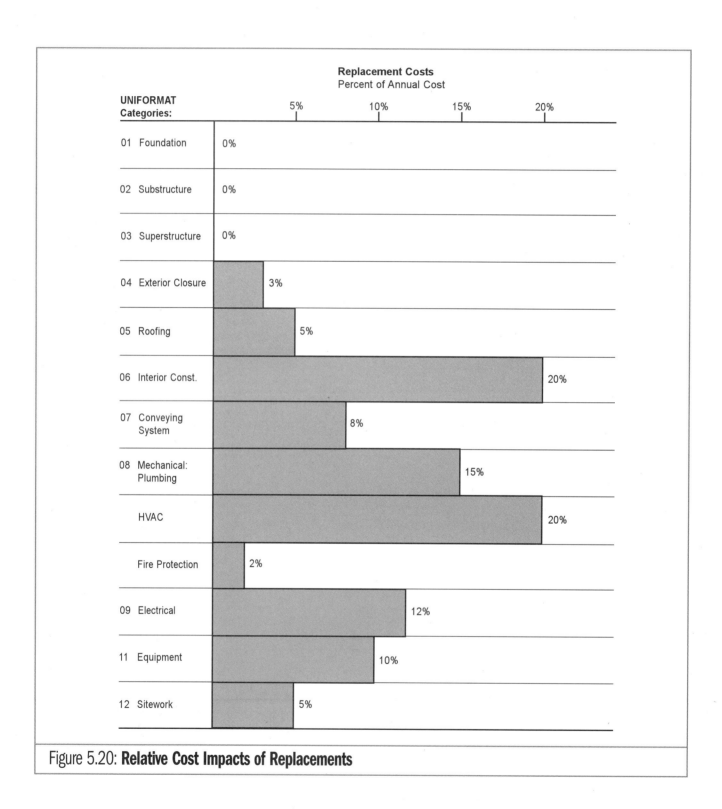

Figure 5.20: **Relative Cost Impacts of Replacements**

The last column of the data in Part 2, titled *Percent Replaced*, refers to the percentage of the economic value of the item being replaced. If, for example, only one component of an item requires replacement at the year listed, that component's percent of the economic value of the total item cost is listed. This allows the designer to estimate the cost of replacement of a particular item. For most cases in this document, 100% of the item is replaced at the year presented.

Alteration and Replacement—Summary

Alteration and replacement costs for office buildings can amount to as much as $5.00 to $9.00 per gross square foot per year. In other words, these costs can be significant as energy consumption and maintenance costs. Therefore, designers should thoroughly analyze and estimate their effect.

Associated Costs

Discussions thus far have focused on: initial project costs; energy costs; maintenance, repair, and custodial costs; and alteration and replacement costs. But what about administrative (building management) costs, interest (debt service), staffing (functional use) costs, denial-of-use costs, and other costs, including security, real estate, taxes, water and sewer, and fire insurance? For example, few life cycle studies touch on personnel salaries—the largest area of expense associated with most facilities, particularly hospitals, correctional institutions, and manufacturing facilities.

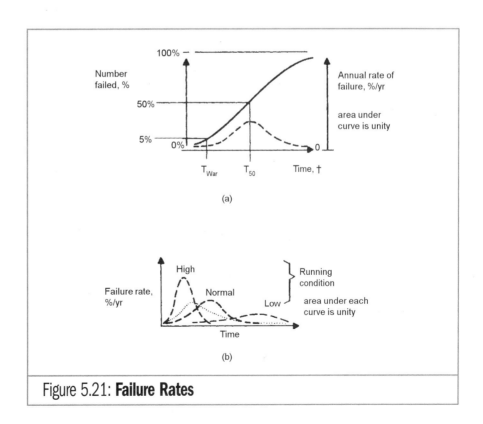

Figure 5.21: **Failure Rates**

The following is a discussion of other life cycle cost items that may be significant in an analysis. Examples are also provided to further explain and justify their significance. The designer must include these costs, in cases where they could influence a decision from one alternative to another.

Administrative Costs

The costs to manage and coordinate the building's maintenance and repair requirements, as well as pay energy and utility bills, can range from $0.60 to $1.20 per GSF/yr. for an office building. Facility managers are professionals who perform these duties on a regular basis. To help control administrative costs, many facility managers will actually coordinate activities for a large number of buildings. Use of computer maintenance management systems has also helped to simplify the record-keeping and management of facilities.

Interest (Debt Service)

Depending on the funding source, the interest rate, and the duration of the loan, the annual interest paid for debt service can be substantial. Consider the following example of an office building:

$$300{,}000 \text{ GSF} \times \$100/\text{GSF} = \$30{,}000{,}000 \text{ construction cost}$$

A 7% interest rate results in a first year interest payment of $2,100,000 or $7.00 GSF. Proper attention should be given to seeking the best financing option and lowest interest rate in order to minimize this significant life cycle cost.

Staffing (Functional-Use) Costs

Functional use includes costs of staff, materials, etc., required to perform the function of the organization using the facility. This is the largest single area of cost in most facilities. For example, a hospital's annual operating budget is often as much as 35% of the entire cost of buying or building the facility that houses these functions. This cost area is also the least subject to scrutiny during the planning stage of many projects, in spite of the significant impact that the facility can have on these costs.

One reason for this, of course, is that functional-use costs are among the most difficult to analyze. The qualitative parameters often make it difficult to quantify and compare alternative decisions. However, functional-use costs are so large that they should be considered in any comprehensive LCC analysis. Care must be taken, though, to compare costs at comparable performance levels. A classic example that the authors experience involved a high-rise jail in New York City. The VE study's LCCA isolated the guards' function as a high-cost area. By revising the layout so that 3 guard-posts could be safely eliminated, some $8,000,000 present worth was saved.

Many facility decisions with functional-use cost implications can be analyzed. Typical decisions include:

1. **Whether a reduction or redefinition of facility capability will be justified by a more economical life cycle cost level.** For example, the decision made by many hospitals to build open-heart surgical capabilities had significant life cycle functional-use cost implications when they could not be fully supported through patient income. Other hospitals have built capabilities such as comprehensive emergency facilities, even though neighboring hospitals' facilities are adequate for the community's full needs. Many hospitals have been living with cost problems as a result of their failure to consider such factors. Another classic example involved a 400-bed hospital, where the functional analysis led to questions regarding the specialized equipment cost and revenue. The VE team member, who worked in a nearby, large hospital complex, told the team that the area was in serious need of "kinder" (children's) dialysis machines, and that no regional hospital was equipped to provide these services for small children. As a result, the maternity ward was revised to free space for the installation of a dialysis machine for the children. Economics indicated less than a one-year break-even, plus an increased service to patients.
2. **Whether to select a design with higher construction costs, but lower functional-use costs.** For example, a recent hospital expansion design called for the construction of two new six-story nursing towers to be connected to a six-story existing facility. It was found that merely expanding the existing nursing units instead would cut the number of nursing stations in half, and the required nursing staff by 30%. (See Case Study 12, in Chapter 8.)
3. **Whether to provide services through contract, concessions, or facility staff.** Such decisions as whether to have in-house laboratories, laundries and food preparation, prepackaged pharmacies, and other support facilities have hinged on this issue.
4. **Whether to build in greater staff efficiency at the price of higher operating costs.** This is critical in such areas as automated patient monitoring and diagnostic equipment.

Denial-of-Use Costs

Denial-of-use includes the extra costs occurring during the construction or occupancy periods, or both, because production (income) is delayed. This may be the result of a process or a facility decision.

The major elements of denial-of-use costs are lost revenues and operating inefficiencies. Where the denial-of-use is attributable to a delay, there are also often considerable increases in construction costs because of inflation, increased interest rates, and other factors. There are also a number of non-quantifiable costs resulting from the non-availability of a facility, such as the loss of doctors or tenants who would otherwise plan to use it.

It is possible to allocate all aspects of denial-of-use costs to the other cost elements: operating inefficiencies to functional-use cost projections, increased cost escalation to initial costs, etc. Denial-of-use cost is really only the difference in total costs between two decisions that lead to different initial occupancy dates. In some analyses, however, the reason for keeping it as an identifiable cost element is to provide a single place for quickly summarizing all the cost implications of a decision that will increase the time required to achieve initial occupancy and start-up of the new facility. Often the major use of these data is to demonstrate to the facility decision-maker the cost of delay. In addition, some typical decisions that hinge on this information include:

1. Whether to employ a construction manager and fast-track construction and thus reduce the time required to complete the facility.
2. Whether to spend the available funds to completely equip several departments and leave others as unfinished shell space until more funds are available, or to partially equip all departments.
3. Whether to use public funding that requires a guaranteed maximum price, thus slowing down a fast-track project until the drawings are sufficiently detailed to obtain the guarantee.
4. Whether to delay construction while reapplying for permission from various agencies for additional space.

Other Costs

Other costs may include any identifiable expense related to a facility decision, not covered in the other categories such as security, real estate taxes, water and sewage, and fire insurance. This special cost compartment accepts quantified estimates of the costs of qualitative areas such as aesthetics, patient convenience, comfort, safety, ecological impact, and increased usable space.

For example, if the design team has a strong preference for a particular exterior skin material, such as bronze-tinted glass, for aesthetic reasons, a special attempt should be made to put a dollar value on that choice to see if it significantly alters the LCC comparison. The regional planner deciding between a single central facility or several community facilities makes a similar attempt to quantify the value or benefit of the added patient convenience of the closer facilities. Other examples are:

1. A value placed on the additional comfort created by designing extra HVAC capacity to meet the maximum load requirements, rather than a lower capacity that would result in some discomfort several days per year.
2. A value placed on the increased safety achieved by installing smoke detectors, sprinkler systems, etc., that are not specifically required by the fire code.
3. A value placed on the added protection of a structural system that far exceeds minimum earthquake design code requirements.

Associated Costs—Summary

In addition to the classical LCC costs such as initial, maintenance and operations, and alteration and replacement, there are many associated costs that may affect a life cycle cost study. These include administrative costs, interest (debt service), staffing (functional use) costs, denial-of-use costs, and other costs. These additional expenditures may range in significance from a few dollars to tens of dollars per gross square foot per year. The owner should require a project's decision-makers to take responsibility for inclusion of those items, together with the usual LCC items considered significant for any alternatives being evaluated.

1. Sizemore, Clark, and Ostrander, Energy Planning for Buildings, American Institute of Architects, 1974.

Chapter 6 Economic Risk Assessment

Take calculated risks.
General George Patton
U.S. Army

It is important to make an assessment of the effects of uncertainties (risks) on the results of life cycle cost analyses (LCCAs). The two leading approaches to uncertainty assessment are the *probabilistic* approach and the *sensitivity* approach. Some owners prefer the probabilistic approach be used wherever possible. The sensitivity approach is recommended as a minimum. These two approaches and the relevant calculations are discussed in this chapter.

The need for uncertainty assessment comes from the fact that input data for LCCAs are based on estimates rather than known quantities. The data input are, therefore, uncertain. There are uncertainties as to:

- **Scope or quantity** of items (e.g., building square footage, number of windows, tonnage of steel, worker-hours of labor).
- **Unit cost** of these items in the marketplace at the time the cost will be incurred.
- **Date** when certain costs will actually be incurred (e.g., when a particular floor covering will have to be replaced).

The uncertainties as to *scope or quantity* and marketplace prices give rise to uncertainties in dollar cost amounts. Uncertainties as *to when costs will be incurred* give rise to uncertainties in the timing of these costs. Both types of uncertainties—in cost amounts and in the timing—contribute to uncertainties in computed PW values. And, because there are uncertainties in the PW values, there is some question as to which of two alternatives really has the lower LCC (PW), even if the LCCA results show that one is lower on the basis of best estimates.

Just how questionable the relative ranking of the two alternatives is—purely in terms of LCC—depends on two factors:

1. The degree of uncertainty inherent in the input data and hence in the PW values, and
2. The difference in PW between the two alternatives, calculated on the basis of best estimates.

Uncertainty assessment helps in judging whether the results of an LCCA can be considered conclusive in light of this uncertainty.

(Note that, for alternatives A and B, if PW_A = $50,000 and PW_B = $300,000, the uncertainties in the input data could be quite significant, but still will not affect the conclusion that PW_B is greater than PW_A. If, however, PW_A = $50,000 and PW_B = $51,000, then the input data could be perfectly certain, and the results of the LCCA would still be uncertain. Because of this latter possibility, the results of an LCCA should automatically be considered inconclusive if the difference between the LCCs of the two lowest-cost alternatives is not more than 5% of the higher of the two LCCs—that is, if $PW_B - PW_A$ is not greater than 0.05 PW_B.)

In the approaches described in this chapter, the input data for two alternatives—usually those with the lowest LCC and the next lowest LCC—are analyzed to determine whether the uncertainties are sufficiently large to render the apparent LCC difference inconclusive. If, after due consideration is given to the uncertainties, there is a clear-cut probability that the alternative with the lower apparent LCC will in fact have the lower LCC, then the results of the LCCA are conclusive. Otherwise, they are not. For example, suppose $PW_B - PW_A$ = $100,000 − $85,000 = $15,000. On the basis of best estimates, alternative A appears to be lower in LCC by $15,000. Since the LCCA was based on "best estimates," the odds are better than even—i.e., the probability is greater than 0.50—that alternative A is, in fact, lower in LCC than alternative B.

The purpose of uncertainty assessment is to categorize the actual probability—i.e., to determine if that probability is in the marginal area (somewhere quite close to 0.50, say 0.52) or in the clear-cut area (say 0.67 or more), or somewhere in between. Neither the accuracy of the $15,000 computed difference nor the exact probability is of importance. All that is required is to place the probability that $PW_B - PW_A$ is greater than 0 in one of three very broad categories. If that category happens to be the one that is most clear-cut (whether by the 0.67 probability rule of thumb or common sense), the results of the LCCA are considered to be conclusive (alternative A has the lower LCC). Otherwise, the results are inconclusive (meaning that it is not sufficiently clear that alternative A has the lower LCC). The general rule of thumb to be clear-cut for economic studies is *about* 0.67, which is equivalent to odds of 2 to 1 (two out of three chances).

The confidence index approach, discussed later in this chapter, is an approximate probabilistic approach to uncertainty assessment. It provides a numerical measure of the degree of uncertainty in the LCCA results. This measure, called the *confidence index*, can be directly related to the 0.67 probability rule of thumb for economic studies. The *sensitivity approach*, also discussed later in this chapter, is used to analyze the changes in PW values that would result from variations in cost data due to uncertainties. The sensitivity approach provides a set of data from which the analyst can judge whether the difference between two PW values is clear-cut according to the 0.67 rule of thumb. It does not, however, provide a direct measure, as does the confidence index approach. Neither approach requires knowledge of probability or statistics, and both are simple and straightforward. Both approaches require the computation of a range of PW values, as discussed in the next paragraph.

Range of Probable Values

As noted earlier, a particular cost element may be uncertain as to both dollar amount and timing. Both types of uncertainty may be expressed as a combination of a best estimate and a range within which the actual value is expected to fall. For example, suppose the dollar amount of a cost is uncertain, but is reasonably expected to be not less than $7,500 and not more than $12,000. This uncertainty in the amount of the cost may be expressed by the best estimate $10,000 and the range $7,500 to $12,000. Similarly, a cyclical cost might be expected to be incurred no more often than once every 4 years but at least once every 10 years; this timing uncertainty could be expressed by the best estimate "every 7 years" and the incurrence range "every 4 years to every 10 years." Present worth values may be calculated based on the best estimate and upper and lower bounds of such ranges. The results are the PW values of the best estimate and upper and lower bounds of a reasonable *range of PW values* that now, together express these uncertainties in terms of present worths. Such ranges of reasonable PW values (expressions of the uncertainty in cost elements) are used in computing the confidence index, and their upper and lower bounds may also be used in the sensitivity approach. The following paragraphs explain in more detail the idea of a reasonable cost estimate for use in uncertainty assessment computations. This section also addresses the computation of PW bounds based on these reasonable estimates.

The *confidence index (CI)* approach to uncertainty assessment, as detailed later in this chapter, makes use of a best estimate of each dollar cost amount or time of incurrence, as well as estimates of reasonable, upper and lower bounds. The estimates of the bounds are reasonable, rather than absolute in the sense that they should not encompass every possible value of the cost amount of every cost timing possibility. Instead, the upper and lower bounds should be determined such that the value of the input data element (cost amount or time of incurrence) is expected to fall within these bounds in roughly 9 out of 10 cases (or, alternatively, such that the value of the element is expected to exceed the upper bound in only one of 20 cases, and to be less than the lower bound in only one of 20 cases). Each pair of upper and lower bounds for an input data element will thus define a range that includes 90% of all possible values of that element. (In what follows, the three estimates for each cost parameter are referred to as the *best estimate* and the *high* and *low 90% estimates*. The range defined by the high and low 90% estimates is called the *90% estimate range*.) All three estimates for a particular parameter should be obtained at the same time and from the same source or sources.

As an example, consider the repaving of a parking lot. The best available estimates indicate that the surface will have to be repaved 12 years after construction at a cost of $8,500. However, experience with similar surfaces indicates that 90% of such surfaces must be replaced between 8 and 14 years after initial use. Then the best estimate of the timing of repaving is every 12 years, and the 90% estimate range is between 8 years and 14 years. A PW calculation based on replacement every 8 years and one based on replacement every 14 years will result in a high and a low PW bound. These PW values will define a PW range that includes the effect of timing uncertainties on the cost of repaving the parking lot.

To compute the PW bounds for an input data element, it is necessary to calculate PW values on its high and low 90% estimates. When only dollar cost amounts are involved, these PW values may be obtained as proportions of the PW of the best estimate of the cost. For example, suppose the best estimate of the annual energy cost for a building is $10,000, and the high and low 90% estimates are $12,000 and $7,500, respectively. Further, suppose that, as is usual, the PW of the $10,000 best estimate has already been calculated and it is $90,770 (10% discount rate, 25-year analysis period). Then the PW of the $12,000 cost may be found as:

$$\text{PW} = \frac{\$12,000}{\$10,000} \times \$90,770 = \$108,924$$

and the PW of the $7,500 annual cost as:

$$\text{PW} = \frac{\$7,500}{\$10,000} \times \$90,770 = \$68,078$$

If the timing of costs is involved in the uncertainty assessment, a separate PW calculation is required for each of the two 90% estimates. Consider, for example, a $10,000 replacement cost that is most likely to be incurred every 10 years. The high and low 90% estimates for the timing of this cost are every 8 years and every 12.5 years, respectively. The analysis period is 25 years, the discount rate is 10%, and the differential escalation rate is 0%. For the best estimate of the cost timing, the PW of this $10,000 cost is based on replacements in years 10 and 20 after occupancy and an artificial salvage value of one-half the cost in year 25. The PW is computed in three parts as shown in Figure 6.1, Paragraph (a):

$$\text{PW (best estimate)} = \$3,855 + \$1,486 + (\$461) = \$4,880$$

The PW of this cost for the high 90% timing estimate is based on replacements in years 8 and 16 after occupancy (under the assumption that the second replacement can be stretched to 9 years use, rather than 8 years, to complete the 25-year economic life of the facility). This PW is computed in Figure 6.1, Paragraph (b):

$$PW \text{ (high 90\%)} = \$4,665 + \$2,176 = \$6,841$$

Similarly, the PW for the low 90% timing estimate is found as shown in Figure 6.1, Paragraph (c):

$$PW \text{ (low 90\%)} = \$3,038$$

Thus, the PW of the best estimate of this replacement cost is $4,880, and its high and low PW bounds are $6,841 and $3,038. These three PW amounts are indicative of the uncertainties in the timing of the replacement cost. These uncertainties may now be assessed with either the CI approach or the sensitivity approach.

Confidence Index (CI) Approach

As noted at the beginning of this chapter, uncertainties in the results of an LCCA are assessed for the purpose of judging whether those results can be considered to be conclusive, given the fact that they were computed from estimated (rather than exact) cost data. More specifically, the input data for two alternatives (usually those with the lowest and next lowest computed LCCs) are analyzed to determine whether there is a clear-cut probability that the alternative with the lower computed LCC will, in fact, have the lower cost. For design economic studies, this probability is considered to be clear-cut if it is about 0.67 or more, and in that case, the LCCA results are considered to be conclusive. Probabilistic approaches to uncertainty assessment provide a direct measure of this probability and, hence, of whether the relative rankings of two alternatives (based on LCCA results) are conclusive.

REPLACEMENT COST/SALVAGE VALUE								
Description	Year	PW Factor	Best Estimate		High 90% Estimate		Low 90% Estimate	
			Est.	PW	Est.	PW	Est.	PW
(a) Replacement, Best Estimate	10.0	0.3855	10,000	3,855				
Replacement, Best Estimate	20.0	0.1486	10,000	1,486				
Salvage Value, Best Estimate*	25.0	0.0923	(5,000)	(461)				
		Total		4,880				
*5 Yr/10 Yr x $10,000 = $5,000								
(b) Replacem't, High 90% Est.	8.0	0.4665			10,000	4,665		
Replacem't, High 90% Est.	16.0	0.2176			10,000	2,176		
		Total				6,841		
(c) Replacem't, Low 90% Est.	12.5	0.3038*					10,000	3,038
* By Interpolation		Total						3,038

Figure 6.1: **Computation of PW Bounds for a Timing Uncertainty**

The confidence index approach is a simplified (approximate) version of the probabilistic approach. The assumptions on which the CI approach is based, the limitations of the approach, and the numerical measure it provides are discussed in the following paragraph. The CI calculation procedure is presented and used to assess uncertainties of various types in later paragraphs. The approach is also applied to several examples.

The CI approach is based on two assumptions: that the uncertainties in all cost data are normally distributed and that the high and low 90% estimates for each cost do, in fact, correspond to the true 90% points of the normal probability distribution for that cost. Fortunately, the results of CI calculations are not very sensitive to either of these assumptions. Even if the probability distributions are not perfectly normal, and even if the 90% estimates do not correspond exactly to the true 90% points of the probability distributions for the costs, the results of CI calculations are still sufficiently accurate for use in uncertainty assessment. As a general rule, the CI approach should be considered valid as long as:

1. the high and low 90% estimates are obtained at the same time and from the same source or sources as the best estimates and are considered to represent knowledgeable judgments rather than guesses, and
2. the differences between the PW of the best estimate for each cost and the PW values of the high and low 90% estimates for that cost are within 25% or so of each other. For example, consider the situation in which:

$$\text{Best estimate} = \$10,000$$
$$\text{Low 90\% estimate} = \$ 7,500$$
$$\text{High 90\% estimate} = \$12,000$$

Here the *low-side* difference ($10,000 – $7,500) is $2,500; the *high-side* difference ($12,000 – $10,000) is $2,000. Since $2,500 is not more than 25% greater than $2,000, the CI approach is considered valid and may be used here.

When the 25% condition does not hold for any cost related to the two alternatives being studied, the CI approach is not valid, and some other approach, such as sensitivity, should be used to assess the effects of uncertainties. When the CI approach is considered valid, the CI is computed as a ratio that takes into account two types of available information concerning the PW values of the alternatives being analyzed. These are:

1. the computed difference $PW_B - PW_A$ between the LCCA of the two alternatives, based on best estimates of their costs, and
2. the uncertainties in the cost data (with regard to both amounts and timing), as embodied in the high and low PW bounds for the costs related to each alternative.

The numerator of this ratio is $PW_B - PW_A$, which is the best estimate of the difference between the LCCA of the two alternatives. The denominator is a direct measure of the uncertainty in $PW_B - PW_A$. The uncertainty in PW_B is—for normally distributed data—found as the square root of the sum of the squares of the uncertainties in all component parts of PW_B. The uncertainty in PW_A is found similarly from uncertainties in the component parts of PW_A. The uncertainty in $PW_B - PW_A$ is the square root of the sum of the squares of the uncertainty in PW_B and in PW_A. The CI approach takes these uncertainties into account directly through its computations. The only judgments required are in estimating the 90% values, and reasonable errors in those judgments are not critical. The calculation of the CI ratio is detailed and illustrated in the following paragraphs.

Confidence Index Calculation

The confidence index for two design alternatives is calculated as follows:

Step 1. Use the high and low 90% estimates and the best estimate to calculate the high-side and low-side differences for each cost.

Step 2. Determine whether these differences are within 25% of each other for each cost. If so, continue. If not, use the sensitivity approach to assess the effects of uncertainties.

Step 3. Determine the difference in the PW values of the two alternatives, based on best estimates. This is the numerator of the CI approach.

Step 4. Compute the PW of the larger of the high-side and low-side differences for each cost, and compute its square. Add the squared PW values of all cost differences for both alternatives, and find the square root of this sum. This is the denominator of the CI ratio.

Step 5. Divide the result of step 3 by the result of step 4 to obtain the confidence index. Use this CI to evaluate the results of the original PW calculations as follows:

- If the CI is below about 0.15, assign a *low* confidence to the results of the LCCA. (A confidence index below 0.15 means that, in the long run, the alternatives with the lower computed LCC will incur the lower actual costs in fewer than 3 out of 5 cases.)
- If the CI is between 0.15 and 0.25, assign a *medium* confidence level to the results of the LCCA. (A confidence index between 0.15 and 0.25 means that, in the long run, the alternative with the lower computed LCC will incur the lower actual costs in about 9 or 10 cases out of 15.)
- If the CI is greater than about 0.25, assign a *high* confidence to the results of the LCCA. (A confidence index above 0.25 means that, in the long run, the alternative with the lower computed LCC will incur the lower actual costs in more than 2 out of 3 cases.)

The paragraphs that follow carry out the steps in the CI computation procedure. The paragraph numbers parallel the step numbers. The data and calculations are organized on a sample worksheet, and results are rounded to an appropriate number of significant figures. The analysis period for this example is 25 years, the discount rate is 10%, and the differential escalation rate is 0%. For the calculations, it is assumed that cost estimates for alternative L—the one with the lower LCC—were obtained as follows:

Cost item	Estimates		
	Low 90%	Best	High 90%
Initial investment	$44,300	$55,400	$66,500
M&R/year	300	500	750

M&R = maintenance and repair.

Further, the results of PW calculations based on best estimates indicate that alternative L has the lowest PW of three feasible alternatives. It is, therefore, one of two alternatives for which a confidence index will be computed. To simplify the illustrations, data and results for the second alternative in the computation—alternative H, the one with the higher LCC—will be shown, but will not be discussed in detail except as needed.

1. *High-side and low-side differences.* Enter a brief description of each cost item for each alternative, as necessary. Enter the low and high 90% estimates and the best estimate for each cost item. For each cost, calculate the low-side cost difference as best estimate minus low estimate, and the high-side difference as high estimate minus best estimate. Enter these differences as in Figure 6.2. For the initial cost of alternative L, for example, the low-side difference is $55.4 − $44.3 = $11.1 (costs in thousands). For the annual M&R cost, the high-side difference is $0.75 − $0.5 = $0.25. Data and differences are similarly entered for alternative H.

2. *Comparison of difference.* Compute the percentage difference between the high-side and low-side differences for each cost. For the initial cost for alternative L, these differences are equal, so the percentage difference is zero. For the annual M&R cost, the percentage difference is 0.05/0.20 × 100 = 25%. The high and low estimates may therefore be assumed to be approximately equidistant from the best estimate. A similar comparison is performed for alternative H (Figure 6.2).

3. *Numerator of CI ratio.* Determine the PW of the best estimate of each cost, and determine the sum of these PW values for each of the two alternatives. In most cases, these individual PW values and sums will have been calculated in the original PW computation for each alternative. For alternative L, the sum of the PW values of the best estimate is $59.9. For alternative H, it is $63.8. Enter these values. Compute the difference between the two sums, or PW values. Here this difference is $63.8 − $59.9 = $3.9. Enter this difference, as in Figure 6.2.

4. *Denominator of CI ratio.* For each cost, compute the PW of the high-side or low-side difference, whichever is greater. Since only cost amounts are involved here, these PW values may be computed as ratios. For the initial cost of alternative L, the PW of the difference is (11.1/55.4) × $55.4 = $11.1. For the M&R cost for alternative L, the larger difference is used to find a PW of (0.25/0.5) × $4.54 = $2.27.

Enter these PW values, and square each of them. For alternative L, $(11,100)^2 = 123,210,000$ and $(2,269)^2 = 5,149,541$. Enter these squares. Similarly, compute and enter the squares of the PW values of the cost differences for alternative H. Add the squared PW values of all the costs for both alternatives, and enter this sum. Here the sum is $210,047,705. Take the square root of this sum to obtain 14,493 as the denominator of the CI ratio. Enter this value, as shown in Figure 6.2.

5. *CI ratio*. Divide the result obtained in step 3 by the result of step 4 to obtain the confidence index. Here, the CI is 3,862/14,493 = 0.266. Assign a confidence level to the PW calculation, based on the CI. Because the CI is greater than 0.25, the results of the LCCA are assigned high confidence. That is, alternative L may be selected for implementation with a high degree of confidence that it is actually the alternative with the lowest life cycle cost.

Confidence Index Examples

In the following paragraphs, the CI approach is applied in two typical examples. The results of the calculations are shown, but the calculations are not discussed in detail. All data and results are presented on CI

LIFE CYCLE COST ANALYSIS (Present Worth Method)
CONFIDENCE INDEX (CI) COMPUTATION
Project/Location: Hospital, Detroit, Michigan

Subject: Example CI Computations
Description: System Selection
Project Life Cycle = 25 YEARS
Discount Rate = 10.00%

COST ITEMS:	ESTIMATES RANGE			DIFFERENCES IN ESTIMATES				PRESENT WORTH		
	LOW	HIGH	BEST	LOW SIDE	HIGH SIDE	DELTA %	ok	BEST ESTIMATE	DELTA	DELTA^2
ALTERNATIVE L(ow)										
Initial Construction Cost	44,300	66,500	55,400	11,100	11,100	0%	ok	55,400	11,100	123,210,000
Energy/ Fuel Annual Cost	0	0	0	0	0	0%	ok	0	0	0
Maintenance & Repair Annual Cost	300	750	500	200	250	25%	ok	4,539	2,269	5,149,541
Totals								59,939		
ALTERNATIVE H(igh)										
Initial Construction Cost	42,200	57,000	50,000	7,800	7,000	11%	ok	50,000	7,800	60,840,000
Energy/ Fuel Annual Cost	200	400	300	100	100	0%	ok	2,723	0	0
Maintenance & Repair Annual Cost	500	1,500	1,000	500	500	0%	ok	9,077	4,539	20,598,164
HVAC Added Cost	1,500	2,400	2,000	500	400	25%	ok	2,000	500	250,000
Totals								63,800		

Note = If high and low 90% estimates > 25%, then use sensitivity analysis

Difference: 3,862   Sum: 210,047,705
(SUM)^1/2: 14,493

Confidence Index = $\dfrac{\text{PW(High)} - \text{PW(Low)}}{(\text{PW Diff(High)}^2 + \text{PW Diff(Low)}^2)^{1/2}} = \dfrac{3{,}862}{14{,}493} = 0.266 = \text{HIGH}$

Confidence Assignment:
Low: CI<0.15
Medium: 0.15 < CI < 0.25
High: CI > 0.25

Figure 6.2: **CI Computations: High-Side and Low-Side Differences**

computation worksheets, as they would be in a complete economic analysis.

Floor Finishes

Figure 6.3 shows the confidence index computation for the two alternatives with the lowest LCCA in an analysis of floor finishes for an administrative facility. The analysis period is 25 years, the discount rate is 10%, and the differential escalation rate is 0%. The computation indicates low confidence in the choice of alternative L for implementation, because of uncertainties in cost amounts. The right-most column in Figure 6.3 shows that almost all the uncertainty is due to the estimates of M&R costs (indicated by the very large denominator amounts for M&R costs). The uncertainty could be mitigated—and the CI improved—if closer estimates of M&R costs could be obtained. In fact, this was done, and the results are shown in Figure 6.4. With new, closer M&R estimates, the confidence index is close to 0.25, and alternative L may be implemented with reasonable assurance that it will have the lowest costs over its lifetime.

Siting and Layout

Figure 6.5 is the confidence index computation for the two alternatives with the lowest LCCA in a study of layouts for a research building. The

LIFE CYCLE COST ANALYSIS (Present Worth Method)
CONFIDENCE INDEX (CI) COMPUTATION
Project/Location: Administrative Facility, Midtown, MI

Subject: Example CI Computations
Description: System Selection
Project Life Cycle = 25 YEARS
Discount Rate = 10.00%

COST ITEMS:	ESTIMATES RANGE			DIFFERENCES IN ESTIMATES				PRESENT WORTH		
	LOW	HIGH	BEST	LOW SIDE	HIGH SIDE	DELTA %	ok	BEST ESTIMATE	DELTA	DELTA^2
ALTERNATIVE L(ow)										
Initial Construction Cost	6,800	7,600	7,200	400	400	0%	ok	7,200	400	160,000
Energy/ Fuel Annual Cost	0	0	0	0	0	0%	ok	0	0	0
Maintenance & Repair Annual Cost	700	2,700	1,800	1,100	900	22%	ok	16,339	9,985	99,695,113
Totals								23,539		
ALTERNATIVE H(igh)										
Initial Construction Cost	13,700	15,300	14,500	800	800	0%	ok	14,500	800	640,000
Energy/ Fuel Annual Cost	0	0	0	0	0	0%	ok	0	0	0
Maintenance & Repair Annual Cost	400	2,000	1,200	800	800	0%	ok	10,892	7,262	52,731,300
Totals								25,392		

Note = If high and low 90% estimates > 25%, then use sensitivity analysis

Difference 1,854
Sum 153,226,413
(SUM)^1/2 12,378

$$\text{Confidence Index} = \frac{PW(High) - PW(Low)}{(PW\,Diff(High)^2 + PW\,Diff(Low)^2)^{1/2}} = \frac{1,854}{12,378} = 0.150 = LOW$$

Confidence Assignment:
Low: CI < 0.15
Medium: 0.15 < CI < 0.25
High: CI > 0.25

Figure 6.3: CI Computation: Floor Finishes, Original Estimates

analysis period is 40 years, the discount rate is 7%, and the differential escalation rate is 0%. The originally computed PW values, based on best estimates, are simply recorded in the *numerator* column on the CI computation worksheet. The PW values of the low-side and high-side differences are computed as proportions of the best estimate PW values. The salvage value is treated as a negative cost in computing both the numerator and the denominator of the CI ratio. The computed CI of 0.134 indicates a low level of confidence in the results of the original LCCA. Although alternative L has a lower computed life cycle cost than alternative H, there is a good chance that its total cost of ownership will be greater than that of alternative H if it is selected for implementation. Here, the initial-cost estimate seems to produce the most uncertainty, but the high-side and low-side estimates are all within about 7% of the best estimate. It may not be possible to refine these estimates further, in which case an alternative should be selected according to a criterion other than lowest LCC.

Sensitivity Approach

The sensitivity approach to uncertainty assessment, like the CI approach, is used to categorize the probability that, of two alternatives, the one with the lower computed LCC will, in fact, have the lower LCC. This probability is considered to be clear-cut if it is about 0.67 or more; in which case then the LCCA results are considered to be conclusive. However, unlike the CI approach, the sensitivity approach does not produce a direct measure of this probability. Instead, it is approximated through judgment based on available cost information, as well as (1) the results of a sensitivity analysis and (2) a computed break-even point. The situations in which the sensitivity approach is used are discussed in the following paragraphs. The required calculations are presented and used to assess uncertainties of various types, and the approach is applied to three LCCA cases.

Sensitivity analysis is, in essence, a method for determining how the value of one parameter is affected by variation in the value of a second parameter on which it depends. These are often called the *output* and *input* parameters, respectively. The method is most useful when the relation between the two parameters cannot be expressed in closed form.

Sensitivity analysis involves:

1. the assignment of several reasonable values to the input parameter,
2. the computation of the output parameter value that corresponds to each input parameter value, and
3. the analysis of the pairs of values of the form (input value, output value) that result from application of the method. These pairs may be tabulated or graphed to indicate the relationship between the two parameters.

The sensitivity method is useful in economic analysis, for example, to determine how variation in the initial cost of a particular HVAC system (as the input variable) affects the LCC of the system (as the output variable).

The LCC of the system may be computed for each of several possible values for the initial cost (including, perhaps, the best estimate of this cost, as well as at least one higher and one lower estimate). The resulting pairs (initial cost, LCC) may then be tabulated or plotted to indicate how the LCC is affected by variation from the best estimate of the initial cost. Since such variation is caused by uncertainties in the initial cost, the table or plot actually indicates how the LCC of this alternative is affected by these uncertainties.

This type of information can be of great value in assessing the effect of uncertainties in cost data. However, additional information, in the form of the *break-even* point, is also available from a sensitivity analysis. In general, the break-even point is a value of the input parameter that results in a particular (or significant) value for the output parameter. In the sensitivity approach to uncertainty assessment, the output parameter is always the LCC of the least-cost alternative; the input parameter is always the input-data element (cost element). The break-even point is then the value of the cost element that causes the LCC of the least-cost alternative to equal the LCC of the next-lowest-cost alternative. The break-even point is of importance because it is the input parameter value for which the least-cost alternative ceases to be the least-cost alternative.

LIFE CYCLE COST ANALYSIS (Present Worth Method)
CONFIDENCE INDEX (CI) COMPUTATION
Project/Location: Administrative Facility, Midtown, MI

Subject: Example CI Computations
Description: System Selection
Project Life Cycle = 25 YEARS
Discount Rate = 10.00%
Present Time = Occupancy Date

	ESTIMATES RANGE			DIFFERENCES IN ESTIMATES				PRESENT WORTH		
	LOW	HIGH	BEST	LOW SIDE	HIGH SIDE	DELTA %	ok	BEST ESTIMATE	DELTA	DELTA^2
ALTERNATIVE L(ow)										
Initial Construction Cost	6,800	7,600	7,200	400	400	0%	ok	7,200	400	160,000
Energy/ Fuel Annual Cost	0	0	0	0	0	0%	ok	0	0	0
Maintenance & Repair Annual Cost	1,300	2,300	1,800	500	500	0%	ok	16,339	4,539	20,598,164
Totals								23,539		
ALTERNATIVE H(igh) COST ITEMS										
Initial Construction Cost	13,700	15,300	14,500	800	800	0%	ok	14,500	800	640,000
Energy/ Fuel Annual Cost	0	0	0	0	0	0%	ok	0	0	0
Maintenance & Repair Annual Cost	600	1,800	1,200	600	600	0%	ok	10,892	5,446	29,661,356
Totals								25,392		

Note = If high and low 90% estimates > 25%, then use sensitivity analysis

Difference 1,854   Sum 51,059,520
(SUM)^1/2 7,146

$$\text{Confidence Index} = \frac{PW(High) - PW(Low)}{(PW\,Diff(High)^2 + PW\,Diff(Low)^2)^{1/2}} = \frac{1,854}{7,146} = 0.259 = \text{HIGH}$$

Confidence Assignment:
Low: CI<0.15
Medium: 0.15 < CI < 0.25
High: CI > 0.25

Figure 6.4: CI Computation: Floor Finishes, Re-estimated Costs

Suppose that in a certain LCCA, Alternative A is found to be the lowest LCC option, and Alternative B is the next lowest. Suppose also that the best estimate of the initial cost of Alternative A is $100,000, and that the break-even point for the initial cost is $117,000. Then, no matter how the initial cost of Alternative A may vary due to uncertainties, it will remain the lowest-cost alternative (that is, PW_A will be less than PW_B) as long as its initial cost is less than $117,000, and nothing else changes. If the chances are good that the initial cost will be below $117,000, then the chances are also good that the results of the LCCA are conclusive—i.e., that Alternative A does, in fact, have the lowest LCC, and Alternative B does not.

This is the essence of the sensitivity approach. The analyst performs a sensitivity analysis for a particular input-data element (either a dollar cost amount or cost timing), computes the break-even point for that element, and then uses all available information—including personal experience with construction costs—to categorize the probability that the actual value of the input parameter will be on the same side of the break-even point as the best estimate. If that probability is 0.67 or more (two out of three chances or better), the results of the LCCA are considered to be conclusive. If that probability is close to 0.5 (one chance out of two), the results are inconclusive.

The sensitivity analysis itself requires no judgment on the part of the analyst, but only the computation of several LCC values. The break-even point is determined directly from these results by solving for the value of the input parameter that yields a particular LCC. There is, however, one important restriction on the application of the approach. As indicated by the phrase "and nothing else changes" above, variations in input-data elements other than the input parameter can affect the results of uncertainty assessments based on the sensitivity approach. For that reason, *the approach should be applied only when the uncertainty in one input-data element is predominant*—and it should be applied with that element as the input parameter. There are essentially two types of situations in which the sensitivity approach should be used in lieu of the confidence index, as follows:

1. When, even before the assessment is undertaken, the uncertainty in one parameter is known to be predominant, *and* the design professional either is unable to determine the high and low 90% estimates required for the CI approach or is sure that they do not satisfy the 25% condition cited previously (so that there is no point in using the CI approach to uncertainty assessment).

2. When the CI approach has been attempted, and initial calculations indicate that the uncertainty in one input parameter predominates, *and* the 90% estimates for that parameter do not satisfy the 25% condition (so that the CI approach cannot be used).

A check to determine whether the uncertainty in one parameter predominates is incorporated in the calculation procedure, described in the next section.

Sensitivity Analysis Calculations

The following section demonstrates the procedure for:
- applying the sensitivity approach,
- determining and verifying that the uncertainty in a particular input-data element is predominant.

The sensitivity approach to uncertainty assessment may be applied in an LCCA as follows:

Step 1. Verify that the uncertainty in one input-data element (the dollar amount or timing of a cost for one of the two lowest LCC alternatives) is the predominant uncertainty. No formal verification is needed when it is known beforehand that uncertainty in a particular input-data element predominates.

Step 2. Conduct a sensitivity analysis of the input element whose uncertainty is predominant, using the best estimate and at least two other values for this element. Tabulate and plot the results of the analysis.

LIFE CYCLE COST ANALYSIS (Present Worth Method)
CONFIDENCE INDEX (CI) COMPUTATION
Project/Location: Research Building, Chicago, IL

Subject: Example CI Computations
Description: Siting and Layout Selection
Project Life Cycle = 40 YEARS
Discount Rate = 7%
Present Time = Occupancy Date

	ESTIMATES RANGE			DIFFERENCES IN ESTIMATES				PRESENT WORTH		
	LOW	HIGH	BEST	LOW SIDE	HIGH SIDE	DELTA %	ok	BEST ESTIMATE	DELTA	DELTA^2
ALTERNATIVE L(ow) COST ITEMS										
Initial Construction Cost	805,000	920,000	865,000	60,000	55,000	9%	ok	865,000	60,000	3,600,000,000
Energy/ Fuel Annual Cost	2,000	3,100	2,500	500	600	0%	ok	33,329	0	0
Distillate Fuel Annual Cost	1,800	3,100	2,400	600	700	17%	ok	31,996	9,332	87,089,886
Maintenance & Repair Annual Cost	3,600	4,800	4,200	600	600	0%	ok	55,993	7,999	63,984,406
Salvage Value	(78,000)	(95,000)	(86,500)	(8,500)	(8,500)	0%	ok	(5,776)	(568)	322,151
Totals								980,543		
ALTERNATIVE H(igh) COST ITEMS										
Initial Construction Cost	805,000	925,000	867,500	62,500	57,500	9%	ok	867,500	62,500	3,906,250,000
Energy/ Fuel Annual Cost	2,400	3,200	2,800	0	400	0%	ok	37,329	0	0
Distillate Fuel Annual Cost	2,200	3,000	2,600	400	400	0%	ok	34,662	5,333	28,437,514
Maintenance & Repair Annual Cost	3,800	5,000	4,400	600	600	0%	ok	58,660	7,999	63,984,406
Salvage Value	(79,500)	(95,000)	(86,800)	(7,300)	(8,200)	11%	ok	(5,796)	(487)	237,609
Totals								992,355		

Note = If high and low 90% estimates > 25%, then use sensitivity analysis

Difference: 11,812
Sum: 7,750,305,972
(SUM)^1/2: 88,036

$$\text{Confidence Index} = \frac{PW(High) - PW(Low)}{(PW\ Diff(High)^2 + PW\ Diff(Low)^2)^{1/2}} = \frac{11{,}812}{88{,}036} = 0.134 = HIGH$$

Confidence Assignment:
Low: CI < 0.15
Medium: 0.15 < CI < 0.25
High: CI > 0.25

Figure 6.5: CI Computation: Siting and Layout

Step 3. Determine the break-even point

Step 4. Categorize the probability that the actual value of the input-data element will lie on the same side of the breakeven point as the best estimate of that element, by estimating whether that probability is expected to be (1) greater than about 0.67, (2) near 0.50, or (3) well above 0.50, but below 0.7. In case 1, the results of the LCCA are considered to be conclusive; in case 2, the results are inconclusive; in case 3, additional cost information may be helpful in determining whether the LCCA results should be used to select an alternative for implementation. The judgment as to the magnitude of this probability should be made by the individual or individuals who provided the best estimate for the input parameter. This judgment should be based on knowledge of the best estimate and break-even point, as well as any other available information concerning the costs at hand.

An example of these sensitivity analysis calculations follows. Either of two types of mechanical equipment may be selected for a certain application. The standard type, Alternative 1, is routinely used by mechanical engineers. A fairly new type, Alternative 2, is more expensive in terms of first cost, but much less expensive to maintain. Both have exactly the same energy requirements. An LCCA was conducted to determine which had the lesser total cost of ownership. Contacts with manufacturers established the present initial cost of Alternative 2 at $100,000. However, industry representatives noted that this price was about to increase substantially—most likely by 50%, but possibly by as little as 10%, or as much as 75%. For use in the LCCA, the best estimate of the initial cost of Alternative 2 was thus taken as $150,000. Complete best-estimate cost data for the two alternatives were as follows:

Cost	Alternative 1	Alternative 2	Years incurred
Initial	$80,000	$150,000	0
M&R/year	$10,000	$1,000	1-25
Energy/year	N/A	N/A	N/A

N/A: not applicable

The PW values (LCC) of the two alternatives, based on these best estimate cost data and 10% discount rate and 25-year analysis period, were found to be $170,770 for Alternative 1 and $159,077 for Alternative 2. To complete the LCCA, the sensitivity approach was used to assess the effect of the uncertainty concerning the impending increase in the initial cost of Alternative 2:

Step 1. It is evident from the nature of the situation, as described above, that the uncertainty in the initial cost is predominant. No further verification is needed.

Step 2. The data and calculations for the sensitivity analysis are organized on sample worksheets in Figures 6.6 and 6.7. On the sensitivity worksheet (Figure 6.6), a brief description of the predominant input-data element (the input parameter) is entered, along with representative values for this parameter, including at least the best estimate and reasonable high and low estimates. (The high and low 90% estimate described in the CI computation may be used here.) Figure 6.6 includes two intermediate values for this parameter as well. Next, the PW (LCC) of each alternative is computed for each representative value of the input parameter. The PW calculations for the best estimate data are shown in Figure 6.7. Similar calculations were performed for five possible initial cost price increases in the range 10% to 75% for Alternative 2. The PW of the M&R cost is added to each of these PW values to obtain the PW of Alternative 2 for each possible initial cost. The resulting pairs (initial cost, PW) are tabulated as they are listed on Figure 6.6. They are also plotted on the axes provided at the lower left, in preparation for determination of the break-even point. Alternative 1 is not affected by this uncertainty or these calculations; its LCC remains fixed at $170,770.

Step 3. The break-even point may be determined graphically by plotting the LCC of Alternative 1 and Alternative 2 on the same graph. The

LIFE CYCLE COST ANALYSIS (Present Worth Method)
SENSITIVITY ANALYSIS (SA) COMPUTATION
Project/Location: Research Building, Chicago, IL

Subject: Example SA Computations
Description: Mechanical Equip. Selection
Project Life Cycle = 25 YEARS
Discount Rate = 10%
Present Time Occupancy Date

Alternative 1: Standard Type Mechanical Equipment
Alternative 2: New Low Maintenance Mechanical Equipment

Notes	Parameter Studied	Representative Values	Alt 1 Parameter Value	Alt 1 PW	Alt 2 Parameter Value	Alt 2 PW	Alt 3 Parameter Value	Alt 3 PW	Alt 4 Parameter Value	Alt 4 PW
SENSITIVITY TEST #1	Initial Cost	Low	80,000	170,770	110,0001	119,077				
		Intermediate	80,000	170,770	125,0001	134,077				
		Best Estimate	80,000	170,770	150,0001	159,077				
		Intermediate	80,000	170,770	160,000	169,077				
		High	80000	170,770	175,0001	184,077				
SENSITIVITY TEST #2		Low								
		Intermediate								
		Best Estimate								
		Intermediate								
		High								

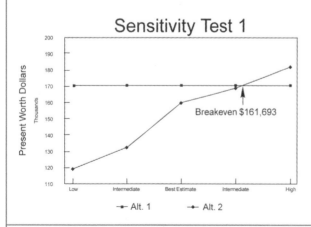

Figure 6.6: **Sensitivity Analysis**

input-parameter value at the point of intersection of the two plots is the break-even point. As Figure 6.6 shows, the horizontal line representing the LCC of Alternative 1 intersects the plot for Alternative 2 at the parameter value of $161,693. This is the break-even point, or the initial cost, (for Alternative 2) which would cause the LCC of Alternative 2 to equal that of Alternative 1. The same result may be obtained, by interpolation, from the tabulation on the sensitivity worksheet.

Step 4. The break-even point of $161,693 and best estimate of $150,000 clarify this step as follows: the analyst and other members of the design team need to categorize the probability that the actual initial cost of Alternative 1 will be *lower* than $161,693 (because the best estimate is lower than the break-even point). Moreover, they need only place that probability in one of three magnitude categories:

a. About 0.67 or higher (that is, 2 to 1 odds or more that the initial cost will be lower than $161,693)

b. Close to 0.5 (that is, about even odds that the initial cost will be lower than the break-even point)

LIFE CYCLE COST ANALYSIS
PRESENT WORTH (PW) COMPUTATION - BEST ESTIMATE
Project/Location: Research Building, Chicago, IL

Subject: Example Sensitivity Computations
Description: Mechanical Equipment Selection
Project Life Cycle = 25 YEARS
Discount Rate = 10.00%
Present Time = Occupancy Date

				Alternative 1 Standard Type Mechanical Equipment		Alternative 2 New Low Maintenance Mechanical Equipment		Alternative 3		Alternative 4	
INITIAL COSTS	Quantity	UM	Unit Price	Est.	PW	Est.	PW	Est.	PW	Est.	PW
Construction Costs											
A. Initial Const Cost (Alt. 1)	1 LS		80,000	80,000	80,000		0		0		0
B. Initial Const Cost (Alt. 2)	1 LS		150,000		0	150,000	150,000		0		0
C.					0		0		0		0
D.					0		0		0		0
E.					0		0		0		0
F.					0		0		0		0
G.					0		0		0		0
Total Initial Cost					**80,000**		**150,000**		**0**		**0**
Initial Cost PW Savings (Compared to Alt. 1)							**(70,000)**		**0**		**0**
REPLACEMENT COST/ SALVAGE VALUE											
Description	Year		PW Factor								
A.			1.0000		0		0		0		0
B.			1.0000		0		0		0		0
C.			1.0000		0		0		0		0
D. Salvage Value (Alt. 1)	40		0.0221	0	0		0		0		0
E. Salvage Value (Alt. 2)	40		0.0221			0	0		0		0
Total Replacement/Salvage Costs					**0**		**0**		**0**		**0**
ANNUAL COSTS											
Description	Escl. %		PWA								
A. Energy/ Fuel Annual Cost	0.000%		9.077	N/A	0	N/A	0		0		0
B. Distillate Fuel Annual Cost	0.000%		9.077	N/A	0	N/A	0		0		0
C. Maintenance & Repair Annual Cost	0.000%		9.077	10,000	90,770	1,000	9,077		0		0
D.	0.000%		9.077		0		0		0		0
E.	0.000%		9.077		0		0		0		0
F.	0.000%		9.077		0		0		0		0
Total Annual Costs (Present Worth)					**90,770**		**9,077**		**0**		**0**
Total Life Cycle Costs (Present Worth)					**170,770**		**159,077**		**0**		**0**
Life Cycle Savings (Compared to Alt. 1)							**11,693**		**0**		**0**

Figure 6.7: **PW Calculations: Best Estimate**

c. Somewhere between the two but closer to 0.67 than to 0.50

They may also decide that:

d. No reasonable judgment is possible.

If they choose category *a*, the results of the LCCA are conclusive, and Alternative 2 should be selected for implementation on the basis of least PW. If they decide the probability is in category *b*, the results of the LCCA are inconclusive and the selection of an alternative must be based on supplementary ranking criteria. If they choose category *c*, they have essentially decided that they cannot decide between categories *a* and *b*; some additional research might provide clarification. In this case, they could contact manufacturers' representatives again, asking them to categorize the probability of the initial cost being less than $161,693. If this additional research leads to the choice of category *a*, the LCCA results are conclusive. If it leads to the choice of category *b*, the results are inconclusive. If it again leads to the choice of category *c*, the LCCA results must be considered inconclusive and supplementary ranking criteria must be used to select an alternative for implementation. If the design team feels that no reasonable judgment of this probability is possible (category *d*), then the entire LCCA must be considered null and void—as if it were never conducted—and an alternative should be selected solely on the basis of initial procurement cost.

Verification of Predominant Uncertainty

In most cases, the analyst performing an LCCA will know, from the input data for the two least-LCC alternatives, whether the uncertainty in a particular input-data element is predominant. In such cases, the sensitivity approach may be used without further verification of predominance. In all other situations, before applying the sensitivity approach, the designer should verify that the uncertainty in one input-data element is predominant. The following procedure, which includes two rule-of-thumb conditions for determining predominance, may be used for this purpose:

Step 1. Compute the present worth of the best estimate and the high and low 90% estimates for each input-data element for the two lowest LCC alternatives.

Step 2. Compute the difference between the present worth of the best estimate and the present worth of the high 90% estimate for each of these input-data elements. Repeat for the low 90% estimate.

Step 3. Determine whether either of the following conditions holds for a particular input-data element; if so, the uncertainty in that element predominates, and that element may be used as the input parameter in the sensitivity approach:

- *Condition 1.* The smaller of the difference between the PW of its best estimate and the PW values of its 90% estimates is at least twice as large as the larger of these differences for any other input-data element and is at least equal to the sum of the larger of these differences for all other cost elements for both alternatives.

- *Condition 2.* The smaller of the differences between the PW of its best estimate and the PW value of its 90% estimates is at least 4 times as large as the next larger such difference for any cost element for both alternatives.

The computation of present worth values for step 1 is covered earlier in this chapter. The remaining steps are self-explanatory. They are illustrated in the following three examples.

Sensitivity Analysis Examples

The following paragraphs describe the sensitivity approach applied to the assessment of uncertainties in three LCCAs. The first example illustrates the use of the approach in connection with an input parameter that affects two costs related to a single alternative, as well as the use of the verification procedure described in the previous paragraph. The second example illustrates the approach applied to uncertainty in cost growth projections that affect both lowest cost alternatives. In the third example, the approach is applied to uncertainty in the economic life of the facility (and thus in the analysis period). The data and computations for these examples give rise to various probability judgment situations. The reasoning by which a particular probability category was chosen is discussed in each case.

Building Exterior Closure

In an LCCA of building exterior closures for a warehouse, the lowest LCC alternative was a conventional concrete block structure (Alternative 1), and the next lowest LCC alternative was an architectural fabric skin (Alternative 2). An uncertainty assessment by means of the CI approach was attempted, but the initial cost for the fabric alternative did not pass the 25% test. Uncertainties were then assessed via the sensitivity approach. The analysis period is 25 years, the discount rate is 10%, and the differential escalation rate is 0%. Complete cost data were as follows:

Cost	Alternative 1	Alternative 2	Year Incurred
Initial	Low $120,000 Best 140,000 High 160,000	Low $70,000 Best 100,000 High 140,000	0
Replacement		Low $70,000 Best 100,000 High 140,000	10
M&R/year	Low $1,000 Best 1,500 High 2,000	Low $1,250 Best 1,750 High 2,250	1–25
Energy loss/year	Baseline	Low $1,600* Best 2,000* High 2,400*	1–25

*Cost of additional annual energy loss relative to conventional enclosure.

Step 1: Verification of predominance. There was considerable uncertainty regarding the initial cost of the fabric enclosure, which suggested that this might be the predominant uncertainty. To determine and verify predominance, the PW values of all high and low 90% bounds and best estimates were first calculated for both alternatives (see Figure 6.8). Then it was noted that the cost of the fabric enclosure affects both the initial cost and the replacement cost for Alternative 2. Therefore the smaller of the PW differences for the initial cost and for the replacement cost for the fabric enclosure were added to obtain $30,000 + $11,566 = $41,566 as the smallest PW difference attributable to the cost of the fabric. This total was compared with the larger difference for every other cost, and was found to be about twice as large as any of these differences (it is actually slightly more than twice the difference of $20,000 between the best estimate and upper bound for the initial cost of Alternative 1). The total is also greater than the sum of the larger differences of all other cost elements and therefore satisfies condition 1 paragraph mentioned earlier in the chapter. The uncertainty in the initial cost of the fabric is predominant. Note that, if the uncertainty in the initial cost of alternative 1 were somewhat larger (that is, if its high and low 90% bounds were somewhat wider apart), then the uncertainty

---

LIFE CYCLE COST ANALYSIS
PRESENT WORTH (PW) COMPUTATION - BEST ESTIMATE
Project/Location: Warehouse, Miami, FL

Subject:	Example Sensitivity Computations		Alternative 1		Alternative 2		Alternative 3		Alternative 4	
Description:	Exterior Closure Selection		Conventional Concrete Block		Architectural Fabric Skin					
Project Life Cycle = 25 YEARS										
Discount Rate = 10.00%										
Present Time = Occupancy Date										

INITIAL COSTS	Quantity	UM	Est.	PW	Est.	PW	Est.	PW	Est.	PW
Construction Costs										
A. Initial Cost (Low Estimate)	1	LS	120,000	120,000	70,000	70,000		0		0
B. Initial Cost (Best Estimate)	1	LS	140,000	140,000	100,000	100,000		0		0
C. Initial Cost (High Estimate)	1	LS	160,000	160,000	140,000	140,000		0		0
D.			*Delta = 20,000 & 20,000*		*Delta = 30,000 & 40,000*			0		0
E.						0		0		0
Total Initial Cost (Best Estimate Only)				**140,000**		**100,000**		**0**		**0**
Initial Cost PW Savings (Compared to Alt. 1)						**40,000**		**0**		**0**

REPLACEMENT COST/ SALVAGE VALUE										
Description	Year	PW Factor								
A. Replacem't Cost (Low Est.)	10	0.3855		0	70,000	26,988		0		0
B. Replacem't Cost (Best Est.)	10	0.3855		0	100,000	38,554		0		0
C. Replacem't Cost (High Est.)	10	0.3855		0	140,000	53,976		0		0
D.		1.0000		0	*Delta = 11,566 & 15,422*			0		0
E.		1.0000			0	0		0		0
Total Replacement/Salvage Costs (Best Estimate Only)				**0**		**38,554**		**0**		**0**

ANNUAL COSTS										
Description	Escl. %	PWA								
A. Maint. & Repair Cost (Low Est.)	0.000%	9.077	1,000	9,077	1,250	11,346		0		0
B. Maint. & Repair Cost (Best Est.)	0.000%	9.077	1,500	13,616	1,750	15,885		0		0
C. Maint. & Repair Cost (High Est.)	0.000%	9.077	2,000	18,154	2,250	20,423		0		0
			Delta = 4,539 & 4,539		*Delta = 4,539 & 4,539*					
D. Energy Loss/Year (Low Est.)*	0.000%	9.077		0	1,600	14,523		0		0
E. Energy Loss/Year (Best Est.)*	0.000%	9.077		0	2,000	18,154		0		0
F. Energy Loss/Year (High Est.)*	0.000%	9.077		0	2,400	21,785		0		0
					Delta = 3,631 & 3,631					
Total Annual Costs (Best Estimate Only)				**13,616**		**34,039**		**0**		**0**
*Cost of additional energy relative to conventional										
Total Life Cycle Costs (Best Estimate Only)				**153,616**		**172,593**		**0**		**0**
Life Cycle Savings (Compared to Alt. 1)						**(18,977)**		**0**		**0**

Figure 6.8: **PW Calculations for Predominance Verification: Exterior Closure**

in the initial cost of Alternative 2 would not satisfy either condition for predominance, and the sensitivity approach would not be valid here.

Step 2: Sensitivity analysis. The PW values required for a sensitivity analysis based on the initial cost of alternative 2 are shown in Figure 6.9.

Step 3: Break-even point. The break-even point is found graphically in Figure 6.9 by plotting, on the same graph, the results of the sensitivity analysis for Alternative 2 and the horizontal line representing the LCC of Alternative 1. The result is a break-even point of $86,304.

Step 4: Probability judgment. To determine whether the LCCA could be considered conclusive, the design team needed to categorize the probability that the cost of the fabric enclosure would be greater than $86,304—that is, on the same side of the break-even point as the best estimate of $100,000. They were unanimous in judging this probability to be greater than 0.67 (2 to 1 odds). Accordingly, the results of the

LIFE CYCLE COST ANALYSIS
SENSITIVITY ANALYSIS (SA) COMPUTATION
Project/Location: Warehouse, Miami, FL

Subject: Example SA Computations
Description: Mechanical Equip. Selection
Project Life Cycle = 25 YEARS
Discount Rate = 10%
Present Time Occupancy Date

Notes	Parameter Studied	Representative Values	Alternative 1 Conventional Concrete Block Parameter Value	PW	Alternative 2 Architectural Fabric Skin Parameter Value	PW	Alternative 3 Parameter Value	PW	Alternative 4 Parameter Value	PW
SENSITIVITY TEST #1	Initial Cost	Low	140,000	153,616	70,000	131,027				
		Intermediate	140,000	153,616	85,000	151,810				
		Best Estimate	140,000	153,616	100,000	172,593				
		Intermediate	140,000	153,616	160,000	200,304				
		High	140,000	153,616	140,000	228,015				
SENSITIVITY TEST #2		Low								
		Intermediate								
		Best Estimate								
		Intermediate								
		High								

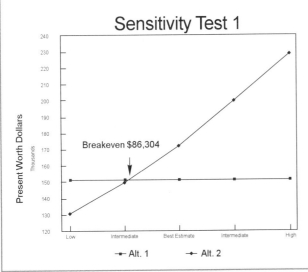

Figure 6.9: **Sensitivity Analysis Results and Break-Even Plot: Exterior Closure**

LCCA were considered to be conclusive, and the conventional, concrete block enclosure (Alternative 1) was ranked economically higher.

Water-Cooling System

During the design of a major new hospital complex, an LCCA was conducted for part of the overall water-cooling system for the complex. The analysis period was 25 years, and the discount rate was 10%. Nothing was unusual about the alternatives, the related costs, or the LCCA results. However, the future growth rate for the cost of electricity was very much in doubt, owing to difficulties that the local utility company had experienced with two nuclear power plants. The utility company had invested heavily in the construction of the plants, but construction delays, licensing problems, and reduced energy demands resulting from conservation had led to doubts as to when (and whether) the plants would be made operational. To remain solvent, the utility company had already raised its demand charges twice in the last two years—in real terms over and above the general rate of inflation. The company was planning further increases in demand charges in each of the next three years, after which no further rate increases were expected. The amount of these increases was not yet determined; current estimates ranged from a 10% to a 75% increase in each of the three years, with a 50% increase most likely to be approved by the public utilities commission. The sensitivity approach was used to assess the uncertainty in the electricity-cost growth rate.

Step 1: Verification of predominance. The situation itself indicates strongly that uncertainty in the growth rate of the electricity demand charge is the predominant uncertainty. In fact, this particular feature (the water-cooling system) was selected for analysis partly because it uses large amounts of electricity and because the future cost of that electricity is so uncertain. Therefore, no formal verification of predominance was necessary.

Step 2: Sensitivity analysis. Figure 6.10 shows the computation of the LCC for the two lowest-LCC alternatives, based on a 50% (best estimate) increase in the electricity demand charge. Since the input parameter for this sensitivity analysis (the electricity-cost growth rate) affects both these alternatives, sensitivity calculations must be performed for both. Calculation of the PW of the electricity demand charge for several possible growth rates from 0% to 100% (including the 50% best estimate) is required for both alternatives. For example, the PWA factor for Alternative 1 demand charge for a 25% annual cost growth for three years would be calculated as follows:

$$\begin{aligned} PWA_1 &= (1+i)^n \times PWA & i = 10\%, n = 25 \\ &= (1 + 0.25)^3 \times 9.077 \\ &= 1.9531 \times 9.077 \\ &= 17.729 \end{aligned}$$

Each of the demand charges must be added to the PW values of all other pertinent costs for each alternative to obtain the tabulation shown at the top of Figure 6.11. For example, to obtain the PW for Alternative 1 corresponding to a 25% growth rate, the PW of $319,115 computed for the alternative growth rate in Figure 6.10 must be added to the PW values of the initial cost, the M&R cost, and the electricity kilowatt hour cost for alternative 1 (See Figure 6.10) to obtain $95,000 + 29,047 + 34,493 + 319,115 = $477,654.

Step 3: Break-even point. The sensitivity data in Figure 6.11 indicate that both alternatives have an LCC of $383,200 when the demand-charge growth rate is 11.2%. The plot of the sensitivity data (also in Figure 6.11) also indicates a break-even point of 11.2%.

Step 4: Probability judgment. The required judgment here is the probability that the demand-charge growth rate (in real dollar terms) will *exceed* 11.2%, since the best estimate *exceeds* the break-even point of 11.2%. The design team decided not to categorize this probability as greater than 0.67 (2 to 1 odds or better), mainly because the PW plots of the two alternatives in Figure 6.11 are so close to each other in the range from 0% to about 60%. Instead, they categorized the probability

LIFE CYCLE COST ANALYSIS
PRESENT WORTH (PW) COMPUTATION - BEST ESTIMATE
Project/Location: Hospital Complex, St. Louis, MO

Subject: Example Sensitivity Computations
Description: Water Cooling System
Project Life Cycle = 25 YEARS
Discount Rate = 10.00%
Present Time = Occupancy Date

				Alternative 1 Conventional Water Cooling System		Alternative 2 Energy Efficient Water Cooling System		Alternative 3		Alternative 4	
INITIAL COSTS	Quantity	UM		Est.	PW	Est.	PW	Est.	PW	Est.	PW
Construction Costs											
A. Initial Construction Cost	1	LS		95,000	95,000	140,000	140,000		0		0
B.					0		0		0		0
C.					0		0		0		0
D.					0		0		0		0
E.					0		0		0		0
Total Initial Cost (Best Estimate)					95,000		140,000		0		0
Initial Cost PW Savings (Compared to Alt. 1)							(45,000)		0		0
REPLACEMENT COST/ SALVAGE VALUE											
Description	Year		PW Factor								
A.			1.0000		0	0	0		0		0
B.			1.0000		0	0	0		0		0
C.			1.0000		0	0	0		0		0
D.			1.0000		0	0	0		0		0
E.			1.0000		0	0	0		0		0
Total Replacement/Salvage Costs					0		0		0		0
ANNUAL COSTS		Differ. Escl. %	PWA								
A. Maintenance & Repair Cost		0.000%	9.077	3,200	29,047	4,200	38,124		0		0
B. Electric (Kwh) Annual Cost		0.000%	9.077	3,800	34,493	3,000	27,231		0		0
C. Electricity Demand Charge: (1)											
Low Estimate	0.0%	0.000%	9.077	18,000	163,387	14,250	129,348		0		0
Intermediate Est.	25.0%	0.000%	17.729	18,000	319,115	14,250	252,632		0		0
Best Estimate	50.0%	0.000%	30.635	18,000	551,430	14,250	436,549		0		0
Intermediate Est.	75.0%	0.000%	48.647	18,000	875,651	14,250	693,223		0		0
High Estimate	100.0%	0.000%	72.616	18,000	1,307,094	14,250	1,034,783		0		0
Total Annual Costs (Best Estimate)					614,969		501,904		0		0
(1) Potential elect. growth rates per year for next 3 years											
Total Life Cycle Costs (Best Estimate Only)					709,969		641,904		0		0
Life Cycle Savings (Compared to Alt. 1)							68,066		0		0

Figure 6.10: **PW Calculations, Best Estimates, Water Cooling Systems**

as being between 0.50 and 0.67 but closer to the latter (or, alternatively, as significantly greater than 0.50 but below 0.67). Following recommended procedure, they then solicited additional knowledgeable judgments of this probability. First, representatives of the utility company were asked to categorize the probability that the real cost growth rate would exceed 11.2% per year over the 3-year period. All the representatives selected the highest category. Contacts were then made with several local and regional economists and several community and industry leaders, and they too were asked to select the appropriate category. Most opted for a probability above 0.67.

On the basis of these findings, the design team reconsidered their original judgment and agreed that the highest probability category was the most appropriate in light of the best information available at the time. Accordingly, the results of the LCCA were considered to be conclusive.

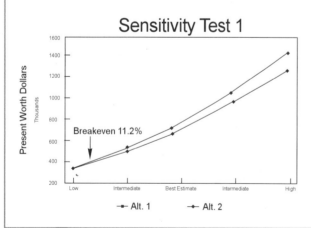

LIFE CYCLE COST ANALYSIS
SENSITIVITY ANALYSIS (SA) COMPUTATION
Project/Location: Hospital Complex, St. Louis, MO

Subject: Example SA Computations
Description: Mechanical Equip. Selection
Project Life Cycle = 25 YEARS
Discount Rate = 10%
Present Time
Occupancy Date

Notes	Parameter Studied	Representative Values	Alternative 1 Conventional Water Cooling System Parameter Value	PW	Alternative 2 Energy Efficient Water Cooling System Parameter Value	PW	Alternative 3 Parameter Value	PW	Alternative 4 Parameter Value	PW
SENSITIVITY TEST #1	Rate of Cost Growth for Electricity Demand Charge	Low	0.00%	321,926	0.00%	334,703				
		Intermediate	25.00%	477,654	25.00%	457,987				
		Best Estimate	50.00%	709,969	50.00%	641,904				
		Intermediate	75.00%	1,034,190	75.00%	898,578				
		High	100.00%	1,465,633	100.00%	1,240,137				
SENSITIVITY TEST #2		Low								
		Intermediate								
		Best Estimate								
		Intermediate								
		High								

Figure 6.11: **Sensitivity Analysis Results and Break-Even Plot: Water Cooling Systems**

Solar-Assisted Domestic Hot Water

A student recreation center including a gymnasium was to be designed for a university in a desert environment. Solar feasibility studies were undertaken for various possible applications. Among these was a study for the domestic hot-water system. At the time this study was being conducted, there was some question concerning the routine use of 25 years for the economic life of solar-energy systems. The university facility manager requested the student recreation center design team conduct a sensitivity study to show the effect of varying the economic life of each solar-energy system considered. The economic life was to be varied from 10 to 25 years, with a best estimate of 18 years.

Moreover, the validity of the results of the study was to take into account the uncertainty in the economic life of the solar-energy system, but ignore all other uncertainties. The projected economic life of the student recreation center was to be taken as 50 years. Uncertainties were assessed by means of the sensitivity approach. Because the study and the uncertainty assessment were both to be performed via the sensitivity approach, they could be (and were) performed simultaneously. The PW computations for the solar-energy system, based on the best-estimate analysis period of 18 years, are shown in Figure 6.12. The analysis was conducted using a discount rate of 7% and the results show PW savings of $6,454. The uncertainty in these results was then assessed.

Step 1: Verification of predominance. Verification was not needed, since the design team was directed to simply ignore all uncertainties other than those related to the analysis period or economic life.

Step 2: Sensitivity analysis. The PW values of the various costs were computed for several analysis periods (from 10 to 25 years). Since the PW values of all four costs are affected by changes in the analysis period, a range of PW values is computed for each cost. (In these computations, the salvage value of the system is assumed to be constant regardless of the economic life of the system. This assumption seems reasonable in light of the fact that the salvage value of the system lies almost entirely in the copper tubing.) The results of the various PW computations are combined and tabulated on the sensitivity/break-even worksheet in Figure 6.13.

Step 3: Break-even point. For an investment-type analysis, the break-even point is the value of the input parameter that results in an LCC of zero. The graph in Figure 6.13 shows that this value is 16 years. The break-even point can also be determined by interpolation from the tabulated sensitivity data as follows:

$$15 + \left(\frac{3,570}{10,024} \times 3\right) = 16 \text{ years}$$

As a check, note that a 16-year economic life provides PW values of $129,000 for initial cost, −$144,723 for liquid-petroleum gas (LPG) cost, $25,695 for M&R costs, and −$9,972 for salvage; their sum is exactly zero.

Step 4: Probability judgment. The design team had to categorize the probability that the economic life of the solar-energy system would exceed 16 years (since the best estimate of that life was greater than 16 years). Their consensus was that this probability is between 0.50 and 0.67—inconclusive. Moreover, since a great deal of effort had already been devoted to this analysis, they agreed that little could be done to reduce the uncertainty.

Summary

While LCCA is not an exact exercise, it improves the decision-making in procurement of facilities and optimization of owning and operating costs. Risk assessments can be challenging to initiate, but their use will lead to further refinement in methodology and data input. *The results are a significant improvement in the facility owner's long-term profits.*

LIFE CYCLE COST ANALYSIS
PRESENT WORTH (PW) COMPUTATION - BEST ESTIMATE
Project/Location: Student Recreation Center, Phoenix, AZ

Subject: Example Sensitivity Computations
Description: Solar-assisted Domestic Hot Water
Project Life Cycle = 18 YEARS
Discount Rate = 7.00%
Present Time = Occupancy Date

Alternative 1: Solar-assisted Domestic Hot Water System

INITIAL COSTS	Quantity	UM	Alt 1 Est.	Alt 1 PW	Alt 2 Est.	Alt 2 PW	Alt 3 Est.	Alt 3 PW	Alt 4 Est.	Alt 4 PW
Construction Costs										
A. Initial Construction Cost	1	LS	129,000	129,000		0		0		0
B.				0		0		0		0
C.				0		0		0		0
D.				0		0		0		0
E.				0		0		0		0
Total Initial Cost (Best Estimate)				129,000		0		0		0
Initial Cost PW Savings (Compared to Alt. 1)						0		0		0

REPLACEMENT COST/ SALVAGE VALUE										
Description	Year	PW Factor								
A.		1.0000		0	0	0		0		0
B.		1.0000		0	0	0		0		0
C.		1.0000		0	0	0		0		0
D.		1.0000		0	0	0		0		0
E. Salvage Value	18	0.2959	(29,440)	(8,710)	0	0		0		0
Total Replacement/Salvage Costs				(8,710)		0		0		0

ANNUAL COSTS	Differ. Escl. %	PWA								
A. Maintenance & Repair Cost	0.000%	10.059	2,720	27,361		0		0		0
B. LPG Fuel Annual Cost (1)	0.000%	10.059	(15,320)	(154,105)		0		0		0
C.				0		0		0		0
D.				0		0		0		0
E.				0		0		0		0
F.				0		0		0		0
G.				0		0		0		0
H.				0		0		0		0
Total Annual Costs (Best Estimate)				(126,744)		0		0		0
(1) Potential energy savings from use of solar-assisted system										
Total Life Cycle Costs (Best Estimate Only)				(6,454)		0		0		0
Life Cycle Savings (Compared to Alt. 1)						0		0		0

Figure 6.12: **PW Calculations, Best Estimates, Solar-Assisted Domestic Hot Water**

LIFE CYCLE COST ANALYSIS
SENSITIVITY ANALYSIS (SA) COMPUTATION
Project/Location: Student Recreation Center, Phoenix, AZ

Subject: Example SA Computations
Description: Solar-assisted Domestic HW
Project Life Cycle = 18 YEARS
Discount Rate = 7.0%
Present Time Occupancy Date

Notes	Parameter Studied	Representative Values	Alternative 1 Parameter Value	PW	Alternative 2 Parameter Value	PW	Alternative 3 Parameter Value	PW	Alternative 4 Parameter Value	PW
			Solar-assisted Domestic Hot Water System 10,024							
SENSITIVITY TEST #1	Economic Life of Solar-assisted Domestic Hot Water System	Low	10	25,538						
		Intermediate	15	3,570						
		Best Estimate	18	(6,454)						
		Intermediate	20	(12,092)						
		High	25	(23,259)						
SENSITIVITY TEST #2		Low								
		Intermediate								
		Best Estimate								
		Intermediate								
		High								

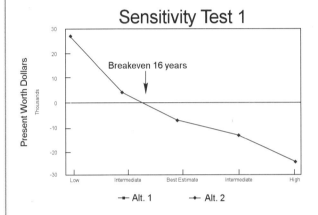

Figure 6.13: **Sensitivity Analysis Results and Break-Even Plot: Solar-Assisted Domestic Hot Water**

Chapter 7 Conducting an LCCA Study

The greatest shortcomings in today's traditional approach to the design of facilities is the use of "uni-discipline"-oriented decision-making. This approach tends to sacrifice total system performance by maximizing subsystem performance. It results, in many cases, in failure to properly consider initial system and total life cycle cost.

A.J. Dell'Isola
Introduction, Life Cycle Costing, ACEC/AIA Seminars

As indicated by discussion in previous chapters, life cycle cost analysis is not simply a matter of performing a few routine calculations. In fact, the availability of compound interest tables, hand calculators, and spreadsheet programs has made the calculations themselves a relatively minor part of the process. Of greater importance is ensuring that the correct economic criteria are followed; all feasible alternatives are studied (along with their associated costs); and the results of the study (including alternative selection), truly reflect both the economic reality of the design situation and the wishes or policies of the organization for which the study is conducted.

This chapter addresses the major steps in conducting an LCCA study, which include:

1. Identifying economic criteria
2. Generating design alternatives
3. Evaluating life cycle costs and benefits
4. Selecting the design alternative

These steps and the key techniques associated with each are illustrated in Figure 7.1.

Identifying Economic Criteria

In both the private and the public sectors, the owners have the prerogative of establishing the criteria under which their facility will be planned, designed, and constructed. The task of the design professional is to provide the desired functional capability within the limits defined by these criteria. The *governing criteria* may affect any of the decisions discussed in Chapter 2—as, for example, when an organizational planning horizon affects the choice of analysis period. Also, the criteria may affect the conduct of the analysis itself—as, for example, when an owner-directed analysis (special investment analysis) requires that particular alternatives be studied. As a first step in conducting an LCCA, it is important that the analyst identify and become familiar with all the relevant criteria. These may be found in a variety of forms: as documented company or agency policy; as unwritten, but generally well-known policy; as less well-known criteria that are implicit in the requirements for the project at hand; or perhaps as informal criteria that can be determined only through discussions between the analyst and the project owners. The governing criteria should be expected to vary for different types of applications, and between some applications that are seemingly quite similar in nature. It is usually necessary, therefore, to determine the particular criteria that govern each individual study, and to seek clarification of these criteria as required.

Figure 7.2 provides a format to help identify and document the required economic criteria for an LCCA study.

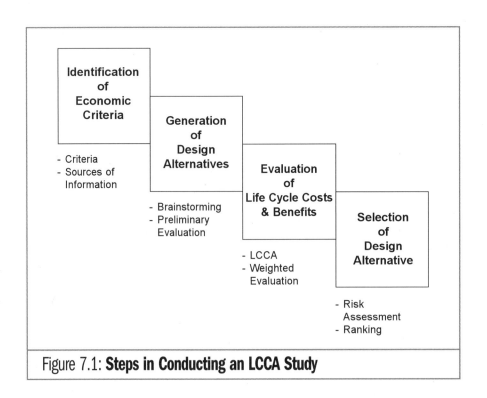

Figure 7.1: **Steps in Conducting an LCCA Study**

Generating Design Alternatives

Generating design alternatives to be studied in an LCCA is essentially a two-part task. The first step is identifying a wide range of possible alternatives. Then each possible alternative is preliminarily evaluated as to whether it meets the minimum functional and technical requirements and falls within the governing criteria. Those that pass this test are included in the analysis. Clearly, the range and quality of the alternatives selected for the study will have a major impact on the design professional's ability to achieve the lowest total cost of ownership. Past experience may, on occasion, seem to indicate that the range of alternatives available for consideration is rather narrow, because current design criteria and standards, performance standards, or master specifications appear to limit the number of alternatives that can be evaluated. Nonetheless, the designer is encouraged to be as creative as practical in the initial identification of possible alternatives. The fact that three particular alternatives qualified for analysis on one project does not mean that the same alternatives (or even the same number of alternatives) are best for another, similar project. Creativity sessions are an effective technique for identifying alternatives. In

Category:	Response:	Examples:	Reference:
Basic Considerations			
Type of Economic Analysis		Pimary-Investment	
Monetary Standard		Constant $, Current $	
Methodology Features			
Basic Equivalence Approach		Present Worth, Annualized	
Cash Flows		Conventional (end-of-year)	
Risk Assessment Approach		Confidence Index, Sensitivity	
Data & Parameters			
Discount Rate		10%, 7%	
Analysis Period		20 to 40 Years	
Present Tie		Occupancy Date	
Escalation Rates			
Energy, Electricity		0% to 5%, per year	
Energy, Natural Gas		0% to 5%, per year	
Energy, Fuel Oil		0% to 5%, per year	
Energy, Electricity		0% to 5%, per year	
Maintenance, Repair		0% to 5%, per year	
Alteration, Replacements		0% to 5%, per year	
Associated, Administrative		0% to 5%, per year	
Associated, Staffing		0% to 5%, per year	
Cost Estimate Sources			
Initial Project Cost		Cost Estimators	
Energy Cost		Mechanical Engineers	
Maintenance, Repair		Facility Manager	
Alteration, Replacements		Facility Manager	
Associated, Administrative		Facility Manager	
Associated, Staffing		Project Director	
Project Description:			

Figure 7.2: **Economic Criteria Documentation Checklist**

the sessions, the design team (or a consultant, preferably representing several disciplines) generates a number of ideas for alternatives that appear to satisfy the needed requirements. It is important to strive for as large a number and range of ideas as possible. Many may be eliminated when they are evaluated, but experience has shown that this free generation of ideas consistently produces superior sets of LCCA alternatives.

There are a number of creativity techniques available for problem-solving situations. The technique most generally applicable to design is brainstorming.

Brainstorming

Brainstorming is a freewheeling type of creativity, usually done in a group. It provides a method or procedure to help the designer generate more solutions to problems. This technique forces mental stimulation and conscious creative thinking. A typical brainstorming session involves four to six people sitting around a table and spontaneously producing ideas designed to solve a specific problem. A group leader, preferably with group dynamics experience, is appointed. During this session, no attempt whatsoever is made to judge or evaluate the ideas. Evaluation takes place after the brainstorming session has ended. Normally, the group leader will open the session by posing a problem and assigning someone to document each idea offered by the group. Before opening the session, the group leader will set the stage by reviewing the following group brainstorming rules:

1. Rule out criticism. Withhold adverse judgment of ideas until later.
2. Generate a large number of possible solutions: set a goal of multiplying the number of ideas produced in the first rush of thinking by 5 or 10.
3. Seek a wide variety of solutions that represent a broad spectrum of possibilities for the problem.
4. Watch for opportunities to combine or improve ideas as they are generated.
5. Before closing the door on possible solutions, allow the time to mull over the problem.

The elimination of judgment from the idea-producing stage allows for the maximum accumulation of ideas. It prevents the premature death of potentially good ideas and conserves the time of the individuals working on the problem by preventing shifts from creation to evaluation of original ideas. Consideration of all ideas encourages everybody to explore new areas, even those that seem impractical. This gives the opportunity to express thoughts to the innovator, who might be reluctant to voice ideas under ordinary conditions for fear of ridicule.

In addition to contributing ideas of their own, participants should suggest how ideas of others can be improved or how two or more ideas can be joined into still another idea. Two or more people working together under these ground rules can generate more ideas than one person working alone. This is mostly because ideas generated by various members of the group can be modified or improved and then offered as possible solutions to the problem. The efficiency of the group goes up as its size increases, until it reaches the point at which its operation becomes too cumbersome, tending to discourage some members' participation. The members of the group should be selected to represent different work backgrounds, and some should have a working familiarity with the subject under study. Group members need not all know one another before the session, but they should all come from similar levels of professional expertise. The group leader needs to obviate the possibility of senior members exerting pressure or dominance on junior members.

The technique and philosophy of brainstorming can also be used by individuals to generate solutions to problems. However, an individual is not usually as productive as a group. It should be emphasized that brainstorming does not always produce a final problem solution or ideas ready for immediate implementation. What it often produces are leads toward the final solution. Figure 7.3 summarizes the rules for brainstorming.

Preliminary Evaluation

The purpose of preliminary evaluation is to select for further analysis and refinement the most promising alternatives from among those generated previously. During the brainstorming creativity session there was a conscious effort to prohibit any judgmental thinking so as not to inhibit the creative process. Now the ideas must be critically evaluated. The first step in the evaluation of alternatives is to compare and rank the ideas. During the initial screening, the ideas are judged according to:

1. Ability to perform the function—ratings. Ratings might be excellent, good, fair, and poor.
2. Ease of implementation, including cost and schedule. Ratings might be:
    a. Simple idea: easy to implement
    b. Moderately complex idea: moderately easy to implement
    c. Complex idea: difficult to implement
3. Magnitude of savings (initial and life cycle)

Another part of the preliminary evaluation process is to judge, as objectively as possible, the advantages and disadvantages of each idea that survives the initial screening mentioned above. All surviving ideas should receive a preliminary evaluation. The initial analysis produces a shorter list of ideas, each of which has passed the evaluation standards set by the group. The selection should be based on an estimate of the relative life cycle cost reduction potential of the alternatives and how well they satisfy the functions required.

The alternatives that remain are developed far enough to obtain more detailed cost estimates. Although the evaluation phase is the responsibility of the study group, others should be consulted for help in developing and estimating the potential costs of these alternatives. Cost estimates must be as complete, accurate, and consistent as possible to minimize errors during the assessment of the alternatives.

Following are some of the questions that should be asked about each alternative:

1. Will the idea work? Can it be modified or combined?
2. What is the life cycle savings potential?
3. What are the chances for implementation? Will it be relatively easy or difficult to make the change?
4. Will it satisfy all the user's needs?

Evaluating Life Cycle Costs and Benefits

Once the most promising design alternatives have been selected, the design professional should then prepare an LCCA (see Chapter 4) and consider non-monetary benefits prior to making a design selection. In LCCAs conducted for public sector facilities, monetary benefits are considered to be negative costs. Thus, for example, salvage values are credited to the life

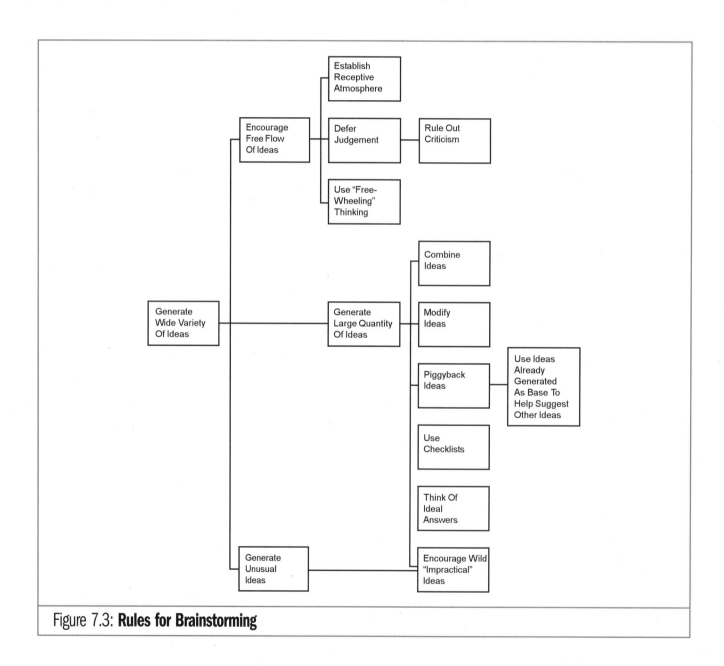

Figure 7.3: **Rules for Brainstorming**

cycle costs of alternatives at the end of the analysis period. Non-monetary benefits (aesthetics, safety, expansion potential, etc.) are normally assumed to be the same for all alternatives that satisfy the minimum functional and technical requirements for inclusion in the analysis. They are, therefore, not considered in the usual LCCA unless special cases are involved. (See the Weighted Evaluation later in this chapter for more on non-monetary benefits.)

Life Cycle Cost Analysis
Estimating Framework

In performing an LCCA, the designer/consultant deals with costs of several types that are expected to be incurred at various times during the analysis period. The analysis may be conducted at any time during the design phase, but can be valuable during the ownership cycle for large owners. An important point in the LCCA is a comparative analysis and, as such, needs only to focus on differing costs amongst alternates. In other words, if all alternates have the same approximate costs for specific line items, these costs can be eliminated from the LCCA. The costs may range in preciseness from actual billed amounts to gross estimates. Moreover, estimates may need to be refined from time to time, as actual costs or more exact estimates become known. Some sort of estimating framework is clearly needed to organize, keep track of, and allow easy access to these costs. Some analysts prefer to use an individual cost format—one that they have devised or are familiar with. Others prefer formats that organize costs according to an industry-recognized classification scheme—either by construction trade or by functional building element. See Chapter 5 for a discussion of recommended estimating format. Whichever type of format is used, it normally includes provisions for organizing the various types of costs: initial, energy, maintenance, etc.

Data Sources

The various types of costs that may enter an LCCA are discussed in Chapter 5. Most design professionals have access to sources of information regarding initial costs, but sources of data concerning other types of costs are not generally well known. Yet information concerning all pertinent costs is necessary for the analysis. Cost data are normally obtained from the most reliable sources, but the accuracy of the data should be commensurate with its use and the level of effort required to obtain it. Part 2 of this book contains life cycle cost data that the authors have assembled.

Initial Project Costs Construction costs are normally developed on the basis of data available from standard estimating sources, staff cost estimators, architect-engineer consultants, construction managers, or cost consultants.

Energy Costs Energy costs may be obtained from energy consumption estimates developed through energy analysis. The various energy analysis methods are discussed in publications of the American Society of Heating, Refrigeration, and Air Conditioning Engineers (ASHRAE). Energy analysis computer programs such as BLAST and DOE-2 (developed in the public sector) and AXCESS and TRACE (developed in the private sector) may also be used to obtain more precise estimates of energy consumption quantities. These software programs help with analysis, selection, and estimating of building products and materials.

DOE-2 and some other software is available free of charge at the U.S. Department of Energy Web site: **www.eren.doe.gov/buildings/ tools_directory**

Maintenance, Repair, and Custodial Costs   Sources are many and varied and include facility managers, colleagues, and other design professionals who have had previous experience with such costs, manufacturers' and distributors' representatives, product literature, handbooks, trade journal articles, and government publications. It is important to ensure that the operating, maintenance, repair, and custodial performance characteristics quoted for the source data are comparable to those expected in the particular facility being analyzed.

Replacement and Alteration Costs   Replacement costs may be obtained from staff cost estimators, facility managers, and design colleagues with experience relative to both the frequency and the cost of replacements. Replacement information may also be obtained from manufacturers' and distributors' representatives, industry associations, and facilities engineers. The facility owner or occupant can provide planned alteration time frames. Once the time frames and the degree of change are established, data regarding the costs of alterations may be obtained from initial cost sources.

Salvage Values   Cost information for salvage, demolition, and disposal normally comes from the designer's and/or user's experience, cost estimators, standard estimating sources, salvage and demolition firms, and industry associations. In many cases, U.S. income tax guidance is utilized.

Associated Costs   Functional use and down-time cost information generally comes from facility managers (in the private sector) and the occupying organizations (in the public sector).

Data Accuracy

Cost estimates are sometimes classified, in ascending order of accuracy and required level of effort, as gross estimates, semi-detailed estimates, and detailed estimates. Each has its use in economic studies.

No matter which level of estimate is being prepared, all the costs that make up the estimate should be consistent as to precision. For example, it is illogical to expend great effort to obtain a very accurate cost estimate for one component (say, to the nearest $10) when other components are estimated only to the nearest thousand dollars. Also, the results of calculations are normally rounded to the proper number of significant digits, so that they do not imply a false and misleading level of accuracy.

Data Collection/Evaluation Effort

The effort expended in the collection and evaluation of cost data at the level for an LCCA can vary widely. At one extreme, little or no effort may be required when the designer or design team can supply, from experience, cost data of the accuracy needed to conduct the study. At the other extreme, the data collection and evaluation effort can be quite substantial, involving contacts with a number of people, literature searches to collect data, and modification of available data to adapt it to the project under study. For this reason, and because cost data are rarely as available or as accurate as expected, the cost data collection/evaluation effort is normally restricted to that justified by potential benefits. Many design professionals begin with a nominal data collection effort, a so-called reconnaissance, involving interviews or telephone conversations with independent sources, e.g., facility managers, colleagues, manufacturers' representatives, other design professionals, and authors of trade journal articles. On the basis of this initial effort, it is possible to make a rational and informed judgment as to the potential effectiveness of more intense data collection efforts. This judgement, along with the credibility, availability, and validity of the data concerning individual alternatives (which affect the uncertainty inherent in the LCCA results), is sufficient to establish the level of effort for additional data collection.

Weighted Evaluation (Evaluating Non-monetary Benefits)

In most LCCAs, monetary benefits are considered to be negative costs. Thus, for example, salvage values are credited to the life cycle costs of alternatives at the end of the analysis period. Non-monetary benefits (aesthetics, safety, expansion potential, environmental friendliness, obsolescence avoidance, etc.) for the public sector are normally assumed to be the same for all alternatives that satisfy the minimum functional and technical requirements for inclusion in the analysis. They are, therefore, not considered in the usual LCCA. However, non-monetary benefits are normally considered for the private sector. Even for the public sector, differences in non-monetary benefits are considered when:

1. Life cycle costs of two or more alternatives are found to be essentially equal
2. The effect of uncertainty is so significant or so interminable that no alternative clearly represents the least cost course of action

When non-monetary benefits are considered, the technique of *weighted evaluation* is helpful in evaluating alternatives.

The procedure for weighted evaluation consists of two processes: (1) criteria identification and weighting and (2) alternatives analysis. The criteria weighting process is designed to isolate important criteria and establish their weights of relative importance. On the criteria scoring matrix, all criteria important in the selection of alternatives are listed. Criteria are compared, one against another. Through this series of comparisons, all criteria receive equal consideration in determining the final weight of importance.

Please note that in conducting a weighted evaluation, the weight assigned should be tempered with owner sensitivities. Having weight assigned by all engineers may tend to lessen sensitivity toward aesthetics, etc., while having all architects assigning weight may tend to lessen sensitivity to initial and life cycle costs. A broad-based team is required to develop more appropriate weights, with owner sensitivity being paramount.

Each benefit is first scored relative to every other benefit, and the results of this paired scoring are used to assign weights to the benefits, on a scale of 1 to 10. As shown in Figure 7.4, the non-monetary benefits (criteria) for selecting a clinic facility layout are scored on the top matrix. Each pair of benefits is compared, and the stronger of the two is scored according to the *how important* scale at the right of the upper matrix. Figure 7.4 shows a minor preference for space relationships over space flexibility (the score of B-2 between the two benefits), and a medium preference (B-3) for space relationships over aesthetic image.

After all possible benefit pairs have been compared and scored, raw scores are developed by adding the scores for the individual criteria. In this case *space relationships* was scored B-2, B-3, B-2, for a raw score of 7. The highest raw score is assigned a weight of 10 in the analysis matrix (lower part of Figure 7.4). The remaining raw scores are then converted to proportional weights. If a criterion does not receive any raw score, and the team feels it still should be considered, the team may arbitrarily assign a weight of, say, 1 or 2 to keep the criterion in the evaluation process.

Figure 7.5 is a checklist of attributes that may be used as criteria in the evaluation of various systems according to the degree to which they are safe, functional, sensible, and practical. This checklist should be reviewed as a part of selecting the non-monetary criteria for use in the weighted evaluation.

Each alternative is then rated on the degree to which it provides each benefit, according to the scale (1 to 5) shown at the bottom of Figure 7.4. A benefit score is found for each alternative-benefit pair by multiplying the alternative rating by the benefit weight. For example, for scheme 1 and *space flexibility,* this score is 2 × 3 = 6 in Figure 7.4. The total benefit score for each alternative is the sum of its individual benefit scores. The alternative with the highest total benefit score is the *best tentative choice* or *recommended alternative*. Weights and ratings do not need to be different.

Two or more benefits may receive the same weight, and two or more alternatives may be rated equally in relation to the same benefit. This technique permits the analyst to evaluate a number of non-monetary benefits that differ considerably in relative importance. The matrix introduces objectivity into the decision-making process, which proceeds through logical steps that can be followed and assessed by reviewers. See Figure 7.4, and the automated spreadsheet on weighted evaluation on the book's Web site: www.rsmeans.com/supplement/67341.asp

Selecting the Design Alternative

Selecting the design alternative for implementation consists of two parts: (1) addressing uncertainty through risk assessment and (2) assigning economic rankings to the alternatives. The alternative with the highest economic ranking, all factors considered, is then recommended for selection and use in the project design.

Risk (Uncertainty) Assessment

An LCCA, by its very nature, involves uncertainty, embodied in assumptions concerning future costs, cost growth, future inflation rates, and facility and element lives. The degree of uncertainty is likely to be substantial, and its effect on the results of the analysis quite significant. An alternative that is lowest in life cycle cost under one set of reasonable assumptions may be highest under another set of reasonable assumptions. Consequently, the results of an LCCA are credible only when uncertainty is taken into account. Therefore an examination (formal or informal) of the prevailing uncertainties must be part of every LCCA. Methods for evaluating the impact of uncertainty vary with the nature of the uncertainty. In particular, two types of uncertainties are distinguished: those that result from assumptions concerning all the alternatives to some degree (called *alternative-independent* uncertainties) and those that result

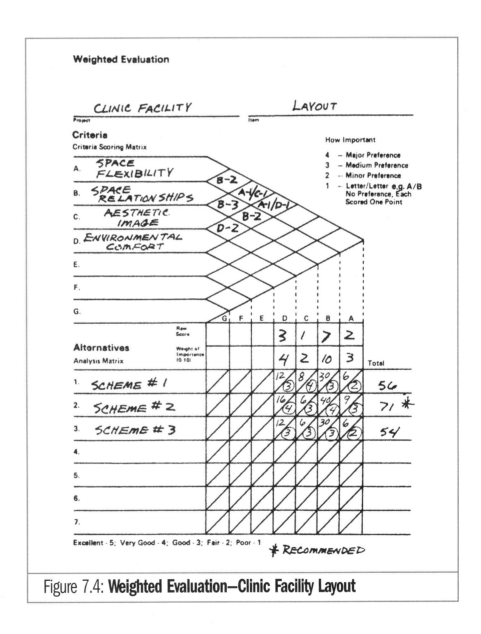

Figure 7.4: **Weighted Evaluation—Clinic Facility Layout**

Attribute	Requirement (R)/Criteria (C)	Test

HEADING 1: SAFETY AND PROTECTION

11 Fire Safety

(01) Fire Areas
 R: Control fire hazard neighboring structures.
 C: Limit distance between structures; limit area within fire barriers; limit ceiling height. — 1

(02) Fire Barriers
 R: Control the spread of fire.
 C: Require fire walls, fire stops, fire-resistance separation between egress openings; require that barrier penetrations maintain rated fire endurance; require fire dampers. — 1

(03) Egress Means
 R: Provide means for emergency evacuation.
 C: State minimum number of exits, maximum travel distance to exits and other means of egress; require minimum width for public corridors and public stairways; limit obstruction by door swing or equipment installation; require exit signs. — 1

(04) Protective Devices
 R: Provide warning devices and automatic fire extinguishing equipment.
 C: State conditions under which automatic fire detection systems, smoke detection systems, sprinkler systems, extinguishing systems, or other protection devices should be provided. — 1,5

(05) Fire Resistance/Combustibility
 R: Maintain integrity for sufficient time to permit evacuation or control of fire.
 C: Require use of noncombustible materials, state minimum hours of fire resistance or level of fire protection; state fire resistance classification. — 1,3

(06) Fire Load/Fuel Contribution
 R: Control fuel contribution of materials.
 C: State maximum potential heat (contribution to fire load) in BTU per hour or BTU per square foot of material. — 3

(07) Surface Spread of Flame
 R: Control surface spread of flame.
 C: State maximum flame spread or flammability of rating. — 3

(08) Flame Propagation
 R: Control propagation of flame through enclosed spaces.
 C: State maximum flame propagation index. — 1,3

(09) Smoke Generation
 R: Control amount and toxic effect of smoke produced.
 C: State maximum smoke development rating; state maximum optical density and maximum time to reach critical density; limit toxicity of smoke, require that smoke be non-noxious. — 3

(10) Smoke Propagation
 R: Control propagation of smoke through enclosed spaces.
 C: Require smoke-tight joints, provide for venting of smoke areas. — 1,3

(11) Accidental Ignition
 R: Protect against accidental ignition of fire.
 C: Design to prevent spark formation; limit equipment overheating; require equipment mounting to permit adequate ventilation. — 1

12 Life Safety (other than fire)

(01) Physical Safety
 R: Protect against physical hazards.
 C: Require guardrails, handrails, protective covers on moving parts; slip-resistant surfaces. — 1

(02) Electrical Safety
 R: Protect against electrical hazards.
 C: Require protective cover, insulation, and grounding; require safety controls and interlocks. — 1

(03) Toxicity
 R: Control dangerous materials and substances.
 C: Limit toxicity of materials; surfaces and finishes; limit toxic emissions below stated temperatures; limit toxic venting and leakage. — 3

Test Codes:
1. Design Drawings 2. Design Calculations 3. Laboratory Certification 4. Prototype Testing 5. Inspection

Figure 7.5: **Criteria Selection Checklist**

Attribute	Requirement (R)/Criteria (C)	Test
(04) Chemical Safety	R: Protect against hazard from chemical substances. C: Identify chemicals and agents, including concentration and anticipated frequency of use, to which the system will be exposed; indicate the level of atmospheric pollution permitted.	1
(05) Azoic Protection	R: Protect against infection from biological sources. C: Identify insects, vermin, fungi, microorganisms and other biological contaminants likely to be encountered and state level of protection to be provided.	1

13 Property Protection

Attribute	Requirement (R)/Criteria (C)	Test
(01) Theft Security	R: Protect equipment and contents against theft. C: Design to control unauthorized entry and access. Design to prevent unauthorized removal of equipment.	1
(02) Security against Vandalism	R: Protect against malicious damage. C: Design to resist malicious damage.	1,5
(03) Resistance to Misuse	R: Protect against accidental or deliberate misuse. C: Design to prevent improper usage. Design for fail-safe operation. Perform factory adjustment. Provide instructions.	1,5

14 Accessibility Considerations

Attribute	Requirement (R)/Criteria (C)	Test
(01) Physical Access	R: Provide for physical access by impaired individuals. C: Design to provide at least one means of ingress and egress for individuals in wheelchairs.	1
(02) Mobility-Impaired Usage	R: Provide for building usage by mobility-impaired individuals, if appropriate. C: Design to permit mobility-impaired individuals access to and use of facilities and equipment, such as restrooms, drinking fountains, vending machines, elevators...	
(03) Vision-Impaired Usage	R: Provide for building usage by vision-impaired individuals, if appropriate. C: Design to permit vision-impaired individuals access to and use of facilities and equipment, such as restrooms, drinking fountains, vending machines, elevators...	1
(04) Hearing-Impaired Usage	R: Provide for building usage by persons with hearing deficiencies, if appropriate. C: Design to permit hearing-impaired full usage of building services, such as, fire alarm systems, door bells, audible signals system...	1

HEADING 2: FUNCTIONAL

21 Strength

Attribute	Requirement (R)/Criteria (C)	Test
(01) Static Loading	R: Sustain gravity loads and superimposed and specified vertical and lateral loads. C: State dead loads to be supported, including forces transmitted from other systems. Specify how and where loads shall be transmitted from other systems.	2
(02) Live Loading	R: Sustain dynamic loads. C: Describe live loads to be supported, including snow load. Identify concentrated loads and state design floor loads.	2
(03) Horizontal Loading	R: Sustain wind loads and other lateral loads. C: For exterior walls, state design wind speeds and other live loads. State typhoon or hurricane conditions. For partitions, state lateral design load per square foot of partition area.	2
(04) Deflection	R: Limit deflection. C: State maximum acceptable deflections.	2
(05) Thermal Loading	R: Sustain loads due to temperature change. C: State the temperature extremes to be used for design.	2

Test Codes:
1. Design Drawings 2. Design Calculations 3. Laboratory Certification 4. Prototype Testing 5. Inspection

Figure 7.5: **Criteria Selection Checklist** *(continued)*

Attribute	Requirement (R)/Criteria (C)	Test
(06) Structural Serviceability	R: Retain serviceability under load and deflection. C: Require structure to sustain design loads without causing local damage.	2
(07) Seismic Loading	R: Sustain earthquake loads. C: State the seismic zone to be used for design.	2
(08) Impact Loading	R: Sustain impact loads and forces. C: Describe the source and magnitude of any impact loads to be sustained.	2
(09) Penetration Resistance	R: Protect against damage from concentrated loads. C: Describe magnitude and location of concentrated loads.	2

22 Durability

Attribute	Requirement (R)/Criteria (C)	Test
(01) Impact Resistance	R: Resist surface degradation due to point impact. C: Limit surface indentation due to specified impact load.	3
(02) Moisture Resistance	R: Resist degradation when exposed to water or water vapor. C: Design for use in specified range of humidity. Limit permanent effect to exposure to water, water retention, and absorption.	3
(03) Thermal Resistance	R: Resist degradation when exposed to temperature ranges expected in normal use. C: Limit physical change when exposed to specified temperature range.	3
(04) Corrosion Resistance	R: Resist degradation when exposed to corrosive agents. C: Limit corrosive effect observed after specified exposure to salt spray or fog; require corrosive-resistant surface treatment; design to avoid contact of dissimilar metals.	3
(05) Chemical Resistance	R: Resist degradation when exposed to chemicals. Resist staining or damage from soluble and insoluble salts, alkali attack, and oxidation. C: Limit changes in appearance or other specified property after exposure to specified chemicals.	3
(06) Weather Resistance	R: Resist degradation when exposed to specified period of simulated weathering. C: Limit changes observed after exposure to specified period of simulated weathering.	3
(07) Ultraviolet Resistance	R: Resist degradation due to exposure to ultraviolet light. C: Limit discoloration after ultraviolet exposure.	3
(08) Surface Serviceability	R: Resist cracking, spalling, crazing, blistering, delaminating, chalking, and fading. C: Limit surface changes observed after exposure to simulated conditions of use.	3
(09) Stain Resistance	R: Resist permanent discoloration when exposed to staining agents and chemicals. C: Limit visual evidence of permanent stains due to treatment with identified agents.	3
(10) Absorbency	R: Resist tendency to absorb and retain water. C: Limit quantity of water retained after specified exposure.	3
(11) Cleanability	R: Resist damage from routine maintenance and cleaning; permit removal of identified stains. C: Limit discoloration or surface change after simulated cleaning with specified cleaning agents.	3
(12) Color Resistance	R: Resist fading over time C: Limit discoloration after stated period.	3
(13) Friability/Frangibility	R: Resist crumbling and brittle fracture. C: Limit damage observed after specified loading.	3
(14) Abrasion Resistance	R: Resist degradation due to rubbing. C: Limit weight loss after specified number of abrasion cycles.	3
(15) Scratch Resistance	R: Resist degradation due to scratching. C: Limit rating on Pencil Hardness Scratch scale.	3
(16) Dimensional Stability	R: Control dimensional changes resulting from changes in environment. C: Limit volume change and movement under specified exposure to moisture and temperature variation.	3

Test Codes:
1. Design Drawings 2. Design Calculations 3. Laboratory Certification 4. Prototype Testing 5. Inspection

Figure 7.5: **Criteria Selection Checklist** (continued)

Attribute	Requirement (R)/Criteria (C)	Test
(17) Cohesiveness/ Adhesiveness	R: Resist peeling and delamination. C: Limit peeling or delamination failures under specified simulated loading.	3
(18) System Life	R: Function properly for identified period. C: Limit failure under accelerated life test. Design life of components consistent with specified life of system.	3,4
23 Transmission Characteristics		
(01) Heat	R: Control heat transmission. C: Design for specified Thermal Transmittance ("U" value).	2
(02) Light	R: Control light transmission. C: Design for specified percentage of light or radiation transmission.	2
(03) Air Infiltration	R: Resist leakage of air. C: Limit infiltration under specified pressure or wind load. Design for specified maximum leakage.	2,3,4
(04) Vapor Penetration	R: Resist vapor penetration. C: Design vapor barrier for minimum vapor permeability.	2,3,4
(05) Water Leakage	R: Resist water leakage. C: Limit infiltration under specified pressure or wind load; design for specified maximum leakage.	2,3,4
(06) Condensation	R: Control admission and condensation of moisture. C: Design to provide moisture barriers and thermal breaks.	2,4
24 Waste Products and Discharge		
(01) Solid Waste	R: Control production of solid waste. Provide for elimination or emission and prevent undesired accumulation. C: Design to accommodate waste produced or accumulated. Require identification of wastes produced.	1,2
(02) Liquid Waste	R: Control production of liquid waste. Provide for elimination or emission and prevent undesired accumulation. C: Design to accommodate waste levels produced, accumulated or omitted. Require identification of waste produced.	2
(03) Gaseous Waste	R: Control production of gases. Provide for elimination and prevent undesired accumulation. C: Design to accommodate levels of gas accumulated or emitted. Require identification of gaseous waste emitted.	1,2
(04) Odor	R: Control formation and persistence of odors. C: Design to prevent odor formation.	1
(05) Particulate Discharge	R: Control production of particulate wastes. Provide for collection of waste and prevent undesired accumulation. C: Design to accommodate amount of particulate waste produced. Limit particulate concentration.	1,2
(06) Thermal Discharge	R: Limit of thermal energy and vibration. Provide for control or reabsorption. C: Design to control thermal discharge produced below specified levels.	2
(07) Radiation	R: Limit emission of radiation. Provide for control or reabsorption. C: Design to control radiation discharge produced below specified levels.	2
25 Operational Characteristics		
(01) Method of Operation	R: Provide operating methods consistent with function. C: List desired operating modes.	1,2,4,5
(02) Results of Operation	R: Provide output consistent with function. C: List desired output quantities and rates.	1,2,5
(03) Cycle Time/Speed of Operation	R: Provide cycle times to accommodate functional requirements. C: List desired repetition rates.	1,2,5

Test Codes:
1. Design Drawings 2. Design Calculations 3. Laboratory Certification 4. Prototype Testing 5. Inspection

Figure 7.5: **Criteria Selection Checklist** (continued)

Attribute	Requirement (R)/Criteria (C)	Test

HEADING 3: SENSIBLE

31 Aesthetic Properties
   (01) Arrangement
      R: Provide order, organization or relationship appealing to visual perception.
      C: Design for pleasing relationships between elements and components.   1

   (02) Composition
      R: Provide unified appearance appealing to visual perception.
      C: Design for pleasing overall appearance.   1

   (03) Texture
      R: Provide surface finishes appealing to tactile perception.
      C: Design surface finishes pleasant to touch and feel.   1,4,5

   (04) Color/Gloss
      R: Provide finishes with pattern or luster appealing to visual perception.
      C: Design surface finishes for pleasing appearance.   1,4,5

   (05) Uniformity/Variety
      R: Provide appropriate consistency or variety of visual environment.
      C: Design to provide pleasing variety of colors, textures, and glosses. Limit visual confusion.   1,4,5

   (06) Compatibility/Contrast
      R: Provide appropriate consistency or variety of visual environment.
      C: Design appearance of elements in a pleasing and harmonious combination.   1,4,5

32 Acoustical Properties
   (01) Sound Generation
      R: Control undesirable sound and vibration generation.
      C: Limit sound generation. Provide specified decibel rating.   2,3,4,5

   (02) Sound Transmission
      R: Control transmission of sound.
      C: Design for specified sound transmission classification. Provide STC or SPP rating.   1,2,3,4,5

   (03) Reflectance
      R: Control reflection, reverberation, and echo production.
      C: Design for specified reverberation time, and sound path length.   1,2,3,4

33 Illumination
   (01) Level
      R: Control quantity of illumination provided.
      C: Design for specified illumination intensity level. Design to provide specified level of natural light.   2,5

   (02) Color
      R: Control color (wavelength) of illumination.
      C: Require lamp color and specified range of correlated color temperature.

   (03) Shadow/Glare
      R: Control illumination uniformity.
      C: Design for specified variation in illumination level over room area.   2,3,4

   (04) Reflection
      R: Control undesirable reflection.
      C: Limit reflected light.   2,5

34 Ventilation
   (01) Air Quality
      R: Control air quality.
      C: Design for specified natural ventilation. Design to control rate of air removal and supply design to control odors.   1,2

   (02) Velocity
      R: Control air movement.
      C: Design to maintain air motion between specified limits.   1,2

   (03) Distribution
      R: Control temperature gradients.
      C: Design to control temperature gradients within specified limits.   1,2

   (04) Pressurization
      R: Control pressure differential.
      C: Design to limit air leakage.   2

Test Codes:
1. Design Drawings 2. Design Calculations 3. Laboratory Certification 4. Prototype Testing 5. Inspection

Figure 7.5: Criteria Selection Checklist *(continued)*

Attribute	Requirement (R)/Criteria (C)	Test
(05) Temperature	R: Control air temperature content.	
	C: State exterior design conditions. Design to control rate of change of mean radiant temperature within specified range.	
(06) Moisture	R: Control air moisture content.	
	C: State exterior design conditions. Design to provide specified range of relative humidity.	2

35 Measurable Characteristics

Attribute	Requirement (R)/Criteria (C)	Test
(01) Levelness	R: Control deviation from identified horizontal.	
	C: Require level installation. Design for ease of level installation.	5
(02) Plumbness	R: Control deviation from identified vertical.	
	C: Require plumb installation within specified tolerance. Design for ease of plumb installation.	5
(03) Dimension/Tolerance	R: Control spatial extent for installation or fit within available space.	
	C: Conform to specified spatial dimensions and tolerances.	5
(04) Volume	R: Control volumetric measure or capacity.	
	C: Conform to specified limits of volume or capacity.	5
(05) Flatness	R: Control planar surface characteristics.	
	C: Limit deviation from flat, smooth, or planar surface.	5
(06) Shape	R: Control surface configuration, contour, or form.	
	C: Conform to specified shape limitations.	5
(07) Weight/Density	R: Control weight or density.	
	C: Conform to specified weight or density limitations.	5

36 Material Properties

Attribute	Requirement (R)/Criteria (C)	Test
(01) Hardness	R: Control resistance to penetration.	
	C: Limit penetration under specified load.	3
(02) Ductility/Brittleness	R: Control capability to shape by drawing. Control tendency to shatter.	
	C: Limit percentage elongation or percent change in cross section before rupture.	3
(03) Malleability	R: Control capability to shape by hammering.	
	C: Limit choice of materials.	
(04) Resilience	R: Control capability to store energy.	
	C: Limit residual deformation after impact load.	3
(05) Elasticity/Plasticity	R: Control capability to retain original shape when load is removed.	
	C: Limit residual deformation after removal of load.	3
(06) Toughness	R: Control capability to change shape without rupture.	
	C: Limit energy absorption before rupture.	3
(07) Viscosity	R: Control fluid resistance to flow.	
	C: Limit coefficient of viscosity.	3
(08) Creep	R: Control permanent change in shape after prolonged exposure to stress or elevated temperature.	
	C: Limit permanent deformation under specified load or temperature conditions.	3
(09) Friction	R: Control tendency of two bodies in contact to resist relative motion.	
	C: Limit coefficient of friction.	3
(10) Thermal Expansion	R: Control change in unit dimension resulting from change in temperature.	
	C: Limit coefficient of thermal expansion.	3

HEADING 4: PRACTICAL

41 Interface Characteristics

Attribute	Requirement (R)/Criteria (C)	Test
(01) Fit	R: Control size and shape of interface elements.	
	C: Design for physical compatibility with specified elements.	1,4,5
(02) Attachment	R: Control physical and electrical connection at interface.	
	C: Design to use specified connections.	1,4,5

Test Codes:
1. Design Drawings 2. Design Calculations 3. Laboratory Certification 4. Prototype Testing 5. Inspection

Figure 7.5: **Criteria Selection Checklist** (continued)

Attribute	Requirement (R)/Criteria (C)	Test
(03) Tolerance	R: Control variation in interface dimension. C: Design to accommodate specified tolerance.	1,4,5
(04) Modularity	R: Control standardized unit dimensions or repeating dimension. C: Design for compatibility with the specified module.	1,4,5
(05) Rotability	R: Control orientation at interface. C: Design to provide or permit specified orientations.	1,4,5
(06) Relocatability	R: Control ability to disassemble, move, or relocate. C: Design to provide specified flexibility to dismount and re-erect.	1,4,5
(07) Erection Sequence	R: Control order of erection or installation. C: Design to provide specified flexibility to dismount or re-erect.	1,4,5

42 Service

Attribute	Requirement (R)/Criteria (C)	Test
(01) Repairability	R: Provide for repair or replacement of damaged or inoperative elements. C: Design for ease of repair. Limit use of special tools, limit amount of labor required.	1,4,5
(02) Interchangeability	R: Provide for interchangeability of elements. C: Design for interchangeability.	1,4,5
(03) Accessibility	R: Provide access for service and maintenance. C: Design with access panels. Avoid placing connections in inaccessible locations.	1,4,5
(04) Replaceability	R: Provide for substitution of equivalent elements. C: Design to permit substitution.	1,4,5
(05) Inconvenience	R: Limit disturbance during maintenance and repair. C: Design to minimize inconvenience. Provide backup or alternate elements.	1,4,5
(06) Extendibility	R: Provide for capability to increase capacity. C: Design to permit or accommodate extension or expansion.	1,4,5
(07) Adaptability	R: Provide for alteration or modification. C: Design to use industry standard connectors and interface elements.	1,4,5
(08) Replacement Sequence	R: Provide for identified order for removal and replacement. C: Design for identified replacement sequence.	1,4,5
(09) Service Frequency	R: Control repair and maintenance frequency. C: Design for identified failure rates and maintenance schedules.	2,4,5

43 Personnel Needs

Attribute	Requirement (R)/Criteria (C)	Test
(01) Maintenance Personnel	R: Control skill levels required for maintenance. C: Design for maintenance by personnel with identified skills.	2
(02) Training	R: Control availability of trained personnel. C: Require provision for training operators and maintenance personnel.	2

Test Codes:
1. Design Drawings 2. Design Calculations 3. Laboratory Certification 4. Prototype Testing 5. Inspection

Figure 7.5: **Criteria Selection Checklist** (continued)

from assumptions concerning specific alternatives (called *alternative-dependent* uncertainties).

Alternative-Independent Uncertainties

Assumptions that give rise to alternative-independent uncertainties include estimates of future inflation rates and the assumptions (such as mission life) under which an analysis period may be chosen. These uncertainties represent future risks; the greater the uncertainty, the greater the risk, and the less attractive is today's spending in terms of future return. Consequently, the effects of alternative-independent uncertainties on the results of an LCCA are usually accounted for by a technique that gives more weight to future costs than to initial costs. The simplest of these weighting or penalty techniques, and the one most commonly employed, is to use a discount rate in the analysis that is greater than the rate that would be used in the absence of uncertainty. The techniques discussed in the next paragraph may also be applied to alternative-independent uncertainties, but they are most often used to assess the effects of alternative-dependent uncertainties.

Alternative-Dependent Uncertainties

Among the assumptions that produce alternative-dependent uncertainties are those concerning:

- *Differential escalation rates:* uncertainty as to whether different alternatives will experience varying rates of cost growth, and whether these rates will be greater than or less than the general rate of inflation.
- *Technological change or obsolescence:* uncertainty as to whether new fuels or faster, more efficient equipment will be developed.
- *Cost-estimate accuracy:* uncertainty as to the reliability and accuracy of cost data, especially as to whether estimates for one alternative are more accurate than those for another.
- *Useful life:* uncertainty as to the relative predictability of the useful lives of the various alternatives.
- *Physical failure:* the failure rate itself is probabilistic and therefore uncertain; in addition, failure rates for various alternatives may be known with varying degrees of uncertainty.

In some cases, a particular source of uncertainty will be pertinent to some of the alternatives, but not all of them. For example, in a study of membrane roofing vs. built-up roofing, uncertainty about how often gravel maintenance will be required is pertinent to only one of the two alternatives. In other cases, a source of uncertainty will pertain to all alternatives, but the magnitude and/or effect of that uncertainty will vary among the different alternatives. For example, in an evaluation of gas-fired, oil-fired, and electric furnaces, uncertainty about the growth of operating costs would pertain to all four alternatives, but different degrees of uncertainty would be involved. Both types of uncertainties are considered to be alternative-dependent; the two techniques most commonly used to account for their effect are the confidence index approach and the sensitivity analysis approach. Both of these techniques are discussed in detail in Chapter 6, "Economic Risk Assessment."

Alternative Ranking/Selection

The very important final step in conducting an LCCA study is the assignment of economic rankings to the design alternatives. The alternative judged to be the most economical, all factors considered, is assigned the highest ranking, and the remaining alternatives are ranked in descending order. The meaning of "most economical, all factors considered" depends very much on the situation at hand—the project, the alternatives, special directives, etc. In the most general situation, the alternative that provides the greatest benefit for the least cost is the most economical. In the straightforward LCCA study for which non-monetary benefits are considered equal, the lowest-cost alternative is assigned the highest ranking; in an LCCA, this is the alternative with the lowest life cycle cost.

In certain situations, specific economic ranking criteria may be indicated, as, for example, the most economical use of energy. In economic studies in general and LCCA in particular, a secondary ranking criterion may be required when there is no obvious *most economical alternative*. This may be the case when uncertainties obscure the differences among life cycle costs. Then, if a study of the uncertainties involved indicates that there is a probability of approximately 0.67 or more that one alternative is lower in life cycle cost than another, that alternative is assigned the higher economic ranking. If this probability is less than 0.67, the alternatives are assumed to be tied, and some means of breaking the tie is normally considered. For the simpler type of uncertainty tie, in which two or more alternatives have essentially the same life cycle costs, one can evaluate their non-monetary benefits using the weighted evaluation technique discussed earlier. If, however, the life cycle costs differ, as in an LCC where minimal energy use is the non-monetary benefit of chief concern, then the higher economic ranking may be assigned to one of two tied alternatives if it:

1. Will consume less energy per annum and will require no additional procurement cost, or
2. Will consume at least 15% less energy per annum *and* will require no more than a 15% additional initial cost, or
3. Will be at least 15% lower in initial cost and will not consume significantly larger quantities of energy per annum

If none of these three conditions applies, then the two alternatives would normally be assigned the same ranking or simply left to the design professional for final selection.

Example LCCA Study

Exterior Wall Study Description

To illustrate a complete LCCA study, this example considers an exterior wall system selection for a university research facility in Minneapolis. The quantities and pricing are based on actual data.

The economic criteria used for this analysis are based on the facility manager's direction to the design professional performing the analysis. This economic data are as follows:

Economic approach	Constant dollars
Project life cycle	25 years

Discount rate	10%
Present time	Occupancy date
Inflation approach	Constant dollars
Differential escalation	0%
Cash flow	End of year

Four alternative exterior wall systems were selected from a variety of possibilities generated during the brainstorming session:

1. Heavy-duty exterior insulation finish on steel studs
2. 4-inch brick with 8-inch concrete masonry
3. Metal panel system (aluminum)
4. Glass curtain wall

These alternatives are illustrated as Figure 7.6.

Life Cycle Cost Analysis

Figure 7.7 shows the LCCA present worth computation for these alternatives using the best estimate of costs prepared by a cost estimator, with input from the university facility manager. This analysis indicates the lowest initial cost is Alternative 1, but the lowest life cycle cost is Alternative 2. The discounted payback period for Alternative 2, compared to Alternative 1, is 2.46 years, as shown at the bottom of Figure 7.7. The equivalent annualized life cycle cost of these alternatives is also calculated at the bottom of the microcomputer spreadsheet.

Since the difference in life cycle cost between Alternatives 1 and 2 was close, the design architect requested the cost estimator to prepare three estimates:

1. High estimate
2. Best estimate
3. Low estimate

The initial costs for these systems are based on design and estimating experience and cost reference books such as Means *Building Construction Cost Data*. The replacement and repair costs are extracted from the report prepared by a maintenance consultant. The average frequency of occurrence was calculated from the data also provided in this report.

Figure 7.8 contains the high, best, and low estimates for each alternative and the resulting total life cycle costs. Note the greater level of cost estimate detail presented on these spreadsheets.

Weighted Evaluation

Non-monetary benefits were also of interest to the university and were therefore assessed by the weighted evaluation technique. Figure 7.9 shows the automated analysis. Seven criteria (benefits) were identified as important in the evaluation of the four alternative wall systems. The criteria scoring matrix and resulting weights are shown on this figure. The lower portion of the weighted evaluation worksheet shows the analysis matrix with resulting scoring. Alternative 2 (masonry wall) scored highest in terms of these non-monetary benefits.

Risk Assessment

Next, an economic risk assessment was made using the confidence index approach. This assessment was based on the two lowest life cycle cost Alternatives (1 and 2). Estimates of the low, best, and high costs were used as a part of the computation shown in Figure 7.10. From this assessment, there is a high confidence the masonry Alternative 2 has the lowest life cycle cost.

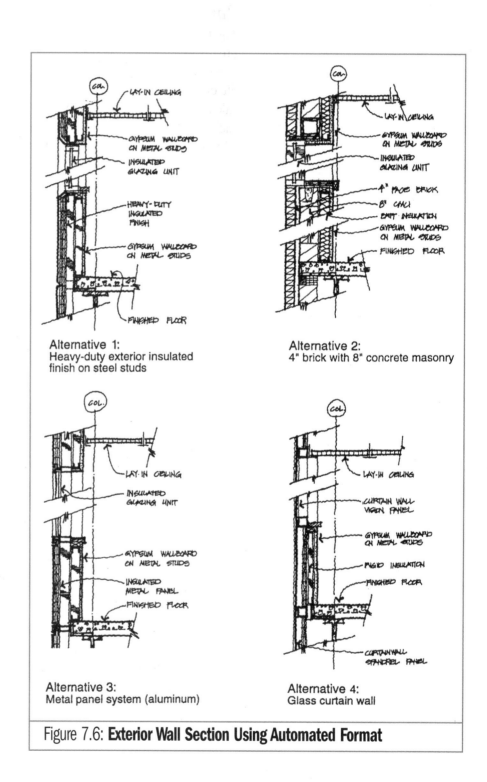

Figure 7.6: **Exterior Wall Section Using Automated Format**

As a further risk uncertainty assessment, the sensitivity approach was used. Figure 7.8 shows the low, best, and high estimated cost, and corresponding present worth value for each of the four alternatives. These costs were the focus of the sensitivity analysis. The data are presented in a graph at the bottom of Figure 7.11. This graph illustrates that under the various estimate possibilities (low to high), Alternative 2 remains the lowest life cycle cost.

Ranking/Selection

The final steps consisted of ranking the four alternatives on the basis of the previous calculations. Figure 7.12 is a spreadsheet that helps to document and summarize this information. The first portion of this sheet summarizes life cycle costs. In addition, the PW savings, compared to Alternative 1, are calculated. The equivalent annualized life cycle costs are also shown for each alternative. The discounted payback is shown (compared to Alternative 1). The results of the economic risk assessment are also listed based on both the CI computation and the sensitivity analysis. In this case, it was assessed as high.

LIFE CYCLE COST ANALYSIS
PRESENT WORTH (PW) COMPUTATION
Project/Location: University Research Facility, Minneapolis, MN

Subject: Example LCCA Study
Description: Exterior Wall System Selection
Project Life Cycle = 25 YEARS
Discount Rate = 10.00%
Present Time : Occupancy Date

			Alternative 1 Heavy-duty Insulation System		Alternative 2 Masonry Wall System (4" Brick w/ 8" Block)		Alternative 3 Metal Panel Wall System		Alternative 4 Glass Curtain Wall System	
INITIAL COSTS	Quantity UM	Unit Price	Est.	PW	Est.	PW	Est.	PW	Est.	PW
Construction Costs (see cost breakdown)		(Approximate)								
A. Heavy-duty Insul. System	83,200 WSF	$18.11	1,506,799	1,506,799		0		0		0
B. Masonry Wall System	83,200 WSF	$19.55		0	1,626,434	1,626,434		0		0
C. Metal Panel Wall System	83,200 WSF	$28.58		0		0	2,377,522	2,377,522		0
D. Glass Curtain Wall	83,200 WSF	$33.65		0		0		0	2,799,675	2,799,675
E.				0		0		0		0
Total Initial Cost				1,506,799		1,626,434		2,377,522		2,799,675
Initial Cost PW Savings (Compared to Alt. 1)						(119,635)		(870,723)		(1,292,876)
REPLACEMENT COST/ SALVAGE VALUE										
Description	Year	PW Factor								
A. Insul. Sealant at Joints	11	0.3505	49,445	17,330		0		0		0
B. Insul. System Major Repairs	15	0.2394	1,176,448	281,632		0		0		0
C. Masonry Coatings/Sealants	13	0.2897		0	231,597	67,085		0		0
D. Masonry Cleaning	15	0.2394		0	163,488	39,137		0		0
E. Masonry Misc. Repair	20	0.1486		0	217,152	32,278		0		0
F. Metal Panel Major Repairs	13	0.2897		0		0	580,464	168,139		0
G. Glass Sealant at Joints	13	0.2897		0		0		0	61,600	17,843
H. Glass Wall Major Repairs	20	0.1486		0		0		0	283,816	42,187
I. Salvage Value	25	0.0923	0	0	0	0	0	0	0	0
Total Replacement/Salvage Costs				298,962		138,500		168,139		60,030
ANNUAL COSTS										
Description	Escl. %	PWA								
A. Energy/ Fuel Annual Cost	0.000%	9.077	884,788	8,031,256	853,743	7,749,459	884,788	8,031,256	977,924	8,876,655
B. Glass Cleaning (2/year)	0.000%	9.077		0		0		0	169,728	1,540,628
C. Metal Panel Cleaning (0.33/year)	0.000%	9.077		0		0	56,576	513,543		0
D.	0.000%	9.077		0		0		0		0
Total Annual Costs (Present Worth)				8,031,256		7,749,459		8,544,799		10,417,283
Total Life Cycle Costs (Present Worth)				9,837,017		9,514,393		11,090,460		13,276,988
Life Cycle Savings (Compared to Alt. 1)						322,624		(1,253,443)		(3,439,971)
Discounted Payback (compared to Alt. 1)		PP Factor				2.46 Years		Never Years		Never Years
Total Life Cycle Costs (Annualized)		0.1102	1,083,725 Per Year		1,048,182 Per Year		1,221,815 Per Year		1,462,700 Per Year	

Figure 7.7: **PW Best Estimate Computation: Exterior Wall System**

The next portion shows the results of the benefits analysis from the weighted evaluation (Figure 7.9). The benefit-to-cost ratio is calculated by dividing the benefit score by the life cycle cost. For example:

$$\text{Alternative 1} = \frac{84 \text{ points}}{\$9,837(000)} = 0.0085$$

From the benefit-to-cost ratios, a final economic ranking was determined (1 = highest rank). Based on this LCCA study, the masonry wall system (Alternative 2) should be selected because of its best value for the project.

Effort is also required in the actual documentation of the best alternatives and their presentation to those having the authority to implement the proposals. This effort usually includes:

1. Preparation and presentation of the LCC proposals.
2. Establishment of a plan of action that will assure implementation according to schedule.
3. Follow-up as required until the recommendations are implemented in the project design.
4. Validation of initial, energy, maintenance, and other life cycle costs once the facility is occupied. This is normally achieved by conducting a post-occupancy evaluation.

Summary

The LCCA study approach can be applied to any facility element. It provides a vehicle to carry the analysis from inception to conclusion. By an organized approach, it ensures that proper consideration has been given to all necessary facets of the study. The methodology divides the study into a distinct set of work tasks. Judgment is required in determining the depth to which each step is performed. Organization for a multi-disciplinary team study should take account of the resources available and the results expected. Typically, a report is produced which (1) summarizes the results of the effort and (2) contains all the detailed backup information.

LIFE CYCLE COST ANALYSIS
PRESENT WORTH (PW) COMPUTATION

Project/Location: University Research Facility, Minneapolis

Subject: Exterior Wall System Selection
Alternative 1: Heavy-duty Insul. Finish, Mtl Studs
- Project Life Cycle = 25 YEARS
- Discount Rate = 10.00%   GSF = 310,452
- Present Time = Occupancy Date

				Alternative 1 High Estimate		Alternative 1 Best Estimate		Alternative 1 Low Estimate			
INITIAL COSTS	Quantity UM	Unit Price (Best Est)		Est. (+10%)	PW	Est. (+0%)	PW	Est. (-10%)	PW	Est.	PW
Construction Costs											
A. Heavy-duty Insul. Finish	83,200 WSF	$12.25		1,121,120	1,121,120	1,019,200	1,019,200	917,280	917,280		0
B. Sealant At Joints	11,000 LF	$1.50		18,150	18,150	16,500	16,500	14,850	14,850		0
C. Insulation (incl. in System)	83,200 WSF	$0.00		0	0	0	0	0	0		0
D. Gyp & Mtl Stud, Sheathing	83,200 WSF	$3.00		274,560	274,560	249,600	249,600	224,640	224,640		0
E. Interior Wall Painting	83,200 WSF	$0.30		27,456	27,456	24,960	24,960	22,464	22,464		0
F. _____											0
General Conditions, OH&P	15.00%			216,193	216,193	196,539	196,539	176,885	176,885		
Total Initial Cost		$18.11			1,657,479		1,506,799		1,356,119		0
Initial Cost PW Difference (Compared to High Estimate)							150,680		301,360		0

REPLACEMENT COST/ SALVAGE VALUE											
Description		Year	PW Factor	Est.	PW	Est.	PW	Est.	PW	Est.	PW
A. Sealant at Joints Ave. Freq		11	0.3505	63,360	22,207	49,445	17,330	35,530	12,453		0
B. Repair Cracks Ave. Freq		15	0.2394	278,720	66,723	247,520	59,254	216,320	51,785		0
C. Replace Surfaces Ave. Freq		15	0.2394	600,704	143,803	517,920	123,985	435,136	104,168		0
D. Coatings		15	0.2394	302,016	72,300	228,384	54,673	154,752	37,046		0
E. Cleaning		15	0.2394	242,112	57,959	182,624	43,718	123,136	29,477		0
Total Replacement/Salvage Costs					362,992		298,960		234,929		0

ANNUAL COSTS	$/GSF/Yr (Best Est)	Escal%	PWA	Est.	PW	Est.	PW	Est.	PW	Est.	PW
A. Energy/ Fuel Cost	$2.85	0.000%	9.077	973,267	8,834,384	884,788	8,031,258	796,309	7,228,132		0
B. Maintenance & Repair (see above)		0.000%	9.077	0	0	0	0	0	0		0
C. _____		0.000%	9.077		0		0		0		0
D. _____		0.000%	9.077		0		0		0		0
E. _____		0.000%	9.077		0		0		0		0
Total Annual Costs (Present Worth)					8,834,384		8,031,258		7,228,132		0

Total Life Cycle Costs (Present Worth)					10,854,855		9,837,017		8,819,180		0
Life Cycle PW Difference (Compared to High Estimate)							1,017,838		2,035,674		0

Figure 7.8: **High, Best, and Low PW Estimates: Exterior Wall Systems** *(continued)*

LIFE CYCLE COST ANALYSIS
PRESENT WORTH (PW) COMPUTATION

Project/Location: University Research Facility, Minneapolis

Subject: Exterior Wall System Selection
Alternative 2: 4" Brick w/ 8" Conc Block
Project Life Cycle = 25 YEARS
Discount Rate = 10.00%   GSF = 310,452
Present Time = Occupancy Date

INITIAL COSTS	Quantity UM	Unit Price (Best Est)	Alternative 2 High Estimate Est. (+10%)	PW	Alternative 2 Best Estimate Est. (+0%)	PW	Alternative 2 Low Estimate Est. (-10%)	PW	Est.	PW
Construction Costs										
A. 4" Brick	83,200 WSF	$8.00	732,160	732,160	665,600	665,600	599,040	599,040		0
B. 8" Conc Block	83,200 WSF	$5.50	503,360	503,360	457,600	457,600	411,840	411,840		0
C. Sealant	2,700 LF	$1.50	4,455	4,455	4,050	4,050	3,645	3,645		0
D. Insulation	83,200 WSF	$1.40	128,128	128,128	116,480	116,480	104,832	104,832		0
E. Interior Gyp & Mtl Stud	83,200 WSF	$1.75	160,160	160,160	145,600	145,600	131,040	131,040		0
F. Interior Wall Painting	83,200 WSF	$0.30	27,456	27,456	24,960	24,960	22,464	22,464		0
General Conditions, OH&P	15.00%		233,358	233,358	212,144	212,144	190,929	190,929		
Total Initial Cost		**$19.55**		**1,789,077**		**1,626,434**		**1,463,790**		**0**
Initial Cost PW Difference (Compared to High Estimate)						162,643		325,287		0

REPLACEMENT COST/ SALVAGE VALUE Description		Year	PW Factor	Est.	PW	Est.	PW	Est.	PW	Est.	PW
A. Coatings	Ave. Freq	13	0.2897	294,528	85,314	217,152	62,901	139,776	40,488		0
B. Sealant at Joints	Ave. Freq	13	0.2897	19,089	5,529	14,445	4,184	9,801	2,839		0
C. Cleaning	Ave. Freq	15	0.2394	203,840	48,797	163,488	39,137	123,136	29,477		0
D. Misc. Repair	Ave. Freq	20	0.1486	294,528	43,779	217,152	32,278	139,776	20,776		0
E. Salvage Value		25	0.0923	0	0	0	0	0	0		0
Total Replacement/Salvage Costs					**183,419**		**138,500**		**93,580**		**0**

ANNUAL COSTS Description	$/GSF/Yr (Best Est)	Escal%	PWA	Est. (+10%)	PW	Est. (+0%)	PW	Est. (-10%)	PW	Est.	PW
A. Energy/ Fuel Cost	$2.75	0.000%	9.077	939,117	8,524,405	853,743	7,749,459	768,369	6,974,513		0
B. Maintenance & Repair (see above)		0.000%	9.077	0	0	0	0	0	0		0
C.		0.000%	9.077		0		0		0		0
D.		0.000%	9.077		0		0		0		0
E.		0.000%	9.077		0		0		0		0
Total Annual Costs (Present Worth)					**8,524,405**		**7,749,459**		**6,974,513**		**0**

Total Life Cycle Costs (Present Worth)					**10,496,901**		**9,514,393**		**8,531,884**		**0**
Life Cycle PW Difference (Compared to High Estimate)							982,508		1,965,018		0

LIFE CYCLE COST ANALYSIS
PRESENT WORTH (PW) COMPUTATION

Project/Location: University Research Facility, Minneapolis

Subject: Exterior Wall System Selection
Alternative 3: Metal Panel System
Project Life Cycle = 25 YEARS
Discount Rate = 10.00%   GSF = 310,452
Present Time = Occupancy Date

INITIAL COSTS	Quantity UM	Unit Price (Best Est)	Alternative 3 High Estimate Est. (+10%)	PW	Alternative 3 Best Estimate Est. (+0%)	PW	Alternative 3 Low Estimate Est. (-10%)	PW	Est.	PW
Construction Costs										
A. Metal Panel System (Alum.)	83,200 WSF	$22.75	2,082,080	2,082,080	1,892,800	1,892,800	1,703,520	1,703,520		0
B. Sealant at Joints	2,700 LF	$1.50	4,455	4,455	4,050	4,050	3,645	3,645		0
C. Insulation (incl. in panel)	83,200 WSF	$0.00	0	0	0	0	0	0		0
D. Interior Gyp & Mtl Stud	83,200 WSF	$1.75	160,160	160,160	145,600	145,600	131,040	131,040		0
E. Interior Wall Painting	83,200 WSF	$0.30	27,456	27,456	24,960	24,960	22,464	22,464		0
F.										0
General Conditions, OH&P	15.00%		341,123	341,123	310,112	310,112	279,100	279,100		
Total Initial Cost		**$28.58**		**2,615,274**		**2,377,522**		**2,139,769**		**0**
Initial Cost PW Difference (Compared to High Estimate)						237,752		475,504		0

REPLACEMENT COST/ SALVAGE VALUE Description		Year	PW Factor	Est.	PW	Est.	PW	Est.	PW	Est.	PW
A. Scrape & Paint	Ave. Freq	13	0.2897	373,568	108,209	300,352	87,001	227,136	65,793		0
B. Coatings	Ave. Freq	13	0.2897	343,616	99,533	264,992	76,758	186,368	53,984		0
C. Sealant at Joints	Ave. Freq	13	0.2897	20,439	5,920	15,120	4,379	9,801	2,839		0
D.		20	0.1486	0	0	0	0	0	0		0
E. Salvage Value		25	0.0923	0	0	0	0	0	0		0
Total Replacement/Salvage Costs					**213,662**		**168,138**		**122,616**		**0**

ANNUAL COSTS Description	$/GSF/Yr (Best Est)	Escal%	PWA	Est.	PW	Est.	PW	Est.	PW	Est.	PW
A. Energy/ Fuel Cost	$2.85	0.000%	9.077	973,267	8,834,384	884,788	8,031,258	796,309	7,228,132		0
B. Panel Cleaning (0.33/year)		0.000%	9.077	74,325	674,654	56,576	513,543	38,827	352,431		0
C.		0.000%	9.077		0		0		0		0
D.		0.000%	9.077		0		0		0		0
E.		0.000%	9.077		0		0		0		0
Total Annual Costs (Present Worth)					**9,509,038**		**8,544,801**		**7,580,563**		**0**

Total Life Cycle Costs (Present Worth)					**12,337,973**		**11,090,460**		**9,842,949**		**0**
Life Cycle PW Difference (Compared to High Estimate)							1,247,513		2,495,025		0

Figure 7.8: High, Best, and Low PW Estimates: Exterior Wall Systems *(continued)*

LIFE CYCLE COST ANALYSIS
PRESENT WORTH (PW) COMPUTATION
Project/Location: University Research Facility, Minneapolis

Subject: Exterior Wall System Selection
Alternative 4: Glass Curtain Wall
Project Life Cycle = 25 YEARS
Discount Rate = 10.00%   GSF = 310,452
Present Time = Occupancy Date

				Alternative 4 High Estimate		Alternative 4 Best Estimate		Alternative 4 Low Estimate			
INITIAL COSTS Construction Costs	Quantity UM	Unit Price (Best Est)		Est. (+10%)	PW	Est. (+0%)	PW	Est. (-10%)	PW	Est.	PW
A. Glass Curtain Wall	83,200 WSF	$28.00		2,562,560	2,562,560	2,329,600	2,329,600	2,096,640	2,096,640		0
B. Sealant At Joints	11,000 LF	$1.50		18,150	18,150	16,500	16,500	14,850	14,850		0
C. Window Gasket	20,800 LF	$4.25		97,240	97,240	88,400	88,400	79,560	79,560		0
D. Mullion Coating	20,800 LF	$0.00		0	0	0	0	0	0		0
E.											0
F.											0
General Conditions, OH&P	15.00%			401,693	401,693	365,175	365,175	328,658	328,658		
Total Initial Cost		$33.65			3,079,643		2,799,675		2,519,708		0
Initial Cost PW Difference (Compared to High Estimate)							279,968		559,935		0

REPLACEMENT COST/ SALVAGE VALUE											
Description		Year	PW Factor	Est.	PW	Est.	PW	Est.	PW	Est.	PW
A. Sealant at Joints	Ave. Freq	13	0.2897	83,270	24,120	61,600	17,843	39,930	11,566		0
B. Gasket Replm't	Ave. Freq	20	0.1486	166,192	24,703	120,328	17,885	74,464	11,068		0
C. Mullion Coating	Ave. Freq	20	0.1486	203,840	30,299	163,488	24,301	123,136	18,303		0
D.		20	0.1486	0	0	0	0	0	0		0
E. Salvage Value		25	0.0923	0	0	0	0	0	0		0
Total Replacement/Salvage Costs					79,122		60,029		40,937		0

ANNUAL COSTS	$/GSF/Yr	Escal%	PWA	Est.	PW	Est.	PW	Est.	PW	Est.	PW
Description	(Best Est)										
A. Energy/ Fuel Cost	$3.15	0.000%	9.077	1,075,716	9,764,319	977,924	8,876,653	880,131	7,988,988		0
B. Glass Cleaning (4/year)		0.000%	9.077	222,976	2,023,962	0	0	0	0		0
C. Glass Cleaning (2/year)		0.000%	9.077		0	169,728	1,540,628	0	0		0
D. Glass Cleaning (1/year)		0.000%	9.077		0		0	116,480	1,057,294		0
E.		0.000%	9.077		0						0
Total Annual Costs (Present Worth)					11,788,281		10,417,281		9,046,282		0
Total Life Cycle Costs (Present Worth)					14,947,045		13,276,985		11,606,926		0
Life Cycle PW Difference (Compared to High Estimate)							1,670,060		3,340,119		0

Figure 7.8: **High, Best, and Low PW Estimates: Exterior Wall Systems**

Weighted Evaluation

Project: Research Facility
Location: University
Study Element: Exterior Wall System
Date:

Criteria Scoring Matrix

Criteria:	Preference		Preference		Preference		Preference		Preference		Preference	
	A or	B	A or	C	A or	D	A or	E	A or	F	A or	G
A. Image/Aesthetics	1	1		2		3		3		4	4	
	B or	C	B or	D	B or	E	B or	F	B or	G		
B. Color Rendition	1	1		2	1	1		3	4			
Environmental	C or	D	C or	E	C or	F	C or	G				
C. Sustainability		2	1	1		2	4					
Obsolescence	D or	E	D or	F	D or	G						
D. Avoidance	1	1	1	1	4							
Operational	E or	F	E or	G								
E. Effectiveness		2	4									
	F or	G										
F. Durability	4											
Future												
G. Expandabilty												

How Important: Major Preference = 4, Medium Preference = 3, Minor Preference = 2, No Preference Each = 1

Analysis Matrix

Criteria	A		B		C		D		E		F		G		
Raw Score	5		7		8		13		10		16		0		
Weight	3		4		5		8		6		10		1		
Alternatives:	Score	WS	Score	WS	Score	WS	Score	WS	Score	WS	Score	WS	Score	WS*	Total
1. Heavy Duty Insulated System	2	6.00	2	8.00	3	15.00	1	8.00	4	24.00	2	20.00	3	3.00	84.00
2. Masonry Wall System	4	12.00	5	20.00	5	25.00	5	40.00	4	24.00	5	50.00	2	2.00	173.00
3. Metal Panel Wall System	3	9.00	4	16.00	4	20.00	4	32.00	4	24.00	4	40.00	4	4.00	145.00
4. Glass Curtain Wall System	4	12.00	5	20.00	4	20.00	4	32.00	3	18.00	4	40.00	5	5.00	147.00
5.		0.00		0.00		0.00		0.00		0.00		0.00		0.00	0.00
6.		0.00		0.00		0.00		0.00		0.00		0.00		0.00	0.00
7.		0.00		0.00		0.00		0.00		0.00		0.00		0.00	0.00
8.		0.00		0.00		0.00		0.00		0.00		0.00		0.00	0.00

Score: Excellent = 5, Very Good = 4, Good = 3, Fair = 2, Poor = 1   * Weight of 1 Scored arbitrarily by the team

Figure 7.9: **Weighted Evaluation: Exterior Wall System**

LIFE CYCLE COST ANALYSIS (Present Worth Method)
CONFIDENCE INDEX (CI) COMPUTATION
Project/Location: University Research Facility, Minneapolis, MN

Subject: Example LCCA Study
Description: Exterior Wall System Selection

Project Life Cycle = 25 YEARS	ESTIMATES RANGE			DIFFERENCES IN ESTIMATES				PRESENT WORTH		
Discount Rate = 10.00%				LOW	HIGH	DELTA		BEST		
Present Time = Occupancy Date	LOW	HIGH	BEST	SIDE	SIDE	%	ok	ESTIMATE	DELTA	DELTA^2
ALTERNATIVE 2 (low) COST ITEMS										
Masonry Wall System Initial Cost	1,463,790	1,789,077	1,626,434	162,644	162,643	0%	ok	1,626,434	162,644	26,453,070,736
Replacement Cost/ Salvage Value (PW)	93,580	183,419	138,500	44,920	44,919	0%	ok	138,500	44,920	2,017,806,400
Energy/ Fuel Annual Cost	768,369	939,117	853,743	85,374	85,374	0%	ok	7,749,459	774,943	600,536,985,725
Totals								9,514,393		
ALTERNATIVE 1 (high) COST ITEMS										
Heavy-duty Insul. System Initial Cost	1,356,119	1,657,479	1,506,799	150,680	150,680	0%	ok	1,506,799	150,680	22,704,462,400
Replacement Cost/ Salvage Value (PW)	234,929	362,992	298,962	64,033	64,030	0%	ok	298,962	64,033	4,100,225,089
Energy/ Fuel Annual Cost	796,309	973,267	884,788	88,479	88,479	0%	ok	8,031,256	803,127	645,013,658,816
Totals								9,837,017		

Note = If high and low 90% estimates > 25%, then use sensitivity analysis

Difference: 322,624   Sum: 1,300,826,209,166

(Sum)^1/2: 1,140,538

$$\text{Confidence Index} = \frac{PW(\text{High}) - PW(\text{Low})}{(PW\,\text{Diff}(\text{High})^2 + PW\,\text{Diff}(\text{Low})^2)^{1/2}} = \frac{322,624}{1,140,538} = 0.283 = \text{HIGH}$$

Confidence Assignment:
Low: CI < 0.15
Medium: 0.15 < CI < 0.25
High: CI > 0.25

Figure 7.10: **CI Computation: Exterior Wall Systems**

LIFE CYCLE COST ANALYSIS
SENSITIVITY ANALYSIS (SA) COMPUTATION

Project/Location: Warehouse, Miami, FL

Subject: Example SA Computations
Description: Mechanical Equip. Selection
- Project Life Cycle = 25 YEARS
- Discount Rate = 10%
- Present Time Occupancy Date

Notes	Parameter Studied	Representative Values	Alternative 1 — Heavy duty Insulation System		Alternative 2 — Masonry Wall System (4" Brick w/ 8" Block)		Alternative 3 — Metal Panel Wall System		Alternative 4 — Glass Curtain Water System	
			Parameter Value	PW	Parameter Value	PW	Parameter Value	PW	Parameter Value	PW
SENSITIVITY TEST #1	Estimated Costs	Low		8,819,180		8,531,864		9,842,949		
		Intermediate								
		Best Estimate		9,837,017		9,514,393		11,090,460		
		Intermediate								
		High		10,854,855		10,496,901		12,337,973		
SENSITIVITY TEST #2		Low								
		Intermediate								
		Best Estimate								
		Intermediate								
		High								

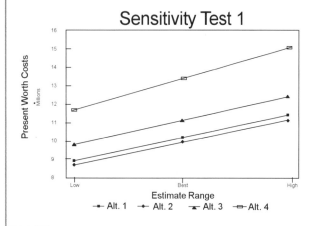

Figure 7.11: Sensitivity Analysis Computation: Exterior Wall Systems

LIFE CYCLE COST ANALYSIS
RANKING / SELECTION
Project/Location: University Research Facility, Minneapolis, MN

Subject: **Example LCCA Study**
Description: **Exterior Wall System Selection**
Project Life Cycle = **25 YEARS**
Discount Rate = **10%**
Present Time = **Occupancy Date**

		Alternative 1 Heavy-duty Insulation System		Alternative 2 Masonry Wall System (4" Brick w/ 8" Block)		Alternative 3 Metal Panel Wall System		Alternative 4 Glass Curtain Wall System	
LIFE CYCLE COST SUMMARY		Est.	PW	Est.	PW	Est.	PW	Est.	PW
Total Initial Cost			1,506,799		1,626,434		2,377,522		2,799,675
Total Replacement/Salvage Value			298,962		138,500		168,139		60,030
Total Annual Costs			8,031,256		7,749,459		8,544,799		10,417,283
Total Life Cycle Costs (Present Worth)			9,837,017		9,514,393		11,090,460		13,276,988
Life Cycle Savings (Compared to Alt. 1)					322,624		(1,253,443)		(3,439,971)
Total Life Cycle Costs (Annualized)		1,083,725	Per Year	1,048,182	Per Year	1,221,815	Per Year	1,462,700	Per Year
Discounted Payback (Compared to Alt. 1)				2.46	Years	Never	Years	Never	Years
Risk/ Sensitivity Confidence			HIGH		HIGH		HIGH		HIGH
WEIGHTED EVALUATION (Non-monetary Benefits)									
Criteria (Benefits)	Weight (1-10)	Evaluation	Score	Evaluation	Score	Evaluation	Score	Evaluation	Score
A. Image/ Aesthetics	3	2	6	4	12	3	9	4	12
B. Color Rendition	4	2	8	5	20	4	16	5	20
C. Environmental Sustainability	5	3	15	5	25	4	20	4	20
D. Obsolescence Avoidance	8	1	8	5	40	4	32	4	32
E. Operational Effectiveness	6	4	24	4	24	4	24	3	18
F. Durability	10	2	20	5	50	4	40	4	40
G. Future Extendability	1	3	3	2	2	4	4	5	5
Total Weighted Evaluation Score			84		173		145		147
BENEFIT TO COST RATIO			0.0085		0.0182		0.0131		0.0111
RANKING (1 = Highest)			4		1		2		3

SELECTION

Based on the results of both the life cycle cost analysis and the weighted evaluation summarized above, Alternative 2 appears to best satisfy the requirements of this project and is therefore recommended for selection and use.

Figure 7.12: **Ranking/Alternative Selection: Exterior Wall Systems**

Chapter 8 Case Studies

The primary criterion to be applied in a choice among alternative proposed investments in physical assets should be selected with the objective of making the best use of limited resources.
Grant and Ireson
Principles of Engineering Economy, 1970

The previous chapters dealt primarily with the principles and concepts necessary for an introduction to life cycle costing (LCC). Although this information is essential, a good understanding of LCC can be acquired only though application of LCC techniques to the development and operation of actual facilities. The user/owner must identify economic criteria, generate a variety of alternatives, evaluate the life cycle costs and benefits, and select the most promising alternative, considering the risks and uncertainties associated with the economic analysis. Once this understanding is attained, advanced computer simulation modeling can be useful in facilitating LCC. *Note: the authors have developed an Excel-based spreadsheet to automate this process. This spreadsheet is available for download at* www.rsmeans.com/supplement/67341.asp

Several individual case studies are presented in this chapter. They represent a variety of applications of life cycle cost analysis (LCCA). The case studies cover different phases, including planning/feasibility, design and construction, and occupancy. They address only key LCC areas. The case studies, numbered 1-16, are taken from actual projects.

The first half of this chapter is comprised of brief narrative explanations of all the case studies. Supporting figures are located in the second half.

The case studies are organized by phase, starting with planning/feasibility, then design, and concluding with construction/occupancy. Only the highlights and critical areas are presented.

LCCA Case Study 1: Office to Museum Renovation

Life cycle costing is a method of analyzing all the costs of ownership over a given period of time and summarizing them using the principles of economic analysis. This process permits comparisons between alternatives for the purpose of selecting the best course of action. LCC is also used to determine the delta costs for budgeting purposes, comparing a current situation and a new use. This was the primary purpose of this study involving conversion of an office building to a museum.

The economic method followed for this estimate is the same as that specified by the U.S. Office of Management and Budget (OMB), circular A-94. All federal government projects follow this guidance, which specifies a 25-year life cycle analysis period and a 10% interest rate (time value of money). *(Note: Recently, this rate has been reduced to 7% to reflect a lower rate of general economy inflation.)* All costs are to be expressed in *constant dollars*, i.e., the general purchasing power of the dollar at the time of the decision. Since the costs would also be used for budgeting, the LCCA needed to be expressed in current dollars.

In addition to the above approach, the delta cost (difference in expenditures between the facility's current use as an office and its proposed use as a museum) has been estimated. These costs are presented in Figure 8.1.1, which begins with the initial (capital) costs. Included in these figures are building renovation construction costs, the architect/engineer (A/E) fee, exhibit display furniture, exhibit designer fee, the prorated cost for a new central plant gas chiller and heater, and the fit-out of leased office space for current staff who will be displaced when the museum is located in the current office building. This total initial cost is approximately $2,176,000.

No major replacement costs or salvage values are estimated after this renovation. Projected annual expenditures are listed next. They include such items as utilities, maintenance, custodial, security, museum staff, and office rental for displaced staff.

Each estimated annual cost is shown first, followed by the present worth figure. The present worth amount is calculated by determining the present worth of annuity factor for a life cycle of 25 years at a 10% interest rate. This factor (taken from Appendix A, Table 4) is 9.077. Multiplying this factor by the annual dollar amount determines the equivalent present worth cost, which is shown in the adjacent column.

LCCA Case Study 1
Continued

The projected annual income (monetary benefits) is shown next. Income is anticipated from renting the ballroom space for special events, profits from gift shop sales, and fund-raising activities. This income is shown in parentheses, since it is positive cash flow, whereas the other figures are negative.

The final two lines on this figure present a *total life cycle cost* (expressed in present worth) for the current use, the proposed use, and the delta, or difference. The first of the two lines only sum the costs without including the income. The last line includes the projected income along with the costs.

Figure 8.1.2 provides the backup to the data used in the life cycle cost calculations. Notes and assumptions include: size and usage of the facility, days and hours of operation, staffing, and economic criteria. The backup for estimated operational costs is also contained in this figure, along with notes regarding how each number was estimated. Notes and assumptions also include methods used to calculate the operational revenue and assumptions regarding the design and construction schedule.

Figures 8.1.3 to 8.1.5 provide cash flow analysis of the estimated costs taken from Figure 8.1.1. Figure 8.1.3 documents the total costs associated with the proposed use. The life cycle costs are identified with the year in which they would occur. These costs are then multiplied by the appropriate cost escalation factor based on a 3% per year compounded inflation rate. The escalated yearly cost (in current dollars) is then presented in a cash flow diagram shown on the right-hand side of Figure 8.1.3.

Figure 8.1.4 presents the cash flow of the delta costs for the proposed museum usage. The costs are taken from Figure 8.1.1 and associated with the year in which they would occur. These costs are then escalated at a rate of 3% per year. The delta cash flow diagram plots both the total dollars (from Figure 8.1.3) and the delta costs.

Figure 8.1.5 presents the delta costs after the anticipated annual museum income has been subtracted. The projected income is taken from Figure 8.1.1. The yearly totals are then escalated at a rate of 3% per year. The cash flow diagram plots these costs, as well as those from Figures 8.1.3 and 8.1.4.

LCCA Case Study 2: District Court Consolidation

District court functions in a major urban area are presently located at several different sites, creating government inefficiencies. This study's objective was to establish the feasibility of consolidating these courts into a historically significant and architecturally rich, old county building.

Several alternatives were proposed during the one week of study effort. Creativity techniques were used to bring out all possible options for the consideration of government officials. After the creativity session was completed, the team began to evaluate various alternatives. The technique of life cycle costing helped select the appropriate course of action. Figure 8.2.1 presents the LCCA that was prepared for this study. Alternative 1 costs were based on the present situation with court functions at various locations that creates inefficiencies for both the city government and litigants. Some renovation and new furnishings are required to continue at the present locations.

Alternative 2 proposed to consolidate the district court functions and locate the district court in the historic county building. Alternative 3 suggested that district court functions be consolidated in a new facility. The old county building would be retained to house non-district-court functions presently in rental space. The final option, Alternative 4, proposed to consolidate district court functions in a new facility similar to Alternative 2, but in this case, the old county building would be demolished, and the existing site used for the new building.

The LCC analysis computer spreadsheet was used to record initial costs, energy, maintenance, replacements/alteration, and associated costs, as appropriate. Estimates were generated for each of these types of costs using the best information available at the time of the study. All estimated costs were converted to present worth using economic criteria of a 40-year life cycle and a 7% discount rate. All costs were expressed in constant dollars.

These present worth costs were then added to arrive at a total life cycle cost for each alternative. From the figures at the bottom of the spreadsheet, it can be seen that Alternative 2 provides the greatest savings to the government, even though the initial cost is considerably higher than the present situation. Efficiencies in staffing and improved public service handling provide a major contribution toward life cycle savings. The old county building was originally designed with LCC in mind, i.e., maintenance-free finishes, long-life roofing system, natural ventilation, and daylighting. Present-day comfort levels for temperature, humidity, acoustics, lighting, and fire safety can be met by the renovation. Non-monetary considerations—including convenience to litigants, historic preservation, environmental sustainability, quality of finishes, and aesthetics—also supported renovation of the existing facility.

LCCA Case Study 2
Continued

The results of this feasibility study indicate clearly that a great court facility can be obtained at minimum cost to the government through court consolidation and the adaptive/creative reuse of the old county building. Its central location, historical significance, and original design as a court facility make this alternative highly desirable. By renovating the existing building, over $30,000,000 in capital cost can be saved, as compared to providing a new court facility. The resulting efficiencies achieved in court consolidation, energy conservation, and historic preservation offer a unique opportunity to develop modern court chambers of a quality that cannot be duplicated in today's marketplace.

LCCA Case Study 3:
Branch Bank Prototype Layout

A design team was asked to develop a prototype branch bank layout that would reflect the initial, energy, and total life cycle cost targets established by the owner from historical data collected from similar facilities.

Several branch banks had been designed and built for the owner over the past several years. To better understand the functional relationships between spaces, the diagram shown as Figure 8.3.1 was prepared. For comparison purposes, a "typical" existing bank was selected for a life cycle cost analysis as Alternative 1. Figure 8.3.2 shows the floor plan for the selected bank.

Numerous alternative design layouts were prepared by the design team. The most promising alternative was selected for life cycle cost analysis. The layout selected (Alternative 2, Figure 8.3.3) is based on the owner's functional requirements. The life cycle cost analysis results for both alternative layouts are presented in Figure 8.3.4. The revised layout was recommended for future branch banks at a significant life cycle cost savings.

LCCA Case Study 4:
Health Care Facility Layout

The original health care facility design consisted of four 1-story, 50-bed units connected by enclosed corridors to a central 2-story support facility. A detailed space function analysis was prepared to further understand and define the scope of the project. From these space planning worksheets, it was determined that a gross area of 18,000 square feet per 50-bed unit, and 21,000 gross square feet for the support facility, would satisfy the functions required. The health care facility was located on a sloping, wooded site to the east of the existing center.

Figure 8.4.1 shows the site and a typical 50-bed unit floor plan (Alternative 1). The grades in the area of the site dropped 20 feet from one side to the other. The grade change resulted in a two-story support

LCCA Case Study 4
Continued

facility with two units connecting to the lower level. The gross area of the project was approximately 99,300 square feet distributed as follows: four 50-bed units at 19,000 square feet each, connecting corridors at 1,500 square feet, and a support facility at 21,810 square feet.

As a result of the layout study and life cycle cost analysis, a major redesign of the 50-bed unit was recommended. Figure 8.4.2 illustrates the following features of the new design (Alternative 2):

- One nursing station in lieu of two per 50-bed unit
- Elimination of the courtyard, with all rooms given a view
- Relocation of the support services to the central core
- Relocation of the bathrooms to the corridor side of the bedrooms

The design also achieves the following beneficial results:

- Reduction of gross area by approximately 1,000 square feet per 50-bed unit
- Reduction of the length of the exterior wall by approximately 450 linear feet
- A more efficient arrangement of mechanical services
- Less ground coverage, resulting in lower site development costs, loss of fewer trees, and improved adaptability on a difficult site
- Significant reduction in travel distance and improved function if the facility is altered in the future
- Virtual elimination of enclosed connecting corridors

The life cycle cost analysis indicated these items provided an initial savings of $515,000. It is estimated that an additional savings of $150,000 would be realized from changes in the site and connecting corridors. A one-level facility as shown provides an additional initial-cost savings of $100,000 in the central support facility. Other savings not quantified included fewer nursing staff with the simplified layout.

Mechanical and Energy Study

The original heating, ventilating, and air-conditioning (HVAC) system for the health care facility consisted of centralized, dual electric rooftop packaged units located at the support facility, which, in turn, distributed chilled water to all bedroom units. High-pressure steam from an outside source was supplied to a centralized mechanical equipment room, where it was converted to hot water for heating. Also, steam was redistributed to each 50-bed unit for domestic water heating.

Each 50-bed unit was provided with a 100% outside air supply rooftop air-handling unit with preheat, cooling, and reheat coils, which conditioned ventilation air to each bedroom and corridor space. The ventilation air offset the exhaust requirements for the bathrooms and toilets within the module. The bathroom air was exhausted to the atmosphere by exhaust fans, with no provisions for energy recovery.

Each 50-bed unit was provided with a perimeter fan coil to compensate for the external environmental changes and interior loads. Also, each

LCCA Case Study 4
Continued

exterior bathroom area was provided with a hot water converter heater, connected to the same forced-hot-water circulating loop serving the fan coil units. Domestic hot water was to be generated in each 50-bed unit on an individual basis with dual steam-to-water converters.

For the required functions, the existing concepts appeared valid, although high in initial and operational cost. To reduce initial and life cycle costs, the mechanical team developed an alternative distribution for the facility. The layout and architectural team studied the proposed HVAC system for the alternative distribution with the goal of maximum cost optimization and energy conservation. The proposed system concept requires a centrally located mechanical equipment room within the support facility, where chilled water for cooling, hot water for heating, and domestic hot water are generated and circulated to all four 50-bed units.

As a result of the new building layout and configuration, the exhaust system may be economically utilized for transfer of building heating and cooling energy to the outside ventilation air by means of a rotary enthalpy exchanger.

Also, to further minimize the outside air intake requirements, the air-conditioning units were arranged so as to simplify the system operation and maintenance. The air-conditioning units serve each bedroom unit with air to balance the bathroom exhausts, and also serve the interior zone areas with conditioned air to satisfy the temperature needs. Air is re-circulated from these areas. Four-pipe fan coil units are arranged to offset the outside environmental changes and satisfy the room temperature control requirements. All the mechanical team suggestions were tested and analyzed using a life cycle cost simulation model. The computer model predicted that, in addition to the savings shown from architectural items, the initial cost of the health care facility would be reduced by $450,000. In energy savings potential alone, the model predicted a 3,711 million Btu reduction per year. Life cycle cost savings amounted to over $800,000 in present worth dollars.

Conclusions

Because of the success achieved with both traditional and new tools and techniques, the health care facility gained improved performance at a reduced total cost of ownership. With the increasing complexity and number of considerations faced by all design professionals, computerized life cycle cost simulation models (such as those shown in Figure 8.4.3) were used to compare the total cost of ownership of competing design alternatives and thereby better satisfy the client's needs.

LCCA Case Study 5: Daylighting

Because of the southwestern U.S. site location, glass exposure of office areas (north and east elevations), and the goal of energy conservation set by the client, an LCCA of various combinations of daylighting schemes and supplemental lighting systems was performed.

Several issues were to be examined in-depth during the course of the study. These included:

1. Fenestration
2. Artificial lighting
3. Energy conservation
4. Mechanical requirements
5. Direct vs. indirect lighting
6. Coffered vs. flat ceiling
7. Lighting reflective devices
8. Fixed vs. operable sash
9. Window washing
10. Emergency ventilation
11. Venetian blinds
12. Initial and life cycle costs

The original daylighting scheme consisted of two 4'-0" high windows in a curtain wall. The ceiling was coffered to allow natural lighting to penetrate into the space to a greater degree. A reflective louver located along the exterior wall served to decrease glare and focus the natural light.

Three alternative daylighting schemes were compared to the original design in terms of both life cycle costs and other important criteria. The first alternative consisted of a coffered ceiling and one 6'-0" high window, with reflective louvers and surfaces added. The second alternative was similar to the original design, i.e., two 4'-0" windows and a coffered ceiling, but without the reflective louvers and surfaces. The final daylighting alternative consisted of a recessed, single 4'-0" high window, slightly sloped, a reflective ceiling panel, and a 9'-0" ceiling height. The wall sections are shown in Figure 8.5.1.

Four means of providing supplemental lighting to the office space (to attain an ambient level of 50 footcandles) were selected for comparison with the schemes. These included a fluorescent, parabolic three-tube, direct 2' × 4' fixture; a fluorescent, two-tube, indirect 6' × 4' fixture; a high-pressure sodium (HPS) 250-watt fixture capable of being dimmed; and an HPS 250-watt fixture with switching capability.

Because of the interaction of the daylighting schemes with the various supplemental lighting means, 16 (4 × 4) possible solutions were evaluated. The team then established criteria, and their corresponding weights of importance. The following criteria were felt to be significant in the evaluation of the various alternatives: glare control, spatial brightness, initial cost, life cycle cost, appearance of device, energy

LCCA Case Study 5
Continued

conservation, color rendition/temperature, fixture response time, and maintainability.

The team then prepared life cycle cost estimates for the various daylighting and lighting fixture alternatives. Figure 8.5.2 is one example of the calculations. Supplemental lighting energy requirements were estimated from daylighting data provided by a computer analysis program, LUMEN 2. A computer analysis of mechanical systems determined the cooling energy requirements based on the alternative configurations.

Once the life cycle costs were established for each of the 16 combinations, the team evaluated these schemes using the technique of weighted evaluation. From the analysis (Figure 8.5.3), the high-pressure sodium fixture appeared most appropriate. Daylighting Alternative 1—using a single, 6'-0" high window—satisfied the required functions at the least life cycle cost. Aesthetics and the benefit of a high upper portion of glass for improved light penetration into the space resulted in final selection of two 3'-0" high windows separated by a solid wall 2' high.

LCCA Case Study 6: *Glass/HVAC Replacement*

Management of a major city hospital requested the replacement of the exterior glass and a modernization of the HVAC system. Four alternatives were suggested by the design firm hired to do the work. These were as follows:

Alternative 1: Insulated clear glass with perimeter induction units. The interior zones would consist of a VAV reheat system.

Alternative 2: Use of low-E (emission) insulated clear glass. HVAC the same as Alternative 1.

Alternative 3: Insulated clear glass with perimeter variable air volume (VAV) reheat system. The interior zones would also consist of a VAV reheat system.

Alternative 4: Use of low-E insulated glass with the same mechanical system as Alternative 3.

The economic criteria given by the hospital facility manager included a project life cycle of 25 years and a discount rate of 10%. All costs were to be expressed in constant dollars.

Figure 8.6.1 presents the life cycle cost analysis for these four alternatives. Alternative 4 proved to have both the lowest initial cost and the lowest life cycle cost. The data are graphically summarized in Figure 8.6.2.

LCCA Case Study 7: Lighting System

The designer of a lighting system was interested in comparing the life cycle cost of direct vs. indirect lighting systems. Two of each type were examined. The alternatives were as follows:

Alternative 1: 2' × 4', two lamp fluorescent, K 12 lens, direct lighting system.

Alternative 2: 2' × 4', two-lamp fluorescent, parabolic louver, direct lighting system.

Alternative 3: Indirect ceiling-/column-mounted two-lamp fluorescent lighting system.

Alternative 4: Indirect floor-/furniture-mounted metal halide (250-watt) lighting system.

The economic criteria used for this LCCA consisted of a 10% discount rate and a 25-year analysis period. All costs were to be expressed in constant dollars. The results of this economic analysis are shown as Figure 8.7.1.

This LCCA indicated that the best choice was Alternative 4. The major savings were due to an assumed employee productivity improvement of 2%, the basis of which was obtained from independent research data.

Since the productivity assumption appeared to be both controversial and highly sensitive to the results of the LCCA, the owner requested a sensitivity analysis. Figure 8.7.2 presents the results of two sensitivity tests that were performed. The first varied the employee productivity from a 0% difference to a 4% improvement. The diagram at the bottom of this figure illustrates the choice changes only in favor of Alternative 1, when employee productivity is assumed not to change at all regardless of the lighting system used.

The second sensitivity analysis varied the assumption on the escalation rate of electricity. The rate was varied from 0–2% increase per year. Under all cases, Alternative 4 remained the lowest life cycle cost alternative. On the basis of these results, the lighting designer recommended Alternative 4.

LCCA Case Study 8: Elevator Selection

This LCCA study concerned itself with the vertical transportation needs of an eight-story computer office building. Over 800 personnel and large quantities of paper required vertical transportation daily.

The original design contained four passenger elevators and two freight elevators. Using a traffic study prepared for the client and standard industry elevator criteria, the team established a requirement of 3.5 elevators for the personnel. Paper transportation could be handled in one of two ways. A passenger elevator could be sized for 6,000-pound loads and used in off-peak personnel travel times. Or a dumbwaiter

LCCA Case Study 8
Continued

could be used to directly transfer paper, etc., from storage to appropriate floors. Both of these suggestions were analyzed by using the life cycle costing approach.

The weighted evaluation process identified the following criteria for analysis of the various alternates with the original design: initial cost, maintenance, efficiency of people/service handling, energy conservation, and aesthetics. The four-passenger elevator scheme, with one rated for 6,000 pounds, was judged by the team to be most desirable. In addition to energy and maintenance savings, the four-elevator scheme reduced the space occupied by elevators by 600 S.F., as compared to the eight-elevator alternative. The life cycle cost analysis for this study is shown as Figure 8.8.1.

LCCA Case Study 9:
Equipment Purchase Examples

LCC can be used to optimize equipment selection and minimize costs by measuring and combining all expenditures and translating them to equivalent dollars. LCC, by putting all costs on a common basis, allows engineers to select the optimum solution.

Example: Baghouse

The application of LCC for selection of housing for industrial filters in a baghouse (a facility used to remove particulates from exhaust gases) is illustrated by the following example.

A compressed-air type of cleaning baghouse must be installed to remove dust from a 40,000 cubic foot per minute air stream. The plant has several existing baghouses, all made by the same manufacturer.

The plant engineer contacted the manufacturer's local representative and requested a quote for four units that could handle the new application. The representative suggested the baghouses described in Figure 8.9.1, Part I. Total installed cost shown includes freight and taxes.

If only installed cost is considered, the 4,200-square-foot baghouse appears to be the best choice. But this information is not sufficient to justify a final decision. Other data must be considered. The equipment is expected to have a 10-year useful life, the cost of electricity is $0.04 per kilowatt-hour and is expected to increase 8% per year, and the total efficiency of the equipment's exhaust fan is 65%. All other factors, such as additional maintenance and bag cleaning costs, will be approximately the same for each baghouse.

For this example, the only maintenance cost that differs with the different sizes of baghouses is the cost of installing the bags. The only significantly differing operating expense is the cost of overcoming the baghouse's resistance to airflow. Depreciation can be allocated equally over the expected 10-year life; therefore, yearly depreciation will be the installed cost less salvage value times 10%.

LCCA Case Study 9
Continued

The cost of capital is determined by expressing all cash flows in present worth terms. Since the primary concern is with the after-tax effect, cash flows must be corrected to an after-tax basis. With an effective tax rate of 50%, the after-tax result of maintenance costs, operating costs, and depreciation is half of the before-tax effect. The after-tax present worth values of each of the four possible selections are shown in Figure 8.9.1, Part II.

The baghouse with the smallest after-tax present worth cost is the best selection. Surprisingly, it is not the 4,200 S.F. unit. The 5,400 S.F. unit, which has an initial installed cost of $6,000 more than the smaller baghouse, has an after-tax present worth cost 7% less than the smaller unit. Figure 8.9.2 illustrates the results.

Example: Refrigerator-Freezer

This example outlines a formal procedure for procurement of equipment (a refrigerator-freezer) using LCC techniques. For this procurement, bidder D was awarded the contract, even though the initial unit cost was $309.50 vs. $231.53 for bidder B. (See Figure 8.9.3 for a summary of bidder LCC costs.) The difference in recurring costs ($357.42 for D vs. $464.91 for B) more than offset (on the basis of present worth analysis) the difference in initial costs. This procurement, based on anticipated demand quantities, provided a projected cost savings over the useful life (15 years) of some $260,000.

The LCC formula used in this procedure is

$$\text{LCC} = A + R$$

where

LCC = life cycle cost in present worth dollars
A = the acquisition cost (bid price)
R = present worth sum of the cost of electrical energy required by the refrigerator-freezer during its useful life

and

$$R = P \times C \times T \times d$$

where

P = computed electrical energy, kilowatt-hours, required during 24 hours of operation
C = cost of 1 kilowatt-hour of electrical energy
T = annual operating time in days
d = discount factor, which will convert the stream of operating costs over the life of equipment to present worth

LCCA Case Study 9
Continued

The discounted cash flow or present worth methodology was used as a decision-making tool to allow direct comparison between different expenditure patterns of alternative investment opportunities. The present worth sum represents the amount of money that would be invested today, at a given rate of interest, to pay the expected future costs associated with a particular investment alternative. For purposes of this procurement, the Federal Supply Service used a discount rate of 8%, and a product life of 15 years.

The value of P in the energy cost equation is a function of net refrigerated volume V of the product being offered and the energy factor EF, which relates refrigerated volume and the electrical energy consumed to maintain the refrigerated volume. Stated in mathematical notation, the value of P is determined as follows:

$$P = \frac{V}{EF}$$

where

$$EF = \frac{\text{Volume of frozen food compartment} \times \text{correction factor} + \text{food compartment}}{\text{Kilowatt-hours of electrical energy consumed in 24 hours of operation}}$$

and the correction factor is a constant of 1.63. Thus the LCC evaluation formula

$$LCC = A + R = A + (P \times C \times T \times d)$$

can be rewritten as

$$LCC = A + \frac{V}{EF} \times \$.04 \times 365 \times 8.559$$

$$= A + \frac{V}{EF} \times \$124.961$$

LCCA Case Study 10: *Construction and Service Contracts*

Saving the owner's money is just as important during the construction and occupancy of a building as in the design phase. During construction, however, an additional party is involved—the contractor. Making changes to the building during construction is much more expensive.

Because of this, a larger cost savings incentive must usually be present before any change will be made. To further compound the situation, the cost incentive must be jointly shared with the two parties involved. This also applies to service contracts during the occupancy phase.

LCCA Case Study 10
Continued

Construction Phase

The General Services Administration (GSA) value incentive clause (VIC) for construction contracts is presented in Figure 8.10.1. It describes how cost savings can be achieved during the construction phase.

The VIC invites the contractor to challenge unrealistic or nonessential owner contract requirements and to profit by doing so. The procurement phase is the last opportunity to save on construction work before it is actually delivered and is very often the last opportunity to have significant impact on ownership costs. These types of clauses are useful when owners desire to benefit from the experience and knowledge of contractors in the areas of cost, new materials, construction techniques, and industry standards.

The VIC rewards contractors who propose changes in contract documents that will result in reducing cost without sacrificing required quality or function. The sharing incentive arrangement is established to encourage contractor initiative and ingenuity in identifying and successfully challenging high-cost areas in contracts. The following types of incentives can be provided in the VIC:

1. *Instant sharing:* Contractors may share in savings realized by modifying or eliminating work required under the terms of the instant contract. All VICs provide for instant sharing.
2. *Collateral sharing:* In addition to instant sharing, all VICs provide that a contractor may share in any savings (resulting from a proposal to change contract requirements) in future government costs of ownership of the work provided under the instant contract. Note that although there need not be a savings in the instant contract, there must be a life cycle savings for the contractor to qualify for collateral sharing.

Example: Chillers

The contractor for a Federal Office Building (FOB) proposed substituting two 380-kilowatt, 600-ton air-conditioning chillers for the specified two 450-kilowatt, 600-ton chillers at an increased cost of $39,970. The GSA regional office indicated that the chillers with 140 kilowatts of reduced energy use were desirable, especially in light of the fact that one chiller would be on 24 hours per day year-round. GSA approved the contract modifications (additional change orders) for the increased chiller cost, which included $5,000 for the contractor's share of the collateral savings.

Occupancy Phase

Clauses similar to those illustrated in Figure 8.10.1 have been developed by GSA for use during the facility occupancy phase. The maintenance contractor (or in-house staff) can provide proposals to reduce maintenance costs, and can provide feedback necessary to design personnel so that future projects will benefit.

LCCA Case Study 10
Continued

Example: Security Service Contract

Figure 8.10.2 provides an example computation for a change in a facility and workforce plan (of a service contract) under the "service and term" contract clause of GSA. The facility, in an as-is condition, is considered a constructive part of the contract, hence is subject to a value change proposal (VCP). Similarly, the workforce plan developed by the contractor not the government) in this example is also part of the contract conditions by the very nature of its use and acceptance. Subcontractor A is a new party to the contract and receives overhead and profit on its new work. In adjusting the contract price there are two possible methods subject to mutual agreement. Method 1 requires the contractor to pay for the new work from the savings accumulated over the contract period, but does change the unit prices. Method 2 advances the contractor money for the new work, while the government accumulates the full savings over the contract period through reduced unit prices.

In either method, the example illustrates the savings potential still present for the owner using the concept of LCC during building occupancy. In this case, the guard service was omitted from the second door, with a video camera and buzzer lock system taking its place.

LCCA Case Study 11: Chemical Stabilization Plant-Dryers vs. Windows

The LCC technique is used quite extensively in the design of process plants. The U.S. Environmental Protection Agency (EPA) requires LCCA for selection of the basic process for any water/wastewater treatment plant (WTP). This example of a proposed chemical stabilization facility for WTP sludge typifies today's problems with sludge disposal—especially its cost. The example was part of a value engineering study that had significant initial and follow-on cost impact on the proposed project.

The VE study was based on the following:

Original Design

The conceptual design report envisioned producing a mix of 56% solids which would be dried to 60% solids by windrowing (drying process). Drying would be accomplished in 7 days for average conditions, and in 3 days for peak conditions. Air removal from the curing area is projected at 1×10^6 cubic feet per minute (CFM) during the summer, and 500,000 CFM during the winter months. This drying step follows the 12-hour heat pulse period. The ambient air must remove 45 tons of water during average conditions and 66 tons of water during peak conditions (based on 7-day-per-week operation).

LCCA Case Study 11 *Continued*

Proposed Design

The VE team proposed using a thermal dryer to remove the water and to eliminate the windrowing step. Two alternatives were presented. Both are the same as the original design, except for the drying process. That is, chemicals are still added to the sludge to raise the pH to 12 or above and to raise the temperature to 52 °C or higher. The temperature is maintained for 12 hours.

Alternative 1 maintains a mix of 56% total solids (TS) entering the dryer. Alternative 2 reduces feed solids to the dryer to 45% TS by reducing the amount of chemicals added. In Alternative 1, the target moisture content is obtained by using the original chemical addition rate and removing the rest of the water in the dryer. In Alternative 2, smaller amounts of chemicals are added, and more water must be removed by the dryer.

Alternative 2 is recommended. While Alternative 1 has higher initial cost savings, Alternative 2 has significantly higher annual cost savings, as it uses fewer chemicals for processing and reduces the amount of material required to be removed.

	Original Design	Alternative 1	Alternative 2
Initial construction costs	$ 56,333,000	$ 13,367,000	$ 28,728,000
O & M annual costs (PW)	$143,522,366	$117,273,091	$ 80,455,788
Total life cycle costs (PW)	$199,855,366	$130,640,091	$109,183,788
Savings compared to original	—	$ 69,215,275	$ 90,671,578

O & M: operation and maintenance
PW: present worth

The life cycle cost analysis is shown as Figure 8.11.1.

LCCA Case Study 12: Campus University Planning Using LCC and Choosing by Advantages

Project Description

Facility leaders of a university in the Detroit metropolitan area were interested in developing a campus framework or master plan. Functions to be achieved included, "maximize four pillars," "create multiple pathways to learning," and "meet global student needs" in order to "facilitate lifelong learning"—to satisfy the university's mission. By answering the question of "how" to satisfy each of these functions, the planning team identified the program of requirements for the campus. This made it possible to explore a series of campus plan alternatives. Later, the functions were used to identify "factors" for the evaluation of

LCCA Case Study 12 *Continued*

alternatives using the decision-making method called "choosing by advantages," or CBA.

Four alternate Campus Plans were developed

Alternative 1
Alternative 2
Alternative 3
Alternative 4

Choosing by Advantages (CBA)

The campus planning team reviewed each of the alternatives and used CBA and LCC for evaluation. CBA assists the stakeholders in considering a variety of "factors" in judging the campus framework plan alternatives. These factors are non-monetary and were selected from the function logic diagram developed earlier. The factors included:

- Education Enhancement
- Religious Culture
- Recreation and Community Relationships
- Campus Life
- Pedestrian Safety
- Pedestrian Circulation Distance and Accessibility
- Campus Image (Face)
- Student Housing Needs
- Operational Effectiveness

Attributes of each factor were listed for each alternative. For example, "campus space functional relationships" were used to measure the factor "education enhancement." After each attribute, a measurement has been listed for each alternative. The alternative with the least preferred attribute measurement was zeroed out. The advantages of the remaining alternatives were listed in comparison with the alternative with no advantage. The best advantage for each factor was highlighted and is shown in bold print. Once the advantages of all factors were listed, the planning team assigned importance points to the advantages, with the "paramount" advantage receiving the most number of points. All importance points are totaled for each alternative. The alternative receiving the highest number of points would be selected if there were no monetary differences.

The campus framework CBA is shown in Figure 8.12.1. The scoring results are as follows:

Alternative 1   430 importance points
Alternative 2   235 importance points
Alternative 3   320 importance points
Alternative 4   620 importance points

Next, the planning team estimated the initial cost and maintenance cost for those items that varied from alternative to alternative, such as the

LCCA Case Study 12
Continued

length of loop roadway. These life cycle costs are shown in Figure 8.12.1. The major initial cost differences included the roadway, the parking lots, and the parking structures. The existing site contained adequate parking for current needs. The major maintenance costs were for the roadway, the parking lots, and the parking structures. All annual maintenance costs were converted to an equivalent "present worth" using a 7% discount rate and 25-year life cycle.

The initial costs were:

Alternative 1	$1.6 million
Alternative 2	$1.0 million
Alternative 3	$2.9 million
Alternative 4	$1.2 million

The life cycle costs were:

Alternative 1	$2.2 million
Alternative 2	$1.4 million
Alternative 3	$3.9 million
Alternative 4	$1.7 million

Two "Importance to Cost" graphs were used to compare alternatives. The first compares the initial costs and benefits for each alternative (see Figure 8.12.2). The preferred alternative is No. 4. The second graph compares the life cycle costs and benefits for each alternative (see Figure 8.12.3). The preferred choice is again No. 4.

Based on the results of this evaluation and the planning team's 100% concurrence, Alternative 4 is recommended. It offers the following advantages:

- Nearly the least separation of classrooms
- Least noise disruption
- Most significant expression of religious culture (green space, paths, etc.)
- Best relationship to enhance community use and awareness (recruiting tool)
- Best views of pond to enhance campus life
- Better pedestrian safety
- Shortest distance from parking to campus buildings
- Shortest distance from academic buildings to cafeteria
- Best campus image (face) with most open campus, best green space, best building visibility from the freeway, and no parking structure view
- Best meets student housing needs, since it is the least institutional in appearance, and has the best functionality.

LCCA Case Study 13: High School Trade-Off Analysis of Initial vs. Staffing Costs

The VE team isolated the staffing costs as the largest expenditure—representing 31% of the total costs. The initial costs were isolated at 18%. The team developed an alternate involving a slight modification to the student-teacher ratio, to see what impact it would have on initial costs. Figure 8.13.1 presents an assessment of how much can be saved in staffing costs in the high school if the student-teacher ratio can be increased from 16 to 17 students per teacher. Assuming all other issues remain constant, the results indicate that nearly $5,000,000 could be invested initially and still achieve a break-even. In other words, if changes in space efficiency, technology, lighting, or other facility aspects could improve teaching efficiency, as much as $5,000,000 could be saved and used for expansion or other investment.

LCCA Case Study 14: Highway for Regional Transportation Authority

The Request For Proposal (RFP) for the project required the design-builders to present the operations, maintenance and replacement cost implications of their submission as a means to evaluate alternate technical approaches and designs. Figure 8.14.1 presents an excerpt from the owner's independent consultant's assessment of a design-builder's proposal for the highway project. The owner analyzed the bidder responses and selected the "optimum" bidder—not necessarily the lowest initial cost.

LCCA Case Study 15: Life Cycle Cost Assessment—HVAC System for a High School

A local school authority put a high school project out for bid, and the resulting bids exceeded their budget. In an effort to realize their project on-time and within budget, they decided on a VE study, with emphasis on total costs. In fact, the bid documents stated, "the basis of HVAC award would be the lowest Life Cycle Costs." Figure 8.15.1 presents the summary of the VE consultant's review of the bidder's HVAC systems as proposed for the school. The analysis indicated that spending $1.7M more initially would yield net savings of $3.2M over the life of the facility and pay back within 5 years. The VE proposal represented a 45% reduction in the school's operations and maintenance costs. The extra initial costs were approved by the school board, and the VE proposal was implemented.

LCCA Case Study 16: Lease vs. Build LCC Analysis

A government owner had a need for additional space. Three choices were considered:

1. To renovate existing space
2. To lease space in another existing government building.
3. To renovate a historic structure

In all cases, the historic facility owned by the agency needed to be upgraded and structurally improved. The owner's consultant was asked to perform an LCCA. The analysis demonstrated that renovating and using the historic structure presented the most favorable life cycle performance. Refer to Figure 8.16.1

Summary

The LCC technique has many uses. The concept of conducting an economic analysis using equivalent costs is an investor's and owner's tool to optimize expenditures or profit, or both. Once the concepts are understood, the applications of LCC are limited only by the user's imagination. With this type of information, owners can realize quality work, while keeping initial costs within their monetary capabilities. This is an achievement not realized under the traditional bidding approach, where the lowest-cost bidder wins regardless and, in too many cases, represents the lower quality option.

LCCA Case Studies:	Figure(s)	Planning/ Feasibility	Design	Construction/ Occupancy
1. Office to Museum Renovation	8-1-1 to 8-1-5	■		
2. District Court Consolidation	8-2-1	■		
3. Branch Bank Prototype Layout	8-3-1 to 8-3-4		■	
4. Health Care Facility Layout	8-4-1 to 8-4-3		■	
5. Daylighting	8-5-1 to 8-5-3		■	
6. Glass/HVAC Replacement	8-6-1 to 8-6-2		■	
7. Lighting System	8-7-1 to 8-7-2		■	
8. Elevator Selection	8-8-1		■	
9. Equipment Purchase Examples	8-9-1 to 8-9-3			■
10. Construction and Service Examples	8-10-1 to 8-10-2			■
11. Chemical Stabilization: Plant-Dryers vs. Windows	8-11-1		■	
12. Campus University Planning Using LCC and Choosing by Advantages	8-12-1 to 8-12-3	■		
13. High School Tradeoff Analysis of Initial vs. Staffing Costs	8-13-1	■		
14. Highway for Regional Transportation Authority	8-14-1	■		
15. LCC Assessment- HVAC System for a High School	8-15-1		■	
16. Lease vs. Build LCC Analysis	8-16-1	■		

Figure 8.1: **LCCA Case Studies**

LIFE CYCLE COST ANALYSIS (Present Worth Method)
Project/Location: Museum, Washington, DC

Description: Office to Museum Renovation
Project Life Cycle = 25 YEARS
Discount Rate = 10.00% "Constant" 1994 Dollars

		Current Use		Proposed Use		Delta Life Cycle Cost	
		Office Only	SF	Museum/ Office	SF		
		Office SF:	18,302	Office SF:	3,302		
		Museum:	0	Museum:	15,000		

INITIAL COSTS

Description	Quantity UM	Unit Price	Est.	PW	Est.	PW	Est.	PW
Construction Costs								
A. E Street Bldg- Existing	18,302 Usable SF	$0.00	0	0	0	0	0	0
B. E Street Bldg- Museum	18,302 Usable SF	$26.48	0	0	484,593	484,593	484,593	484,593
C. Museum A/E Fee	15.00% Pct	$3.97	0	0	72,689	72,689	72,689	72,689
D. Exhibit Display Furniture	3,500 Exhibit SF	$300.00	0	0	1,050,000	1,050,000	1,050,000	1,050,000
E. Exhibit Designer Fee	25.00% Pct	$75.00	0	0	262,500	262,500	262,500	262,500
F. Gas Chiller/Heaters	1 Lump Sum	$474,000	0	0	81,370	81,370	81,370	81,370
Note: Pro rata @ 17.17%								
G. Staff Displacem't Office Fitout	15,000 Usable SF	$15.00	0	0	225,000	225,000	225,000	225,000
H.			0	0		0	0	0
I.			0	0		0	0	0
J.			0	0		0	0	0
K.			0	0		0	0	0
Total Initial Cost				0		2,176,152		2,176,152
Initial Cost Difference (Compared to Current Use)						(2,176,152)		(2,176,152)

REPLACEMENT COST/ SALVAGE VALUE

Description	Year	PW Factor						
A.	5	0.6209	0	0		0		0
B.	5	0.6209		0		0		0
C.	10	0.3855		0		0		0
D.	10	0.3855		0		0		0
E.	15	0.2394		0		0		0
F.	15	0.2394		0		0		0
G.	20	0.1486		0		0		0
H.	20	0.1486		0		0		0
I. Salvage Value	25	0.0923	0	0	0	0		0
Total Replacement/Salvage Costs				0		0		0

Description: Office to Museum Renovation

ANNUAL COSTS

Description	Est. Notes	Escl. %	PWA	Current Use Est.	PW	Proposed Use Est.	PW	Delta Life Cycle Cost Est.	PW
A. Utilities	(5)	0.000%	9.077	55,000	499,237	64,015	581,070	9,015	81,833
B. Maintenance	(5)	0.000%	9.077	66,000	599,085	76,818	697,285	10,818	98,200
C. Custodial	(5)	0.000%	9.077	24,000	217,849	31,868	289,267	7,868	71,418
D. Security	(5)	0.000%	9.077	17,000	154,310	57,000	517,391	40,000	363,082
E. HVAC	(5)	0.000%	9.077	4,500	40,847	5,238	47,542	738	6,695
F. Landscaping	(5)	0.000%	9.077	3,400	30,862	3,400	30,862	0	0
G. Fire/ Safety	(5)	0.000%	9.077	4,300	39,031	4,300	39,031	0	0
H. Elevator	(5)	0.000%	9.077	2,200	19,969	2,921	26,516	721	6,547
I. Trash Removal	(5)	0.000%	9.077	1,500	13,616	1,992	18,079	492	4,464
J. Other	(5)	0.000%	9.077	1,500	13,616	1,500	13,616	0	0
K. Director	(5)	0.000%	9.077	0	0	65,000	590,008	65,000	590,008
L. Admin. Assistant	(5)	0.000%	9.077	0	0	21,500	195,156	21,500	195,156
M. Curator	(5)	0.000%	9.077	0	0	47,500	431,159	47,500	431,159
N. Registrar	(5)	0.000%	9.077	0	0	30,500	276,850	30,500	276,850
O. Museum Educator	(5)	0.000%	9.077	0	0	26,000	236,003	26,000	236,003
P. Museum Technician	(5)	0.000%	9.077	0	0	21,500	195,156	21,500	195,156
Q. Mus. Shop Manager	(5)	0.000%	9.077	0	0	28,000	254,157	28,000	254,157
R. Museum Operations	(5)	0.000%	9.077	0	0	104,000	944,012	104,000	944,012
S. Staff Displacement	(5)	0.000%	9.077	0	0	390,000	3,540,046	390,000	3,540,046
Total Annual Costs (Present Worth)				$1,628,421	$983,052	$8,923,207	$803,652	$7,294,786	

ANNUAL PROJECTED INCOME

Description	Est. Notes	Escl. %	PWA	Est.	PW	Est.	PW	Est.	PW
A. Museum Income	(6)	0.000%	9.077	0	0	0	0	0	0
B. Ballroom Income	(6)	0.000%	9.077	0	0	(52,000)	(472,006)	(52,000)	(472,006)
C. Gift Shop Income	(6)	0.000%	9.077	0	0	(25,000)	(226,926)	(25,000)	(226,926)
D. Grant Income	(6)	0.000%	9.077	0	0	0	0	0	0
E. Fund Raiser Income	(6)	0.000%	9.077	0	0	(30,000)	(272,311)	(30,000)	(272,311)
Total Annual Income (Present Worth)					$0	($107,000)	($971,243)	($107,000)	($971,243)

Total Life Cycle Costs (Present Worth)	$1,628,421	$11,099,358	$9,470,938
Total Life Cycle Costs - Projected Income (PW)	$1,628,421	$10,128,115	$8,499,694

Figure 8.1.1: **LCCA Case Study 1: Office to Museum Renovation**

LIFE CYCLE COST ANALYSIS (Present Worth Method)
Project/Location: Museum, Washington, DC
Description: Office to Museum Renovation

NOTES & ASSUMPTIONS:

Description	Current Use	Proposed Use	Delta to Current Use
1. Building Usage and Areas	Office Only SF Office SF: 18,302 Museum: 0	Museum/ Office SF Office SF: 3,302 Museum: 15,000	SF Office SF: (15,000) Museum: 15,000
2. Facility Operation Assumptions			
A. Days of Operation	Monday thru Friday	Monday thru Saturday	Saturday Museum Use
B. Hours of Operation	Office: 8:30 a.m. - 5:00 p.m.	Office: 8:30 a.m. - 5:00 p.m. Museum: 10:00 a.m. - 6:00 p.m.	Office: No Time Change Museum: 10:00 a.m. - 6:00 p.m.
3. Staffing			
A. Security Staff	1 Security Guard /Shift (Total Site)	1 Security Guard /Shift (Total Site) 1 Security Guard Dedicated to Museum During Operating Hours	1 Security Guard Dedicated to Museum During Operating Hours
B. Museum Staff	None Required	Director Administrative Assistant Curator Registrar Museum Educator Museum Technician Museum Shop Manager	
4. Economic Criteria			
A. Analysis Period (Life Cycle)	25 Years	25 Years	As Per OMB Circular A-94
B. Interest Rate (Discount Rate)	10 Percent	10 Percent	As Per OMB Circular A-94
C. Escalation Rate	3 Percent	3 Percent	Based on current inflation rate
D. Economic Approach	Present Worth	Present Worth	As Per OMB Circular A-94

Description					Current Use	Proposed Use		Delta to Current Use
5. Operational Costs	Current Office (*)	Proposed Use Office	Museum	Total	Office Cost/ SF	Office Cost/ SF	Museum Cost/ SF	
Utilities	$55,000	$9,923	$54,092	$64,015	$3.01	$3.01	$3.61	Extra Day Usage (+20%)
Maintenance	$66,000	$11,908	$64,911	$76,818	$3.61	$3.61	$4.33	Extra Day Usage (+20%)
Custodial	$24,000	$4,330	$27,538	$31,868	$1.31	$1.31	$1.84	High Traffic & Extra Day (+40%)
Security	$17,000	$3,067	$53,933	$57,000	$0.93	$0.93	$3.60	Extra Guard: $40,000/yr
HVAC	$4,500	$812	$4,426	$5,238	$0.25	$0.25	$0.30	Extra Day Usage (+20%)
Landscaping	$3,400	$613	$2,787	$3,400	$0.19	$0.19	$0.19	No Change
Fire/ Safety	$4,300	$776	$3,524	$4,300	$0.23	$0.23	$0.23	No Change
Elevator	$2,200	$397	$2,524	$2,921	$0.12	$0.12	$0.17	High Traffic & Extra Day (+40%)
Trash Removal	$1,500	$271	$1,721	$1,992	$0.08	$0.08	$0.11	High Traffic & Extra Day (+40%)
Other	$1,500	$271	$1,229	$1,500	$0.08	$0.08	$0.08	No Change
Director	$0	$0	$65,000	$65,000	$0.00	$0.00	$4.33	Staffing Costs normalized @ Year 5
Admin. Assistant	$0	$0	$21,500	$21,500	$0.00	$0.00	$1.43	Staffing Costs normalized @ Year 5
Curator	$0	$0	$47,500	$47,500	$0.00	$0.00	$3.17	Staffing Costs normalized @ Year 5
Registrar	$0	$0	30,500	$30,500	$0.00	$0.00	$2.03	Staffing Costs normalized @ Year 5
Museum Educator	$0	$0	26,000	$26,000	$0.00	$0.00	$1.73	Staffing Costs normalized @ Year 5
Museum Technician	$0	$0	21,500	$21,500	$0.00	$0.00	$1.43	Staffing Costs normalized @ Year 5
Mus. Shop Manager	$0	$0	28,000	$28,000	$0.00	$0.00	$1.87	Staffing Costs normalized @ Year 5
Museum Operations	$0	$0	104,000	$104,000	$0.00	$0.00	$6.93	Operations Costs normalized @ Year 5
Staff Displacement	$0	$390,000	$0	$390,000	$0.00	$26.00	$0.00	Office Rental @ $25-$27/SF

(*) From Regression Analysis Using Historical Data

6. Operational Revenue								Projected Income normalized @ Year 5
Museum Income	$0	$0	$0	$0	$0.00	$0.00	$0.00	Free Admission to Museum
Ballroom Income	$0	$0	($52,000)	($52,000)	$0.00	$0.00	($3.47)	Ballroom Income(1 night/ wk @ $1000)
Gift Shop Income	$0	$0	($25,000)	($25,000)	$0.00	$0.00	($1.67)	Gift Shop Income
Grant Income	$0	$0	$0	$0	$0.00	$0.00	$0.00	Grant Income
Fund Raiser Income	$0	$0	($30,000)	($30,000)	$0.00	$0.00	($2.00)	Fund Raiser Income(1 maj. & 15 small)

7. Design & Construction Schedule

Design   Jan-Dec 1999
Const.   Jan-Dec 2000
Occup.   Jan 2001+

Figure 8.1.2: **LCCA Case Study 1: Backup Data**

LIFE CYCLE COST ANALYSIS (Present Worth Method)
Project/Location: Museum, Washington, DC
Description: Office to Museum Renovation

CASH FLOW ANALYSIS (TOTAL COST): *CURRENT DOLLARS*

Yr. Description	Date	1994 Amount	Escal Rate/yr 3.00%	Escal Amount
0	1994	0	1.0000	$0
1	1995	0	1.0300	$0
2	1996	0	1.0609	$0
3 Gas Chiller/Heaters	1997	81,370	1.0927	$88,915
4 Office Fitout (Displ)	1998	225,000	1.1255	$253,239
5 A/E Fee & Designer	1999	335,189	1.1593	$388,576
6 Const. & Display	2000	1,534,593	1.1941	$1,832,384
7 O&M Cost	2001	983,052	1.2299	$1,209,031
8 O&M Cost	2002	983,052	1.2668	$1,245,301
9 O&M Cost	2003	983,052	1.3048	$1,282,661
10 O&M Cost	2004	983,052	1.3439	$1,321,140
11 O&M Cost	2005	983,052	1.3842	$1,360,775
12 O&M Cost	2006	983,052	1.4258	$1,401,598
13 O&M Cost	2007	983,052	1.4685	$1,443,646
14 O&M Cost	2008	983,052	1.5126	$1,486,955
15 O&M Cost	2009	983,052	1.5580	$1,531,564
16 O&M Cost	2010	983,052	1.6047	$1,577,511
17 O&M Cost	2011	983,052	1.6528	$1,624,836
18 O&M Cost	2012	983,052	1.7024	$1,673,581
19 O&M Cost	2013	983,052	1.7535	$1,723,789
20 O&M Cost	2014	983,052	1.8061	$1,775,502
21 O&M Cost	2015	983,052	1.8603	$1,828,767
22 O&M Cost	2016	983,052	1.9161	$1,883,630
23 O&M Cost	2017	983,052	1.9736	$1,940,139
24 O&M Cost	2018	983,052	2.0328	$1,998,343
25 O&M Cost	2019	983,052	2.0938	$2,058,294
26 O&M Cost	2020	983,052	2.1566	$2,120,042
27 O&M Cost	2021	983,052	2.2213	$2,183,644
28 O&M Cost	2022	983,052	2.2879	$2,249,153
29 O&M Cost	2023	983,052	2.3566	$2,316,628
30 O&M Cost	2024	983,052	2.4273	$2,386,126
31 O&M Cost	2025	983,052	2.5001	$2,457,710

Figure 8.1.3: **LCCA Case Study 1: Cash Flow Analysis (Total Cost)**

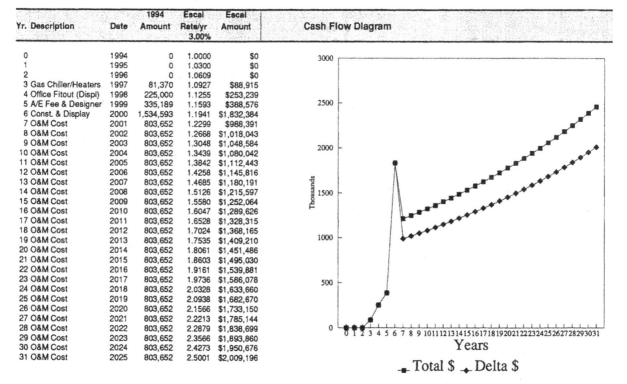

Figure 8.1.4: **LCCA Case Study 1: Office to Museum Renovation Cash Flow Analysis (Delta Costs)**

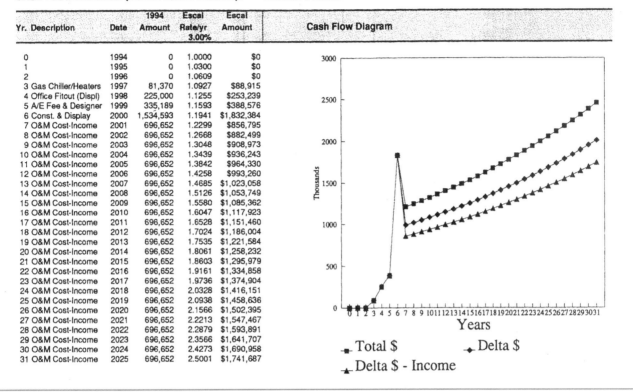

Figure 8.1.5: **LCCA Case Study 1: Cash Flow Analysis (Delta Costs–Income)**

LIFE CYCLE COST ANALYSIS
PRESENT WORTH (PW) COMPUTATION

Project/Location: District Courts Consolidation, Lansing, MI

Subject: Economic Feasibility Study Description: Renovate, Consolidate, New Project Life Cycle = 40 YEARS Discount Rate = 7.00% Present Time = Occupancy Date			**Alternative 1** Renovate - Court Functions at Various City Locations with a Variety of Inefficiencies		**Alternative 2** Consolidate & Locate District Courts in Historic Old County Building		**Alternative 3** Consolidate & Locate District Courts in New Facility, Retain Old County Building		**Alternative 4** Consolidate & Locate District Courts in New Facility, Demolish Old County Building	
INITIAL COSTS			Est.	PW	Est.	PW	Est.	PW	Est.	PW
Construction Costs										
A. Rental of Temporary Space, Setup Cost			0	0	950,000	950,000		0	950,000	950,000
B. Staff Moves				0	450,000	450,000	225,000	225,000	450,000	450,000
C. Old County Bldg Renovation			5,000,000	5,000,000	14,500,000	14,500,000	1,500,000	1,500,000		0
D. District Court Furnishings			1,500,000	1,500,000	1,500,000	1,500,000	1,500,000	1,500,000	1,500,000	1,500,000
E. New Courts Bldg Construction						0	42,500,000	42,500,000	42,500,000	42,500,000
F. New Courts Bldg Land Acquisition				0		0	2,000,000	2,000,000		0
G. County Courts Building Demolition				0		0		0	1,200,000	1,200,000
H. Planning, Programming & Design Service Fees			525,000	525,000	1,615,000	1,615,000	1,875,000	1,875,000	2,075,000	2,075,000
Total Initial Cost				**7,025,000**		**19,015,000**		**49,600,000**		**48,675,000**
Initial Cost PW Savings (Compared to Alt. 1)						(11,990,000)		(42,575,000)		(41,650,000)
REPLACEMENT COST/ SALVAGE VALUE										
Description	Year	PW Factor								
A. Building Alterations	10	0.5083	1,900,000	965,863	1,650,000	838,776	3,300,000	1,677,552	1,650,000	838,776
B. Major Systems Replacem't	15	0.3624	2,500,000	906,115	1,300,000	471,179	2,700,000	978,604	1,350,000	489,302
C. Building Alterations	20	0.2584	1,900,000	490,996	1,650,000	426,391	2,700,000	852,782	1,650,000	426,391
D. Major Systems Replacem't	30	0.1314	1,350,000	177,345	1,350,000	177,345	2,700,000	354,691	1,350,000	177,345
E. Building Alterations	30	0.1314	1,900,000	249,597	1,650,000	216,755	3,300,000	433,511	1,650,000	216,755
Total Replacement/Salvage Costs				**2,789,916**		**2,130,446**		**4,297,140**		**2,148,569**
ANNUAL COSTS										
Description	Escl. %	PWA								
A. Energy/ Fuel Annual Cost	0.000%	13.332	185,000	2,466,366	165,000	2,199,732	350,000	4,666,098	165,000	2,199,732
B. Maintenance Annual Cost	0.000%	13.332	345,000	4,599,440	280,000	3,732,878	575,000	7,665,733	290,000	3,866,196
C. Space Rental	1.000%	15.160	2,500,000	37,899,118	2,500,000	37,899,118	0	0	2,500,000	37,899,118
D. Staffing Cost	2.000%	17.392	8,000,000	139,135,551	6,400,000	111,308,441	6,400,000	111,308,441	6,400,000	111,308,441
E.	0.000%	13.332		0		0		0		0
Total Annual Costs (Present Worth)				**184,100,475**		**155,140,170**		**123,640,272**		**155,273,487**
Total Life Cycle Costs (Present Worth)				**193,915,391**		**176,285,616**		**177,537,412**		**206,097,056**
Life Cycle Savings (Compared to Alt. 1)						17,629,776		16,377,980		(12,181,665)
Discounted Payback (compared to Alt. 1)	PP Factor					5.40 Years		9.63 Years		Never Years
Total Life Cycle Costs (Annualized)		0.0750	14545427 Per Year		13223032 Per Year		13316928 Per Year		15459163 Per Year	

Figure 8.2.1: **LCCA Case Study 2: District Court Consolidation**

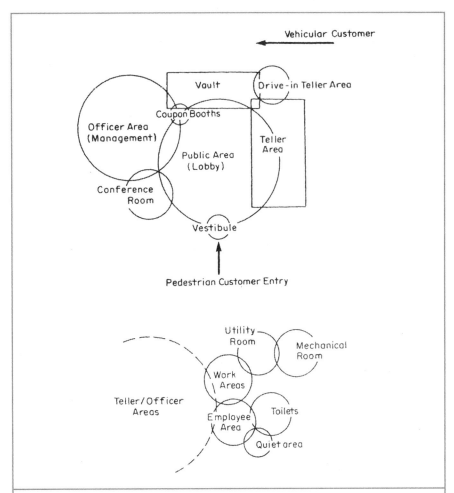

Figure 8.3.1: **Space Affinity Diagrams. Top: Public Service Spaces. Bottom: Bank Support Spaces.**

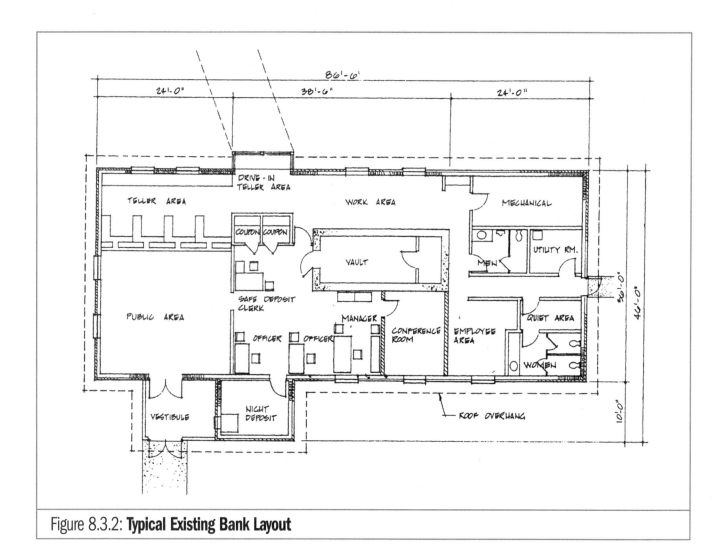

Figure 8.3.2: **Typical Existing Bank Layout**

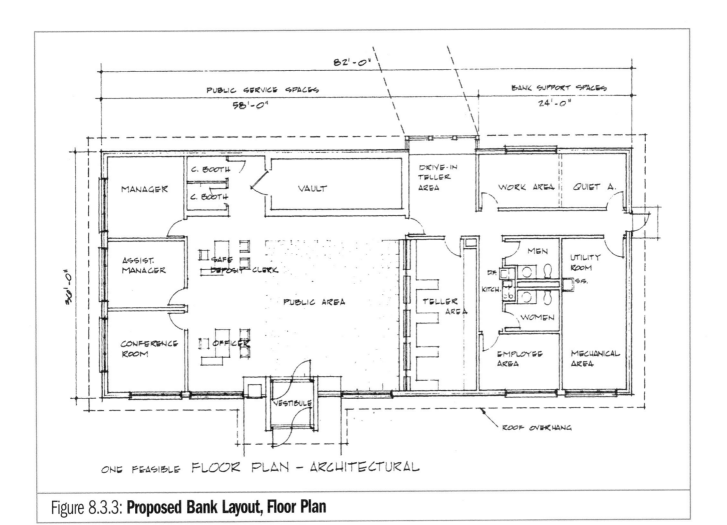

Figure 8.3.3: **Proposed Bank Layout, Floor Plan**

LIFE CYCLE COST ANALYSIS
PRESENT WORTH (PW) COMPUTATION
Project/Location: Branch Bank, MI

Subject: Building Layout Study
Description: Current Standard Vs New
Project Life Cycle = 30 YEARS
Discount Rate = 10.00%
Present Time : Occupancy Date

			Alternative 1 Current Typical Standard Layout (See Floor Plan)		Alternative 2 Revised New Bank Standard Layout (See Floor Plan)		Alternative 3		Alternative 4	
INITIAL COSTS			Est.	PW	Est.	PW	Est.	PW	Est.	PW
Construction Costs										
A. Current Typical Standard Layout			460,000	460,000		0		0		0
B. Revised New Standard Layout				0	350,000	350,000		0		0
C.				0		0		0		0
D.				0		0		0		0
E.				0		0		0		0
F.				0		0		0		0
G.				0		0		0		0
H.				0		0		0		0
Total Initial Cost				460,000		350,000		0		0
Initial Cost PW Savings (Compared to Alt. 1)						110,000		0		0
REPLACEMENT COST/ SALVAGE VALUE										
Description	Year	PW Factor								
A. Building Alterations	10	0.3855	46,000	17,734	35,000	13,494		0		0
B. Major Systems Replacem'ts	15	0.2394	92,000	22,024	70,000	16,757		0		0
C. Building Alterations	20	0.1486	46,000	6,837	35,000	5,202		0		0
D.		1.0000		0		0		0		0
E. Re-Sale Value	30	0.0573	(115,000)	(6,590)	(262,500)	(15,043)		0		0
Total Replacement/Salvage Costs				40,005		20,410		0		0
ANNUAL COSTS										
Description	Escl. %	PWA								
A. Energy/ Fuel Annual Cost	3.000%	12.667	6,220	78,792	5,300	67,138		0		0
B. Maintenance Annual Cost	1.000%	10.355	6,800	70,417	5,600	57,990		0		0
C.	0.000%	9.427		0		0		0		0
D.	0.000%	9.427		0		0		0		0
E.	0.000%	9.427		0		0		0		0
Total Annual Costs (Present Worth)				149,208		125,128		0		0
Total Life Cycle Costs (Present Worth)				649,213		495,538		0		0
Life Cycle Savings (Compared to Alt. 1)						153,676		0		0
Discounted Payback (compared to Alt. 1)		PP Factor				0.00 Years		0.00 Years		0.00 Years
Total Life Cycle Costs (Annualized)		0.1061		68,868 Per Year		52,566 Per Year		0 Per Year		0 Per Year

Figure 8.3.4: Branch Bank Alternative: LCCA Results

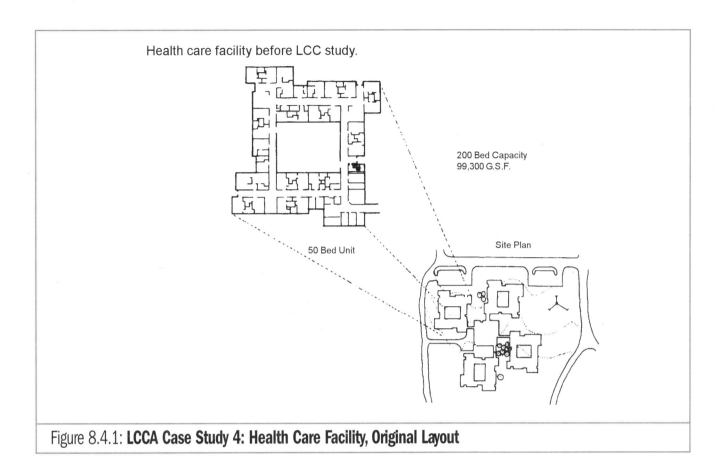

Figure 8.4.1: **LCCA Case Study 4: Health Care Facility, Original Layout**

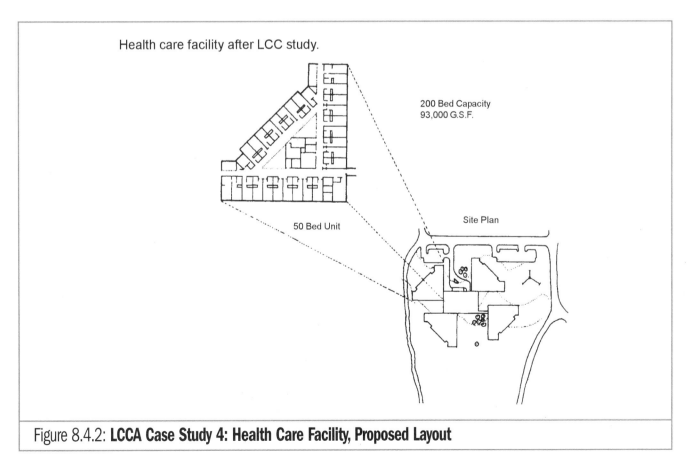

Figure 8.4.2: **LCCA Case Study 4: Health Care Facility, Proposed Layout**

LIFE CYCLE COST ANALYSIS
PRESENT WORTH (PW) COMPUTATION
Project/Location: 200 Bed Health Care Facility, Dayton, Ohio

Subject: 50 Bed Unit Layout Study
Description: Courtyard Vs Triangular Shape
Project Life Cycle = 25 YEARS
Discount Rate = 8.00%
Present Time = Occupancy Date

			Alternative 1		Alternative 2		Alternative 3		Alternative 4	
			Single Story Courtyard Layout (See Floor Plan) 4 - 50 Bed Units		Single Story Triangular Layout (See Floor Plan) 4 - 50 Bed Units					
INITIAL COSTS			Est.	PW	Est.	PW	Est.	PW	Est.	PW
Construction Costs										
A. Single Story, Courtyard Layout			13,500,000	13,500,000		0		0		0
B. Single Story, Triangular Layout				0	11,300,000	11,300,000		0		0
C.				0		0		0		0
D.				0		0		0		0
E.				0		0		0		0
F.				0		0		0		0
G.				0		0		0		0
H.				0		0		0		0
Total Initial Cost				13,500,000		11,300,000		0		0
Initial Cost PW Savings (Compared to Alt. 1)						2,200,000		0		0
REPLACEMENT COST/ SALVAGE VALUE										
Description	Year	PW Factor								
A. Building Alterations	10	0.4632	1,350,000	625,311	1,130,000	523,408		0		0
B. Major Systems Replacem't	15	0.3152	2,700,000	851,152	2,260,000	712,446		0		0
C. Building Alterations	20	0.2145	1,350,000	289,640	1,130,000	242,439		0		0
D.		1.0000		0		0		0		0
E.		1.0000		0		0		0		0
Total Replacement/Salvage Costs				1,766,103		1,478,293		0		0
ANNUAL COSTS										
Description	Escl. %	PWA								
A. Energy/ Fuel Annual Cost	2.000%	12.928	397,200	5,134,810	325,500	4,207,907		0		0
B. Maintenance Annual Cost	0.000%	10.675	297,900	3,180,016	255,750	2,730,074		0		0
C.	0.000%	10.675		0		0		0		0
D.	0.000%	10.675		0		0		0		0
E.	0.000%	10.675		0		0		0		0
Total Annual Costs (Present Worth)				8,314,825		6,937,981		0		0
Total Life Cycle Costs (Present Worth)				23,580,928		19,716,274		0		0
Life Cycle Savings (Compared to Alt. 1)						3,864,655		0		0
Discounted Payback (compared to Alt. 1)		PP Factor				0.00 Years		0.00 Years		0.00 Years
Total Life Cycle Costs (Annualized)		0.0937		2,209,033 Per Year		1,846,996 Per Year		0 Per Year		0 Per Year

Figure 8.4.3: Health Care Facility: LCCA Results

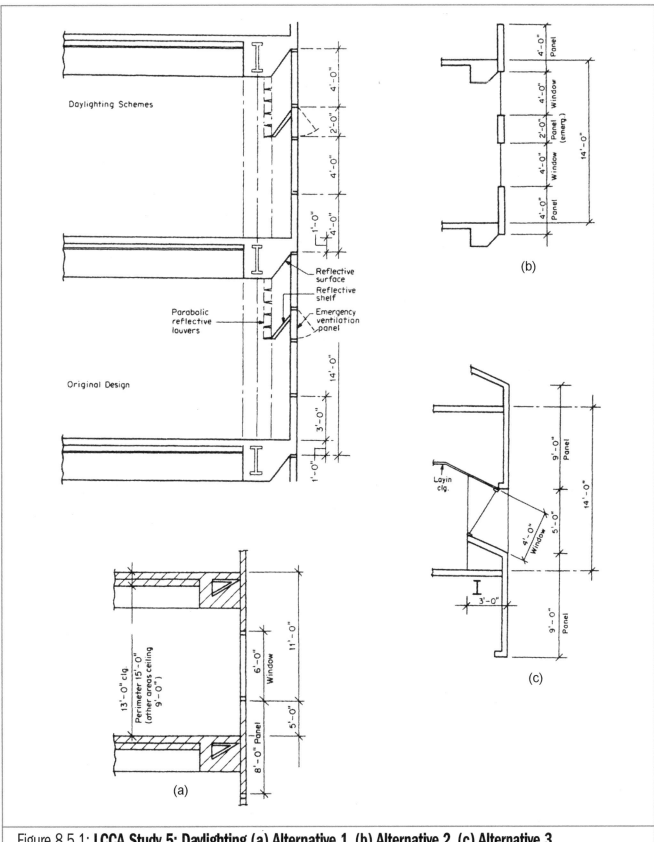

Figure 8.5.1: **LCCA Study 5: Daylighting (a) Alternative 1. (b) Alternative 2. (c) Alternative 3.**

Life Cycle Costing Estimate
General Purpose Work Sheet

Study Title: DAY LIGHTING SCHEMES
HIGH PRESSURE SODIUM *FIXTURES-SWITCHED ½ BAY
Discount Rate: 10% **Economic Life:** 40 YEARS
*NEW (W) LAMP 250W H S 7500 HRS/LIFE

Original Describe: TWO 4'-0" WINDOWS, REFLECTIVE LOUVERS AND COFFERED CEILING FIXED SASH

Alternative 1 Describe: COFFERED CEILING w/ONE 6'-0" WINDOW FIXED SASH

Alternative 2 Describe: TWO 4'-0" HIGH WINDOWS NO REFLECTIVE LOUVERS FIXED SASH

Alternative 3 Describe: RECESSED SINGLE SLOPED 4'-0" WINDOWS FIXED SASH

			Original Est. Cost	Original PW	Alt 1 Est. Cost	Alt 1 PW	Alt 2 Est. Cost	Alt 2 PW	Alt 3 Est. Cost	Alt 3 PW
Initial Costs										
A. Alum. & Glass Curtain Wall - N&E Elev.			22.00/WSF	800,800	19.00/WSF	691,580	22.00/WSF	800,800	22.00/WSF	800,800
B. Venetian Blinds Vertical Reflective Backing			10.00/WSF	16,000	10.00/WSF	12,000	10.00/WSF	16,000	10.00/WSF	8,000
C. Parabolic Reflective Louvers 4' High E. Elev.			50.00/LF	60,000	—	—	—	—	—	—
D. Baseboard Convector HVAC Syst. (Differentials)			25.00/LF	63,000		0	25.00/LF	63,000		0
E. Reflective Surface - Alzak 24 GA & Furring			3.00/SF	18,000	—	—	—	—	—	—
F. Layin Ceiling System (22½ LF)			—	—	—	—	—	—	1.04/SF	44,550
G. Acoustical Tile on Furring			1.31/SF	67,746	1.31/SF	59,000	1.31/SF	67,746	—	—
H. Ceiling Furring - 5/8" Gyp & Paint (30,000 SF)			1.20/SF	36,000	1.70/SF	—	1.20/SF	36,000	1.20/SF	15,120
I. H.P. Sodium Fixtures-Switched Distribution, Controls				354,040		354,040		354,040		354,040
J. Contingencies 5%			14,158	70,779	1,116,620	55,831	1,337,586	66,879	1,222,510	66,126
K. Escalation 5%				70,779		55,831		66,879		66,126
Total Initial Cost				1,557,144		1,228,282		1,476,344		1,344,762
Operations (Annual) ENERGY	Diff. Escal. Rate	PWA W/Escal.								
A. High Pressure Sodium	0%	9.779	7,032	68,766	7,695	75,249	6,833	66,820	8,889	86,926
B. HVAC Space Cooling	0%	9.779	13,589	132,887	11,000	107,569	13,589	132,887	11,000	107,569
C. High Pressure Sodium (Remaining Area)	0%	9.779	14,742	144,162	14,742	144,162	14,742	144,162	14,742	144,162
Total Annual Operations Costs				345,815		326,980		343,869		338,657
Maintenance (Annual)										
A. Alum. Panel Clean ($0.09/SF/YR)	0%	9.779	1,400	13,692	1,750	17,115	1,400	13,692	2,800	27,384
B. Window Washing ($.17/SF/YR)	0%	9.779	3,543	34,650	2,657	25,958	3,543	34,650	1,772	17,324
C. Parabolic Louver ($.25/SF/YR)	0%	9.779	1,200	11,735	—	—	—	—	—	—
D. Baseboard HVAC ($.32/LF/YR)	0%	9.779	806	7,885	—	—	806	7,885	—	—
E. Reflective Surface ($.25/SF/YR)	0%	9.779	1,500	14,669	—	—	—	—	—	—
F. Venetian Blinds ($.20/SF/YR)	0%	9.779	3,200	31,293	2,400	23,470	3,200	31,293	1,600	15,646
G. H.P. Sodium Fixt. ($6.00/FIXT./YR)	0%	9.779	12,822	125,386	12,822	125,386	12,822	125,386	12,822	125,386
Total Annual Maintenance Costs				139,310		191,956		212,906		185,740
Replacement/Alterations (Single Expenditure)	Year	PW Factor								
A. Parabolic Reflective Louvers	10	.386	60,000	23,160	—	—	—	—	—	—
B. Parabolic Reflective Louvers	20	.149	60,000	8,940	—	—	—	—	—	—
C. Parabolic Reflective Louvers	30	.057	60,000	3,520	—	—	—	—	—	—
D. Venetian Blinds	10	.386	16,000	6,176	12,000	4,632	16,000	6,176	8,000	3,088
E. Venetian Blinds	20	.149	16,000	2,384	12,000	1,788	16,000	2,384	8,000	1,192
F. Venetian Blinds	30	.057	16,000	912	12,000	684	16,000	912	8,000	456
G. Baseboard Convector HVAC	20	.149	20,000	2,980	—	—	20,000	2,980	—	—
H. High Pressure Sodium Fixture	20	.149	354,040	52,752	354,040	52,752	354,040	52,752	354,040	52,752
Total Replacement/Alteration Costs				100,724		59,856		65,204		57,488
Tax Elements	Diff. Escal. Rate	PWA W/Escal.								
A. Parabolic Reflective Louvers	1	.909	(6,000)	(5,454)	—	—	—	—	—	—
B. Reflective Surface	1	.909	(1,800)	(1,636)	—	—	—	—	—	—
C. High Pressure Sodium										
D. Depreciation over 7 Years				(22,457)		(22,457)		(22,457)		(22,457)
Total Tax Elements				(29,547)		(22,457)		(22,457)		(22,457)
Associated (Annual)	Diff. Escal. Rate	PWA W/Escal.								
A. Denial of Use (Space) Loss	0%	9.779	—	—	—	—	—	—	75,600	739,300
B. ($10.00/SF/YR) x 7560 SF										
Total Annual Associated Costs				—		—		—		739,300
Total Owning Present Worth Costs										
Salvage At End Of Economic Life	Year	PW Factor								
Building (Struc., Arch., Mech., Elec., Equip.)	40	.022	(141,557)	(3,114)	(111,648)	(2,456)	(133,759)	(2,502)	(122,251)	(2,689)
Other 10% of Initial Cost										
Sitework										
Total Salvage				(3,114)		(2,456)		(2,502)		(2,689)
Total Present Worth Life Cycle Costs				2,210,332		1,782,161		2,068,364		2,640,801
Life-Cycle Present Worth Dollar Savings				—		428,171		141,968		(430,469)

PW — Present Worth   PWA — Present Worth Of Annuity

Figure 8.5.2: **Life Cycle Costing Estimate**

Lighting Fixture Description:	Criteria:	Daylighting Schemes			
		Original Two 4'-0" High Windows Refl. Louvers & Coffered Ceiling	Alt. 1: Coffered Ceiling w/one 6'-0" Window	Alt. 2: Two 4'-0" High Windows, No Reflectors, Coffer Ceiling	Alt. 3: Recessed Single 4'-0" Window, 9'-0" Ceiling
Fluorescent Parabolic 3-tube, Direct	Energy	40	45	45	40
	Life Cycle Cost	50	70	55	30
	Other*	153	147	140	146
	Total	243	262 (4)	240	216
	Initial Cost	$1,613,200	$1,284,338	$1,527,400	$1,400,816
	Life Cycle Cost	2,203,482	1,778,704	2,046,933	2,644,575
	Energy Cost	339,175	323,783	323,088	341,923
Fluorescent 2-tube, Indirect 6" × 4' Fixture	Energy	15	15	20	10
	Life Cycle Cost	35	55	45	15
	Other*	190	184	177	183
	Total	240	254	242	208
	Initial Cost	$1,723,677	$1,394,815	$1,637,877	$1,511,293
	Life Cycle Cost	2,484,842	2,064,873	2,319,894	2,938,525
	Energy Cost	456,131	445,498	431,645	472,189
High Pressure Sodium Fixture, Dimmed	Energy	40	50	50	45
	Life Cycle Cost	45	65	50	20
	Other*	171	165	158	164
	Total	256 (7)	280 (2)	258 (6)	229
	Initial Cost	$1,720,384	$1,391,522	$1,634,584	$1,508,002
	Life Cycle Cost	2,381,953	1,936,710	2,206,244	2,800,842
	Energy Cost	332,410	296,499	297,164	313,681
High Pressure Sodium Fixture, Switched	Energy	35	40	40	40
	Life Cycle Cost	50	70	55	30
	Other*	179	173	166	172
	Total	264 (3)	283 (1)	261 (5)	242
	Initial Cost	$1,557,144	$1,228,282	$1,471,344	$1,344,762
	Life Cycle Cost	2,210,332	1,782,161	2,068,364	2,640,801
	Energy Cost	345,815	326,980	343,869	338,657

*Criteria: Glare control, spatial brightless, initial costs, Appearance of device, color rendition/temp., maintainability.

Figure 8.5.3: **Summary Matrix**

LIFE CYCLE COST ANALYSIS (LCCA)

Project/Location: City Hospital, Detroit, Michigan

Subject: Energy Study
Description: Glass & HVAC System Selection
Project Life Cycle = 25 Years
Discount Rate = 10.00%
Present Time = Date of Occupancy

				Alternative 1 Insulated Clear Glass Perim: Induction Units Inter: VAV Reheat Sys.		Alternative 2 Low E Insul Clr Glass Perim: Induction Units Inter: VAV Reheat Sys.		Alternative 3 Insulated Clear Glass Perim: VAV Reheat Sys. Inter: VAV Reheat Sys.		Alternative 4 Low E Insul Clr Glass Perim: VAV Reheat Sys. Inter: VAV Reheat Sys.	
INITIAL COSTS	Quantity UM	Unit Price		Est.	PW	Est.	PW	Est.	PW	Est.	PW
Construction Costs											
A. Insulated Clear Glass	56,816 WSF	$35.00		1,988,560	1,988,560		0	1,988,560	1,988,560		0
B. Low E Insul Clr Glass	56,816 WSF	$35.75			0	2,031,172	2,031,172		0	2,031,172	2,031,172
C. HVAC System, Alt. 1	1,320 TONS	$4,620		6,098,400	6,098,400		0		0		0
D. HVAC System, Alt. 2	1,104 TONS	$4,620			0	5,100,480	5,100,480		0		0
E. HVAC System, Alt. 3	1,368 TONS	$3,627			0		0	4,961,736	4,961,736		0
F. HVAC System, Alt. 4	1,272 TONS	$3,627			0		0		0	4,613,544	4,613,544
Total Initial Cost					8,086,960		7,131,652		6,950,296		6,644,716
Initial Cost PW Savings (Compared to Alt. 1)							955,308		1,136,664		1,442,244

REPLACEMENT COST / SALVAGE VALUE											
Description	Year	PW Factor									
A. HVAC System, Alt. 1	12	0.3186		1,524,600	485,784		0		0		0
B. HVAC System, Alt. 2	12	0.3186			0	1,275,120	406,292		0		0
C. HVAC System, Alt. 3	12	0.3186			0		0	1,240,434	395,240		0
D. HVAC System, Alt. 4	12	0.3186			0		0		0	1,153,386	367,504
E. Salvage Value		1.0000		0	0	0	0	0	0	0	0
Total Replacement/Salvage Costs					485,784		406,292		395,240		367,504

ANNUAL COSTS											
Description	Escl. %	PWA									
A. Electrical Consumption	0.000%	9.077		1,937,328	17,585,204	1,856,616	16,852,578	1,790,256	16,250,225	1,752,192	15,904,717
B. Gas Consumption	0.000%	9.077		385,920	3,503,011	343,584	3,118,726	262,896	2,386,318	242,424	2,200,492
C. Maintenance Cost: HVAC Alt. 1	0.000%	9.077		42,240	383,414		0		0		0
D. Maintenance Cost: HVAC Alt. 2	0.000%	9.077			0	35,328	320,674		0		0
E. Maintenance Cost: HVAC Alt. 3	0.000%	9.077			0		0	41,040	372,522		0
F. Maintenance Cost: HVAC Alt. 4	0.000%	9.077			0		0		0	38,160	346,380
Total Annual Costs (Present Worth)					21,471,629		20,291,977		19,009,065		18,451,589

Total Life Cycle Costs (Present Worth)					30,044,373		27,829,921		26,354,601		25,463,809
Life Cycle Savings (Compared to Alt. 1)							2,214,452		3,689,773		4,580,564
Discounted Payback (Compared to Alt. 1)		PP Factor					0.00 Years		0.00 Years		0.00 Years
Total Life Cycle Costs (Annualized)		0.1102		3,309,931	Per Year	3,065,969	Per Year	2,903,436	Per Year	2,805,299	Per Year

Figure 8.6.1: **LCCA Case Study 6: Glass/HVAC System**

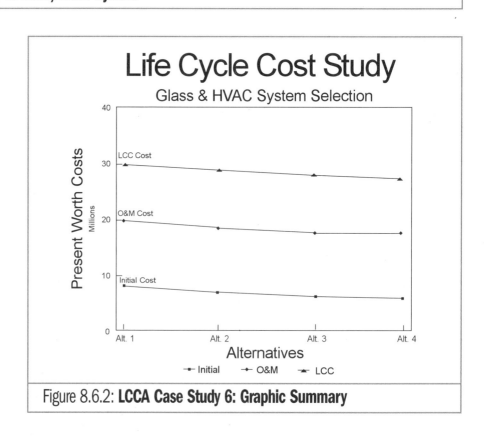

Figure 8.6.2: **LCCA Case Study 6: Graphic Summary**

Life-Cycle Costing Estimate
General Purpose Work Sheet

Study Title: Lighting Systems
Discount Rate: 10% **Date:** ___
Economic Life: 25 years

Life-Cycle Cost Analysis Using Present-Worth Costs

				Alternative 1 Describe: 2'x4' 2 lamp fluorescent, k-12 lens		**Alternative 2** Describe: 2'x4' 2 lamp fluorescent, parabolic louver		**Alternative 3** Describe: indirect ceiling/column mounted 2 lamp fluorescent		**Alternative 4** Describe: indirect floor/furniture mntd metal halide (250W)	
				Estimated Costs	Present Worth	Estimated Costs	Present Worth	Estimated Costs	Present Worth	Estimated Costs	Present Worth
Initial/Collateral Costs	A. 2'x4' 2-lamp fluorescent, k-12 lens			380 fixtures $42,500							
	B. 2'x4' 2-lamp fluorescent, parabolic					380 fixtures $46,400					
	C. indirect ceiling/col. mnt'd 2 lamp fluor.							560 fixtures $100,800			
	D. indirect floor/furn. mnt'd metal halide									150 fixtures $60,000	
	E.										
	F.										
	G.										
	Total Initial/Collateral Costs										
Replacement/Salvage Costs (Single Expenditure)		Year	PW Factor								
	A. Lighting System	10	0.386	42,500	16,400	46,400	17,900	100,800	38,900	60,000	23,200
	B. Lighting System	20	0.149	42,500	6,300	46,400	6,900	100,800	15,000	60,000	8,900
	C.										
	D.										
	E.										
	F.										
	G.										
	H.										
	Salvage (Resale Value)	25	0.092		NIC		NIC		NIC		NIC
	Total Replacement/Salvage Costs				22,700		24,800		53,900		32,100
Annual Costs		Diff. Escal. Rate	PWA W/Escal.								
	A. Electrical Energy	1%	9.894	14,500	143,500	14,500	143,500	21,400	211,700	17,400	172,200
	B. Maintenance	0%	9.077	1,400	12,700	1,400	12,700	2,100	19,100	3,200	29,000
	C. Employee Productivity	0%	9.077	40,000	363,100	40,000	363,100	0	0	0	0
	D. (assumed 2% improvmnt.										
	E. for indirect systems)										
	F.										
	G.										
	Total Annual Costs				519,300		519,300		230,800		201,200
LCC	**Total Present-Worth Life-Cycle Costs**				584,500		590,500		385,500		293,300
	Life-Cycle Present-Worth Dollar Savings				—		(6,000)		199,000		291,200

PW – Present Worth   PWA – Present Worth Of Annuity   ( ) Additional Cost

Figure 8.7.1: LCCA Case Study 7: Lighting System, PW Computation

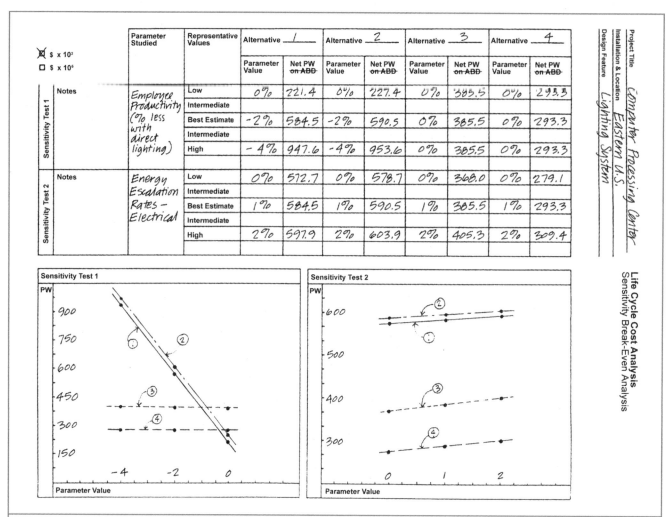

Figure 8.7.2: **LCCA Case Study 7: Sensitivity Analysis**

			Original Describe: 4 PASSENGER & 2 FREIGHT ELEVATORS		Alternative 1 Describe: 4 PASSENGER ELEV. INCL. ONE AS FREIGHT (6000)		Alternative 2 Describe: 4 PASSENGER INCL. ONE AS FREIGHT + D.W.		Alternative 3 Describe:	
Study Title: CONVEYING SYSTEM: ELEVATORS (8 FLOORS) Discount Rate: 10% Economic Life: 40 YEARS			Estimated Costs	Present Worth	Estimated Costs	Present Worth	Estimated Costs	Present Worth	Estimated Costs	Present Worth
Initial Costs										
A. PASSENGER ELEVATORS			260,000		195,000		195,000			
B. 6000 LB. PASSENGER ELEVATOR			—		85,000		85,000			
C. FREIGHT ELEVATORS			172,000		—		—			
D. DUMBWAITERS (D.W.)			—		—		22,000			
E. SHAFTS - STRUCTURAL			94,000		60,000		70,500			
F. PITS			8,000		5,500		6,000			
G. ELECTRICAL			15,000		10,000		10,500			
H. HEATING & VENTILATION			6,000		4,000		4,000			
I. SUB-TOTAL:			555,000		359,500		393,000			
J. Contingencies ___%										
K. Escalation 21%			116,000		75,000		82,000			
Total Initial Cost			671,000		434,500		475,000			
Operations (Annual)	Diff. Escal. Rate	PWA W/Escal.								
A. ENERGY - ELEVATORS (25 HP)	0%	9.779	12,220	119,500	8,146	79,660	8,146	79,660		
B. ENERGY - DUMBWAITERS (5 HP)	0%	9.779	—	—	—	—	611	5,975		
C.										
D.										
E.										
F.										
Total Annual Operations Costs				119,500		79,660		85,635		
Maintenance (Annual)	Diff. Escal. Rate	PWA W/Escal.								
A. CLEAN, ETC. ($550/ELEV.)	0%	9.779	2,200	20,534	1,400	13,691	1,400	13,691		
B. CLEAN, ETC. ($135/D.W.)	0%	9.779	—	—	—	—	135	1,320		
C. INSPECT, REPAIR ($450/STOP)	0%	9.779	21,600	211,226	14,400	140,818	14,400	140,818		
D. INSPECT, REPAIR D.W. ($170/S.)	0%	9.779	—	—	—	—	1,360	13,299		
E.										
F.										
G.										
Total Annual Maintenance Costs				231,762		154,509		169,128		
Replacement/Alterations (Single Expenditure)	Year	PW Factor								
A. PASSENGER ELEVATORS	20	.1486	260,000	38,636	195,000	28,977	195,000	28,977		
B. 6000 LB. ELEVATORS	20	.1486	—	—	85,000	12,631	85,000	12,631		
C. FREIGHT ELEVATORS	20	.1486	172,000	25,559	—	—	—	—		
D. DUMBWAITERS	20	.1486	—	—	—	—	22,000	3,269		
E.										
F.										
G.										
H.										
I.										
J.										
Total Replacement/Alteration Costs				64,195		41,608		44,877		
Tax Elements	Diff. Escal. Rate	PWA W/Esc.								
A. NOT CONSIDERED										
B.										
C.										
D.										
E.										
F.										
G.										
Total Tax Elements										
Associated (Annual)	Diff. Escal. Rate	PWA W/Esc.								
A. DENIAL OF USE COST - SPACE				N/A	4,800 SF	(216,000)	4,444 SF	(200,000)		
B. (8 FLOORS AT $45/SF)										
C.										
Total Annual Associated Costs				N/A		(216,000)		(200,000)		
Total Owning Present Worth Costs				415,457		59,777		99,640		
Salvage At End Of Economic Life	Year	PW Factor								
Building (Struc., Arch., Mech., Elec., Equip.)	40	.022	(43,200)	(950)	(28,000)	(616)	(30,200)	(664)		
Other (10% OF INITIAL COST)										
Sitework										
Total Salvage				(950)		(616)		(664)		
Total Present Worth Life Cycle Costs				1,085,507		493,661		574,640		
Life-Cycle Present Worth Dollar Savings						591,899		510,867		

PW – Present Worth   PWA – Present Worth Of Annuity

Figure 8.8.1: **Life Cycle Costing Estimate in General-Purpose Worksheet**

Part I Economic comparisons of proposed baghouses				
Baghouse size, S.F. filter area	4,200	4,800	5,400	6,000
Purchase cost, dollars	21,000	23,200	25,400	27,600
Replacement filter cost, dollars	1,680	1,920	2,160	2,400
Air-to-cloth ratio	9.5:1	8.3	7.4:1	6.7:1
Total anticipated resistance to air flow, inches w.g.	6.5	5.8	5.4	5.2
Total installed cost, dollars	33,600	36,600	39,600	42,600

Part II LCC comparisons of proposed baghouses				
Baghouse size, S.F.	4,200	4,800	5,400	6,000
Installed cost, dollars	33,600	36,600	39,600	42,600
Maintenance cost, dollars	5,162	5,899	6,636	7,374
Operated cost, dollars	73,800	65,853	61,311	59,040
Depreciation, dollars	−9,291	−10,120	−10,950	−11,779
Salvage value, dollars	−1,295	−1,430	−1,565	−1,700
Total after-tax present value, dollars	101,976	96,802	95,032	95,535

Figure 8.9.1: **Comparisons of Proposed Baghouses**

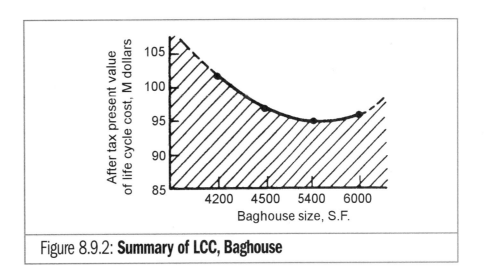

Figure 8.9.2: **Summary of LCC, Baghouse**

Summary of Life Cycle Costs—Top-Mounted Freezer

Zone	Type Cost	Bidders A	B	C	D	E	F
1	A	242.21	231.53	263.45	309.50	252.90	248.36
	R	518.01	464.91	431.24	357.42	486.96	493.40
	LCC	760.22	696.44	694.69	666.92	739.86	741.76
2	A	243.33	230.37	263.45	309.50	244.95	248.36
	R	518.01	464.91	431.24	357.42	486.96	493.40
	LCC	761.34	695.28	694.69	666.92	731.91	741.76
3	A	250.84	232.98	263.45	309.50	251.69	248.36
	R	518.01	464.91	431.24	357.42	486.96	493.40
	LCC	768.85	697.89	694.69	666.92	738.65	741.76
4	A	272.09	245.04	257.45	309.50	267.25	248.36
	R	518.01	464.91	431.24	357.42	486.96	493.40
	LCC	790.10	709.95	688.69	666.92	754.21	741.76

Figure 8.9.3: **Summary of Life Cycle Costs, Top-mounted Freezer**

52.248-3 Value Engineering--Construction

As prescribed in 48.202, insert the following clause:
Value Engineering -- Construction
(March 1989)

(a) *General.* The Contractor is encouraged to develop, prepare, and submit value engineering change proposals (VECPs) voluntarily. The Contractor shall share in any instant contract savings realized from accepted VECPs, in accordance with paragraph (f) below.

(b) *Definitions.* "Collateral costs," as used in this clause, means agency costs of operation, maintenance, logistic support, or Government-furnished property.

"Collateral savings," as used in this clause, means those measurable net reductions resulting from a VECP in the agency's overall projected collateral costs, exclusive of acquisition savings, whether or not the acquisition cost changes.

"Contractor's development and implementation costs," as used in this clause, means those costs the Contractor incurs on a VECP specifically in developing, testing, preparing, and submitting the VECP, as well as those costs the Contractor incurs to make the contractual changes required by Governmental acceptance of a VECP.

"Government costs," as used in this clause, means those agency costs that result directly from developing and implementing the VECP, such as any net increases in the cost of testing, operations, maintenance, and logistical support. The term does not include the normal administrative costs of processing the VECP.

"Instant contract savings," as used in this clause, means the estimated reduction in Contract cost of performance resulting from acceptance of the VECP, minus the allowable Contractor's development and implementation costs, including subcontractors' development and implementation costs (see paragraph (h) below).

"Value engineering change proposal (VECP)" means a proposal that --

  (1) Requires a change to this, the instant contract, to implement; and

  (2) Results in reducing the contract price or estimated cost without impairing essential functions or characteristics; *provided*, that it does not involve a change --

    (i) In deliverable end item quantities only; or

    (ii) To the contract type only.

(c) *VECP preparation.* As a minimum, the Contractor shall include in each VECP the information described in subparagraphs (1) through (7) below. If the proposed change is affected by contractually required configuration management or similar procedures, the instructions in those procedures relating to format, identification, and priority assignment shall govern VECP preparation. The VECP shall include the following:

  (1) A description of the difference between the existing contract requirement and that proposed, the comparative advantages and disadvantages of each, a justification when an item's function or characteristics are being altered, and the effects of the change on the end item's performance.

  (2) A list and analysis of the contract requirements that must be changed if the VECP is accepted, including any suggested specification revision.

  (3) A separate, detailed cost estimate for (i) the affected portions of the existing contract requirements and (ii) the VECP. The cost reduction associated with the VECP shall take into account the Contractor's allowable development and implementation costs, including any amount attributable to subcontracts under paragraph (h) below.

  (4) A description and estimate of costs the Government may incur implementing the VECP, such as test and evaluation and operating and support costs.

  (5) A prediction of any effects the proposed change would have on collateral costs to the agency.

  (6) A statement of the time by which a contract modification accepting the VECP must be issued in order to achieve the maximum cost reduction, noting any effect on the contract completion time or delivery schedule.

  (7) Identification of any previous submissions of the VECP, including the dates submitted, the agencies and contract numbers involved, and previous Government actions, if known.

(d) *Submission.* The Contractor shall submit VECPs to the Resident Engineer at the worksite, with a copy to the Contracting Officer.

(e) *Government Action.*

  (1) The Contracting Officer shall notify the Contractor of the status of the VECP within 45 calendar days after the contracting office receives it. If additional time is required, the Contracting Officer shall notify the Contractor within the 45-day period and provide the reason for the delay and the expected

Table 8.10.1: **Value Incentive Clause (Construction Contract)** *(continued)*

52.248-3 Value Engineering--Construction

date of the decision. The Government will process VECPs expeditiously; however, it shall not be liable for any delay in acting upon a VECP.

(2) If the VECP is not accepted, the Contracting Officer shall notify the Contractor in writing, explaining the reasons for rejection. The Contractor may withdraw any VECP in whole or in part, at any time before it is accepted by the Government. The Contracting Officer may require that the Contractor provide written notification before undertaking significant expenditures for VECP effort.

(3) Any VECP may be accepted, in whole or in part, by the Contracting Officer's award of a modification to this contract citing this clause. The Contracting Officer may accept the VECP, even though an agreement on price reduction has not been reached, by issuing the Contractor a notice to proceed with the change. Until a notice to proceed is issued or a contract modification applies a VECP to this contract, the Contractor shall perform in accordance with the existing contract. The Contracting Officer's decision to accept or reject all or any part of any VECP shall be final and not subject to the Disputes clause or otherwise subject to litigation under the Contract Disputes Act of 1978 (41U.S.C.601-613).

(f) *Sharing.*

(1) *Rates.* The Government's share of savings is determined by subtracting Government costs from instant contract savings and multiplying the result by

(i) 45 percent for fixed-price contracts or

(ii) 75 percent for cost-reimbursement contracts.

(2) *Payment.* Payment of any share due the Contractor for use of a VECP on this contract shall be authorized by a modification to this contract to--

(i) Accept the VECP

(ii) Reduce the contract price or estimated cost by the amount of instant contract savings; and

(iii) Provide the Contractor's share of savings by adding the amount calculated to the contract price or fee.

(g) *Collateral savings.* If a VECP is accepted, the instant contract amount shall be increased by 20 percent of any projected collateral savings determined to be realized in a typical year of use after subtracting any Government costs not previously offset. However, the Contractor's share of collateral savings shall not exceed (1) the contract's firm-fixed-price or estimated cost, at the time the VECP is accepted, or (2)$100,000, whichever is greater. The Contracting Officer shall be the sole determiner of the amount of collateral savings, and that amount shall not be subject to the Disputes clause or otherwise subject to litigation under 41U.S.C.601-613.

(h) *Subcontracts.* The Contractor shall include an appropriate value engineering clause in any subcontract of $50,000 or more and may include one in subcontracts of lesser value. In computing any adjustment in this contract's price under paragraph (f) above, the Contractor's allowable development and implementation costs shall include any subcontractor's allowable development and implementation costs clearly resulting from a VECP accepted by the Government under this contract, but shall exclude any value engineering incentive payments; *provided,* that these payments shall not reduce the Government's share of the savings resulting from the VECP.

(i) *Data.* The Contractor may restrict the Government's right to use any part of a VECP or the supporting data by marking the following legend on the affected parts:

"These data, furnished under the Value Engineering--Construction clause of contract........., shall not be disclosed outside the Government or duplicated, used, or disclosed, in whole or in part, for any purpose other than to evaluate a value engineering change proposal submitted under the clause. This restriction does not limit the Government's right to use information contained in these data if it has been obtained or is otherwise available from the Contractor or from another source without limitations."

If a VECP is accepted, the Contractor hereby grants the Government unlimited rights in the VECP and supporting data, except that, with respect to data qualifying and submitted as limited rights technical data, the Government shall have the rights specified in the contract modification implementing the VECP and shall appropriately mark the data. (The terms "unlimited rights" and "limited rights" are defined in Part 27 of the Federal Acquisition Regulation.)

(End of clause)

Alternate I (APR 1984). When the head of contracting activity determines that the cost of calculating and tracking collateral savings will exceed the benefits to be derived in a construction contract, delete paragraph (g) from the basic clause and redesignate the remaining paragraphs accordingly.

Source: *Value Engineering Program Guide for Design and Construction,* PBS-PQ251, May 10, 1993, Vol. 2, p. 4-7.

Table 8.10.1: **Value Incentive Clause (Construction Contract)**

Situation: Guard service contract where below described change will be implemented in the last six months of a one year contract. Contract contains a one year renewal option.

Description of change:

Revise the contractors approved manpower plan to delete all guard service from door #2 of building 21 and install a video camera and buzzer lock system to be controlled at door #1 of the building.

INSTANT CONTRACT COMPUTATIONS

	Contract Cost		Proposal Cost	Difference
Prime contractor:				
direct labor 1/2 year				
normal hours	$ 2,000	VCP processing	$ 100	
overtime	750			
holidays, weekends	1,200			
fringe benefits	800			
	$ 4,750		$ 100	($ 4,650)
Subcontractor A:				
	N.A.	monitoring sys.	$ 3,000	
		overhead & prof.	300	$ 3,300
			$3,300	
Prime contractor:				
			net saving 1st year	($ 1,350)
optional year				
normal hours	$ 4,000			
overtime	1,500			
holidays, weekends	2,400			
fringe benefits	1,600		0	($ 9,500)
	$ 9,500		0	
			net saving 2nd year	($10,850)

COLLATERAL COMPUTATIONS

	Costs Before		Costs After	Difference
10 years labor	$91,000	instand contract increase	0	
		10 yrs system maintenance	$5,000	
		10 yrs electrcity	1,000	
totals	$91,000		$6,000	($ 85,000)
		average one years savings		($ 8,500)

ADJUSTMENT TO CONTRACT PRICE

Method 1:

Under Option 1 reduce contract price by 50% of $1,350 or, $675 and,
Under Option 2 increase contract price by 20% of $8,500 or, $1,700.

Method 2:

Under Option 2 increase contract price by $3,300 for monitoring system plus $1,700 for contractor collateral share and,
Under Option 5 reduce the full quantities of labor from the manpower plan to reflect an annual savings of $9,500 and make a lump sum payment of $675.

Figure 8.10.2: **Service Contract Example**

LIFE CYCLE COST ANALYSIS (LCCA)

Project/Location: Chemical Stabilization Facility, Boston, MA

Subject: Solid Waste Drying System
Description: Value Engineering Study
Project Life Cycle = 25 Years
Discount Rate = 8.00%
Present Time = Date of Occupancy

				Original Design Windrows Solid Waste Drying System		Alternative 1 Mechanical Dryer Type 1 Solid Waste System (56% Total Solids Entering Dryer, Chemicals Added)		Alternative 2 Mechanical Dryer Type 2 Solid Waste System (45% Total Solids Entering Dryer, Less Chemicals)		Alternative 3	
INITIAL COSTS	Quan.	UM	Unit Price	Est.	PW	Est.	PW	Est.	PW	Est.	PW
Construction Costs											
A. Windrows Bldg & Equipm't	1	LS	$56,333,000	56,333,000	56,333,000		0		0		0
B. Mech Dryer (1) Bldg & Eqmt	1	LS	$13,367,000		0	13,367,000	13,367,000		0		0
C. Mech Dryer (2) Bldg & Eqmt	1	LS	$28,728,000		0		0	28,728,000	28,728,000		0
D.					0	0	0		0		0
E.					0		0		0		0
Total Initial Cost					56,333,000		13,367,000		28,728,000		0
Initial Cost PW Savings (Compared to Orig. Design)							42,966,000		27,605,000		0

REPLACEMENT COST/ SALVAGE VALUE											
Description	Year	PW Factor									
A. No Change in LCC Outcome	—	1.0000		Similar	0	Similar	0	Similar	0		0
B.		1.0000			0		0		0		0
C.		1.0000			0		0		0		0
D. Salvage Value		1.0000		0	0	0	0	0	0	0	0
Total Replacement/Salvage Costs					0		0		0		0

ANNUAL COSTS											
Description	Escl. %	PWA									
A. Water/ Sewage	0.00%	10.675		1,548,000	16,524,554	812,000	8,667,918	1,828,000	19,513,491		0
B. Fuel	0.00%	10.675		242,000	2,583,296	628,000	6,703,759	1,593,000	17,004,918		0
C. Power	0.00%	10.675		1,865,000	19,908,458	1,024,000	10,930,971	1,611,000	17,197,064		0
D. Labor	0.00%	10.675		4,071,000	43,457,014	3,481,000	37,158,896	4,071,000	43,457,014		0
E. Chemicals	0.00%	10.675		4,264,000	45,517,246	3,977,000	42,453,585	2,477,000	26,441,421		0
F. Savings in Product Hauling	0.00%	10.675		0	0	0	0	(5,338,000)	(56,981,955)		0
G. Equipment Maintenance	0.00%	10.675		1,455,000	15,531,799	1,064,000	11,357,962	1,295,000	13,823,835		0
H.	0.00%	10.675			0		0		0		0
Total Annual Costs (Present Worth)					143,522,366		117,273,091		80,455,788		0

Total Life Cycle Costs (Present Worth)					199,855,366		130,640,091		109,183,788		0
Life Cycle Savings (Compared to Orig. Design)							69,215,275		90,671,578		0
Discounted Payback (Compared to Orig.) PP Factor							0.00 Years		0.00 Years		0.00 Years
Total Life Cycle Costs (Annualized)		0.0937		18,722,207	Per Year	12,238,204	Per Year	10,228,204	Per Year	0	Per Year

Figure 8.11.1: **LCCA Case Study 11: Plant-Dryers vs. Windows**

LIFE CYCLE COST ANALYSIS (LCCA)

Project/Location: Case Study 12- Campus Planning
Subject: Site & Buildings Layout
Description: Conceptual Schemes
Project Life Cycle = 25 Years
Discount Rate = 7.0%

					Alternative 1		Alternative 2		Alternative 3		Alternative 4	
					Contained Campus Framework Plan		Street Campus Framework Plan		Vertical Campus Framework Plan		Reconsideration Campus Framework Plan	
INITIAL COSTS		Quantity UM		Total Price	Est.	PW	Est.	PW	Est.	PW	Est.	PW
A.	Alternative 1 Roads	2,700	LF	$270	729,000	729,000	0	0	0	0	0	0
B.	Alternative 2 Roads	2,960	LF	$270	0	0	799,200	799,200	0	0	0	0
C.	Alternative 3 Roads	3,400	LF	$270	0	0	0	0	918,000	918,000	0	0
D.	Alternative 4 Roads	3,400	LF	$270	0	0	0	0	0	0	918,000	918,000
E.	Alternative 1 Parking Lot	500	Cars	$1,750	875,000	875,000	0	0	0	0	0	0
F.	Alternative 2 Parking Lot	130	Cars	$1,750	0	0	227,500	227,500	0	0	0	0
G.	Alternative 3 Parking Lot	55	Cars	$4,000	0	0	0	0	220,000	220,000	0	0
G.	Alternative 3 2 Level Deck	1,000	Cars	$1,750	0	0	0	0	1,750,000	1,750,000	0	0
H.	Alternative 4 Parking Lot	160	Cars	$1,750	0	0	0	0	0	0	280,000	280,000
Total Initial Cost						**1,604,000**		**1,026,700**		**2,888,000**		**1,198,000**

REPLACEMENT COST/ SALVAGE VALUE												
Description		Year		PW Factor								
A.	_____	0		1.0000	0	0	0	0	0	0	0	0
B.	_____	0		1.0000	0	0	0	0	0	0	0	0
C.	_____	0		1.0000	0	0	0	0	0	0	0	0
Total Replacement/Salvage Costs						**0**		**0**		**0**		**0**

ANNUAL COSTS			Diff.									
Description		Cost	Escl. %	PWA								
A.	Road Maint.	$10.00 LF	0.00%	11.654	27,000	314,647	29,600	344,946	34,000	396,222	34,000	396,222
B.	Parking Maint.	$50.00 Car	0.00%	11.654	25,000	291,340	6,500	75,748	52,750	614,727	8,000	93,229
C.			0.00%	11.654	0	0	0	0	0	0	0	0
Total Annual Costs (Present Worth)						**605,986**		**420,694**		**1,010,948**		**489,450**

| **Total Life Cycle Costs (Present Worth)** | | | | | | **2,209,986** | | **1,447,394** | | **3,898,948** | | **1,687,450** |
| **Total Life Cycle Costs (Annualized)** | | | | PP Factor 0.0858 | 189,640 Per Year | | 124,202 Per Year | | 334,571 Per Year | | 144,801 Per Year | |

Figure 8.12.1: Life Cycle Cost Analysis: Campus Planning

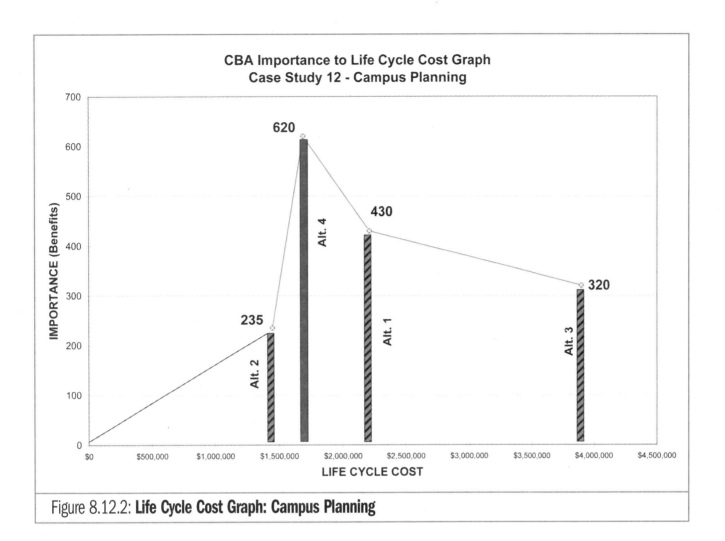

Figure 8.12.2: **Life Cycle Cost Graph: Campus Planning**

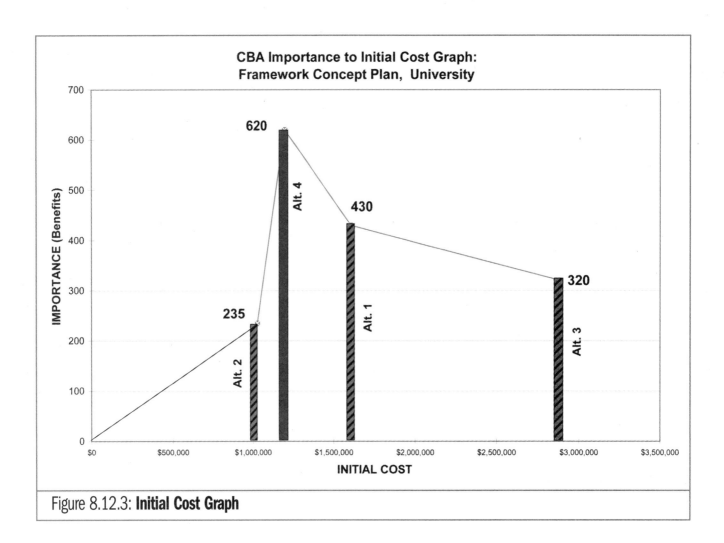

Figure 8.12.3: **Initial Cost Graph**

Statistics:			
Number of Years for Period		30	
Discount Rate (Interest)		8%	
Escalation Rate (Inflation)		3%	
Gross Area	sf	244,000	
Students	Each	1,600	
Student per Teacher		17.0	(Modified from 16)
Staff Support per Teacher		2.2	
Benefits/Overhead	%	32%	

ITEM	Measure	Units	Unit Cost	Current Cost	Factor	Present Value	Percent
Capital Costs							
Construction	$/sf	244,000	$133.00	$ 32,452,000	1	$ 32,452,000	18%
Furnishings/Fitout	$/sf	244,000	$26.50	$ 6,466,000	1	$ 6,466,000	4%
Fees	$/sf	244,000	$7.00	$ 1,708,000	1	$ 1,708,000	1%
Other Project Costs	$/sf	244,000	$7.00	$ 1,708,000	1	$ 1,708,000	1%
Construction Contingency	$/sf	244,000	$6.50	$ 1,586,000	1	$ 1,586,000	1%
Sub-Total - Initial Capital Costs	$/sf	244,000	$180.00	$ 43,920,000	1	$ 43,920,000	24%
Major Capital Replacements	$/sf/yr	244,000	$ 2.50	$ 610,000	15.63	9,534,894	5%
Grand Total Capital Costs						$ 53,454,894	30%
Operations & Maintenance							
Maintenance	$/sf/yr	244,000	$ 1.40	$ 341,600	15.63	$ 5,339,541	3%
Energy	$/sf/yr	244,000	$ 0.80	$ 195,200	15.63	$ 3,051,166	2%
Sub-Total				$ 536,800		$ 8,390,707	5%
Functional Operation							
Educational Staffing	Teachers	94	$ 38,000	$ 3,572,000	15.63	$ 55,833,838	31%
Support Staffing	Staff	43	$ 31,000	$ 1,333,000	15.63	$ 20,836,088	12%
Benefits/Overhead	%	32%		$ 1,569,600	15.63	$ 24,534,376	14%
Textbooks	Student	1,600	$ 90	$ 144,000	15.63	$ 2,250,860	1%
School Allotment	Student	1,600	$ 65	$ 104,000	15.63	$ 1,625,621	1%
Other Support Costs	Student	1,600	$ 135	$ 216,000	15.63	$ 3,376,290	2%
Transportation	Student	1,600	$ 360	$ 576,000	15.63	$ 9,003,441	5%
				$ 7,514,600		$ 117,460,515	66%

Grand Total			
Present Value Cost		$	179,306,115
Equivalent Annual Cost	0.0640	$	11,471,206
Equivalent Annual Cost per Student	1,600	$	7,170
Equivalent Annual Cost per Student (Excluding Capital)	1,600	$	5,032
Original Approach =		$	184,010,413
Investment Potential =		$	4,704,298

Figure 8.13.1: **Life Cycle Cost Calculations High School Trade Off Analysis of Initial Costs vs. Staffing**

Regional Transportation Authority
Maintenance Cost Assessment

Proposer:	Panel Assessment - Proposer No. 1
Period =	50 Yrs.
Discount Rate =	6%
Inflation Rate =	3%

September 13, 1999

Present Value Factors	
50 Year annual present worth =	26.162
5 Year Replacement Cycle =	4.698
10 Year Replacement Cycle =	2.053
15 Year Replacement Cycle =	1.109
20 Year Replacement Cycle =	0.642
25 Year Replacement Cycle =	0.488

Maintenance Category	Yearly Cycle	Labor Hours	Labor Cost @$75/Hr.	Materials Cost	Equipment Cost	Total Cost	PV Factor	Present Value Total
I. ANNUAL MAINTENANCE								
A. Trench Structural System								
1 Inspection								
a. Trench Walls		120	$ 9,000		$ 200	$ 9,200	26.162	$ 240,692
b. Bridges		186	$ 13,950		$ 3,410	$ 17,360	26.162	$ 454,175
c. Ladders		400	$ 30,000		$ 1,000	$ 31,000	26.162	$ 811,027
d. Fence		160	$ 12,000			$ 12,000	26.162	$ 313,946
e. Slabs		20	$ 1,500			$ 1,500	26.162	$ 39,243
f. Struts		1,613	$ 120,975		$ 48,390	$ 169,365	26.162	$ 4,430,956
2 Surface Maintenance			$ -					
a. Graffiti removal		2,352	$ 176,400	$ 98,270	$ 11,760	$ 286,430	26.162	$ 7,493,630
b. Cracks & spalls		960	$ 72,000	$ 15,000	$ 4,800	$ 91,800	26.162	$ 2,401,687
c. Weeds		32	$ 2,400	$ 2,000	$ -	$ 4,400	26.162	$ 115,114
d.			$ -			$ -	26.162	$ -
e.			$ -			$ -	26.162	$ -
3 Other Maintenance								
a. Trench			$ -			$ -	26.162	$ -
b. Bridges		576	$ 43,200	$ 3,500	$ 2,880	$ 49,580	26.162	$ 1,297,120
c. Ladders		200	$ 15,000	$ 500	$ 1,000	$ 16,500	26.162	$ 431,676
d. Fence		320	$ 24,000	$ 8,505	$ 4,800	$ 37,305	26.162	$ 975,980
e. Litter removal		128	$ 9,600	$ -	$ 640	$ 10,240	26.162	$ 267,901
f. Lighting system		67	$ 5,025	$ 750	$ 1,675	$ 7,450	26.162	$ 194,908
g.			$ -			$ -	26.162	$ -
B. Drainage System								
1 Inspection								
a. Trench & grate		40	$ 3,000			$ 3,000	26.162	$ 78,487
b. Pump Stations		64	$ 4,800		$ 320	$ 5,120	26.162	$ 133,950
c. Water treatment		32	$ 2,400		$ 160	$ 2,560	26.162	$ 66,975
d.			$ -			$ -	26.162	$ -
2 Annual Operating & Maintenance								
a. Debris removal		96	$ 7,200		$ 1,440	$ 8,640	26.162	$ 226,041
b. Pump Stations		192	$ 14,400	$ 500	$ 2,880	$ 17,780	26.162	$ 465,163
c. Power			$ -	$ 5,955		$ 5,955	26.162	$ 155,796
d. Water treatment		64	$ 4,800	$ 2,000	$ 320	$ 7,120	26.162	$ 186,275
e.			$ -			$ -	26.162	$ -
TOTAL ANNUAL MAINTENANCE		7,622	$ 571,650	$ 136,980	$ 85,675	$ 794,305		$ 20,780,742
II. CAPITAL MAINTENANCE								
A. Trench Structural System								
1 Repair & replacement								
a. Trench Walls						$ -		$ -
b. Bridges						$ -		$ -
c. Ladders	25		$ -	$ 400,000		$ 400,000	0.488	$ 195,139
d. Fence	25		$ -	$ 268,500		$ 268,500	0.488	$ 130,987
e. Lighting System	25		$ -	$ 100,000		$ 100,000	0.488	$ 48,785
f. General Bridge Repair			$ -			$ -		$ -
B. Drainage System								
1 General Work								
a. Joints & leaks	10	2,148	$ 161,100	$ 71,600	$ 10,740	$ 243,440	2.053	$ 499,866
b. Trench & grate	25		$ -	$ 107,400		$ 107,400	0.488	$ 52,395
c. Pump Stations	25		$ -	$ 100,000	$ -	$ 100,000	0.488	$ 48,785
d. Water treatment	15	240	$ 18,000	$ 20,000	$ 3,600	$ 41,600	1.109	$ 46,152
e.			$ -			$ -		$ -
TOTAL CAPITAL MAINTENANCE		2,388	$ 179,100	$ 1,067,500	$ 14,340	$ 1,260,940		$ 1,022,109
GRAND TOTAL PRESENT VALUE		N/A	N/A	N/A	N/A	N/A		$ 21,802,850

Figure 8.14.1: **LCC Assessment of a Highway Project**

General Purpose Worksheet

					Alternative 1 Individual Rooftop Units		Alternative 2 Central Plant with 4 - Pipe Fan Coil Units	
Study Title: HVAC System Analysis					Estimated Costs	Present Worth	Estimated Costs	Present Worth
Discount Rate: 8.0% Date: 12/17/97								
Life Cycle (Yrs.) 30								

INITIAL / COLLATERAL COSTS

					Estimated Costs	Present Worth	Estimated Costs	Present Worth
Initial/Collateral Costs								
A.	Equipment				1,212,354	1,212,354	2,803,000	2,803,000
B.	Screening				55,501	55,501	40,000	40,000
C.	Plant Space						128,000	128,000
D.								
E.								
F.								
G.								
H.								
I.								
J.								
Total Initial/Collateral Costs					$1,267,855	$1,267,855	$2,971,000	$2,971,000
Difference								($1,703,145)

REPLACEMENT / SALVAGE COSTS

		Year	Inflation/ Escal. Rate	PW Factor	Estimated Costs	Present Worth	Estimated Costs	Present Worth
Replacement/Salvage **(Single Expenditures)**								
A.	Rooftop Units (70%)	8		0.540	770,000	416,007		
B.	Rooftop Units (70%)	16		0.292	770,000	224,756		
C.	Rooftop Units (70%)	24		0.158	770,000	121,428		
D.	Fan Coils (100%)	15		0.315			408,888	128,899
E.	Central Plant Equipment	20		0.215			600,000	128,729
F.								
G.								
H.								
I.								
J.								
Total Replacement/Salvage Costs						$762,191		$257,627

ANNUAL COSTS

			Inflation/ Escal. Rate	PW Factor	Estimated Costs	Present Worth	Estimated Costs	Present Worth
Annual Costs								
A.	Maintenance - Rooftops		1%	12.496	75,360	941,693		
B.	Maintenance - Fan Coils		1%	12.496			37,680	470,847
C.	Maintenance - Central Plant & Distrib.		1%	12.496			28,800	359,883
D.	Energy		3%	15.631	468,000	7,315,296	192,000	3,001,147
E.				11.258				
F.								
G.								
H.								
I.								
J.								
Total Annual Costs					$543,360	$8,256,989	$258,480	$3,831,876
Sub-Total Replacement/Salvage + Annual Costs (Present Worth)						$9,019,180		$4,089,504
Difference								$4,929,676

LIFE CYCLE COSTS

Total Life Cycle Costs (Present Worth)						$10,287,035		$7,060,504
Life Cycle Cost PW Difference								$3,226,531
Payback - Simple Discounted (Added Cost / Annualized Savings)								3.9 Yrs.
- Fully Discounted (Added Cost+Interest / Annualized Savings)								4.8 Yrs.
Total Life Cycle Costs - Annualized					Per Year:	$913,771	Per Year:	$627,166

Figure 8.15.1: **Life Cycle Cost Assessment—Design-Builder's HVAC System**

Life Cycle Cost Analysis
General Purpose Worksheet

Study Title: Lease Build Analysis - Government Office Bldg.
Discount Rate: 6.3% Date: 1/28/98
Life Cycle (Yrs.): 30

				Alternative 1 Renovate Historic Facility		Alternative 2 Lease Space From GSA		Alternative 3 Renovate Current Space	
				Estimated Costs	Present Worth	Estimated Costs	Present Worth	Estimated Costs	Present Worth
Initial/Collateral Costs									
A.	Work in MEB								
B.	Hazardous Materials Removal & Structural Rehab			5,509,000	5,509,000	5,509,000	5,509,000	5,509,000	5,509,000
C.	Historic Preservation			2,255,258	2,255,258				
D.	Basic Program Needs			6,930,742	6,930,742				
E.	Lease Space - Interior Fitout- Basic Program Needs					500,000	500,000		
F.	Work in Current Space (3-Buildings)								
G.	Hazardous Materials Removal & Structural Rehab			2,490,000	2,490,000	2,490,000	2,490,000	2,490,000	2,490,000
H.	Historic Preservation							1,170,000	1,170,000
I.	Basic Program Needs - Construct., Fees, Sup. & Mods.							8,610,000	8,610,000
J.	- Moving & Swing Space Cost							1,450,000	1,450,000
	Total Initial/Collateral Costs			$17,185,000	$17,185,000	$8,499,000	$8,499,000	$19,229,000	$19,229,000
	Difference						$8,686,000		($2,044,000)
Replacement/Salvage (Single Expenditures)		Year	PW Factor						
A.	Major Replacements	10	0.543	358,750	194,742			323,000	175,336
B.	Major Replacements	15	0.400	717,500	286,961			646,000	258,365
C.	Major Replacements	20	0.295	861,000	253,710			775,200	228,428
D.	Major Replacements	25	0.217	1,076,250	233,658			969,000	210,374
E.									
F.									
G.									
H.									
I.									
J.									
	Total Replacement/Salvage Costs				$969,072				$872,502
Annual Costs		Differential Escal. Rate	PW Factor						
A.	Lease annual cost per OSF								
B.	27,000 OSF @ $26.25	2.7%	18.380			708,840	13,028,274		
C.	O&M Costs								
D.	30,000 OSF @ $5.00	2%	16.848	150,000	2,527,207				
E.									
F.	30,000 OSF @ $4.43	2%	16.848					132,900	2,239,106
G.	(equivalent over 28 years)								
H.									
I.									
J.									
	Total Annual Costs			$150,000	$2,527,207	$708,840	$13,028,274	$132,900	$2,239,106
Life Cycle Costs									
	Total Life Cycle Costs (Present Worth)				$20,681,279		$21,527,274		$22,340,608
	Life Cycle Cost PW Difference						($845,996)		($1,659,329)
	Discounted Payback (Alt. 2,3,4 vs. Alt. 1)						12.2		70.9
	Total Life Cycle Costs - Annualized			Per Year:	$1,551,016	Per Year:	$1,614,462	Per Year:	$1,675,459

Figure 8.16.1: **Lease versus Build Analysis**

Chapter 9
Management Considerations

If, in a business enterprise or in government, many important decisions that in the aggregate can have a major influence on the success (and sometimes on the survival) of the enterprise are badly made by persons of modest incompetence, these bad decisions are not primarily the fault of those persons, they are the fault of the management.

Grant and Ireson
Principles of Engineering Economy, 1970

A basic objective of life cycle costing is to ensure that maximum benefit is realized at the lowest cost. This same objective can normally be applied to the overall economic study effort, and to the individual design decisions related to life cycle cost analysis (LCCA) studies. Numerous decisions are made during project design. Use of an LCCA to support every decision would obviously be impractical. To achieve the lowest total cost of ownership for the project while minimizing the cost of the economic study effort itself requires careful planning and control—in short, effective management. This chapter focuses on the management of the overall economic study effort through the planning and control of the individual LCCAs. The discussions are directed primarily toward project managers, but they are also of interest to design supervisors and designers. Major topics include planning of the level of effort, the selection of study areas, contingency planning, and the documentation of LCCA.

Strengths and Weaknesses of Life Cycle Costing

The need for cost-effective management—of both the overall economic study of the project design, and the individual LCCA—becomes apparent when the strengths and weaknesses of the study process are examined.

Strengths

Life cycle costing contributes the following principal strengths to an overall economic study effort:

- It satisfies owner demands for projects with the best value for the money.
- It reduces the total cost of ownership.
- It assists in selecting alternatives that are environmentally sustainable, while minimizing premature building obsolescence.
- It creates cost awareness and visibility.
- It provides the most appropriate yardstick for measuring the overall economy of design alternatives.
- It develops data that can be valuable for future studies.
- It develops data that a higher authority such as project designer and/ owner can review to validate the selection of alternatives.
- It is systematic, rational, and objective and thus imparts credibility to the design decisions.

Weaknesses

The process of life cycle costing possesses the following principal weaknesses:

- It can be costly and time-consuming, primarily as a result of the need for multiple designs and life cycle cost data acquisition.
- The validity of results may be questionable because of the difficulty of obtaining valid cost data, especially regarding future costs.
- The study effort may be rendered ineffective by the imposition of a criterion other than long-term economy, such as a limit on initial project costs that would preclude the implementation of the most economical life cycle cost alternative.

Careful attention should be given to several management aspects of the LCC process if its inherent weaknesses are to be offset, and the cost of the study effort is to be commensurate with expected savings. The cost of the study effort may be controlled by:

- Establishing at an early stage the level of effort for conducting the study. This level of effort is normally planned to ensure the availability of sufficient funds and personnel to conduct and complete the analysis.
- Allocating resources at the early design phases (schematic and design development phases).
- Carefully selecting those design features on which the bulk of the effort should be concentrated.
- Planning for the contingency in which the construction or procurement of a more economical design may require funds exceeding those budgeted for the project.
- Ensuring that an appropriate level of effort is devoted to data collection, evaluation, and documentation of the study effort, i.e., neither inadequate nor excessive for the particular situation at hand.

The LCCA is one of several tools that are available to the designer. As is true of any tool, it is useful only to the extent that it is effective. Managerial control helps ensure that the LCCA is used properly and effectively.

Level of Effort

Managing the level of effort for the overall economic study implies first determining the appropriate level of effort, including funding, and then allocating personnel and monetary resources to the individual studies that will compose the overall effort. Both of these managerial tasks are discussed in the following paragraphs.

General Considerations

Timely planning and control of the level of effort for both the overall economic study and individual LCCA is critical to the success of life cycle costing. To this end, a series of decisions is required, beginning at the pre-schematic design phase and continuing throughout design. Initially, the main characteristics of the project are examined to determine where the most savings are likely to be produced with a reasonable expenditure of life cycle costing effort, and the level of effort that will be involved. Based on this broad survey, necessary resources are allocated to ensure the availability of personnel and design fee hours when the LCCA are conducted. At the schematic design phase, the design features or areas with the greatest potential for cost savings are selected for study. Normally, a higher percentage of resources is apportioned to the schematic design phase, and a smaller percentage to the design development phase. This allocation is based on the authors' experience that indicates greater opportunities for savings from LCCA during the schematic phase.

Planning the Overall Study Effort

The planning for these resources normally is based on a number of project-related factors, mainly initial investment cost, size, type, complexity, uniqueness, and projected annual costs of energy, maintenance, etc. The *initial project cost* can be a significant factor in determining levels of effort, since the amount of money available for LCCA studies tends to be greater for projects with larger initial costs. The *size* of the project, expressed in a unit, such as square feet, also influences the level of study effort. Because of the effects of scale, even a small unit cost saving for, say, a 400,000 square foot building can result in substantial savings. In addition, the magnitude of *energy* and *maintenance costs* may justify intensive study. Size often correlates with initial cost, but not in every case. Two buildings may both have budgeted costs of $10 million, but one may contain considerably more space as the other. The planned study efforts for the two buildings may be about equal or vastly different, depending on *building type,* relative *complexity,* and *projected energy* and *maintenance costs.* Complex facilities normally incorporate features that are high in cost, with intricate interrelationships that provide many opportunities for LCCA. Complex projects, such as hospitals and health care facilities, research and development laboratories, and manufacturing facilities, may warrant a much higher level of effort than relatively low-complexity projects, such as warehouses and parking structures.

Uniqueness can affect the planned level of effort in about the same way as complexity; *lack of uniqueness* also may be a consideration. Sometimes an owner may be planning a number of comparable or identical projects in the near future. A greater-than-usual proportion of the available study resources might be assigned to one or a few of these projects, and the study results applied to the remainder of the projects.

For project managers and organizations with *limited LCCA experience*, an effective approach to the allocation of dollar resources to the overall study effort begins with the setting of limits. A reasonable upper limit is 0.5% of the budgeted project cost. Except in unusual cases, it is unlikely that a greater amount could be either allocated or justified. The lower limit might be two tenths of 1% of the budgeted project cost, or not less than $10,000. Such an amount is normally sufficient to determine the areas most likely to produce savings, select at least one design feature for study, organize a team if necessary, conduct the study, and document the results.

For a $10 million corporate headquarters office building, this approach would result in a funding range of $20,000–$50,000 for life cycle cost studies. It is likely that these funds would be available within the design fee for a project with this initial cost, yet the funding range just described is quite wide. A specific choice of funding level within the range is normally based on the nature of the other project-related factors discussed above: size, type, complexity, etc. Depending on whether these factors are indicative of high, low, or average costs, the funding level is located higher, lower, or fairly centrally within the funding range. In the corporate headquarters office building example, a check with designers and/or facilities personnel might show that energy and maintenance costs are expected to be quite high, the building is somewhat more complex than average, and other factors are about average. The project manager might then select a funding level that is above the middle of the funding range—perhaps $40,000.

Project managers and organizations with *some experience* in LCCA usually develop and continually update a set of estimating criteria that take into account the project-related elements discussed above. Figure 9.1 shows the hours of effort recommended by the authors for performing life cycle costing.

Apportioning Resources to Schematic and Design Development Phases

Once the funding level for the overall economic study has been determined, the study funds should be apportioned between the schematic phase and the design development phase. Because earlier studies tend to effect greater savings, the larger part of the funding (and, thus of the study effort) is usually apportioned to the studies that will be conducted during the schematic phase. However, the relative amounts assigned to the two phases depend very much on the design situation, including the project-related factors discussed above, the availability of cost data, and scheduling. For the example corporate office building, the overall funding level of $40,000 might be apportioned as $25,000 for schematic phase studies and $15,000 for design development phase studies, in the absence of unusual circumstances.

Selecting the Study Areas

The resources assigned to the schematic phase and to the design development phase are normally allocated to studies of design areas and/or features (*study areas*) in the manner that provides for the most effective use of those resources. A practical approach to selecting the study areas that should receive the bulk of the analysis effort consists of two steps:

- Develop a preliminary list of high-cost areas or features.
- Establish a priority list of these areas/features.

There is no one best way to develop a list of study areas and establish priorities within that list. Essentially, each project is a special case. However, the following are some general guidelines.

Listing High-Cost Areas and Features

The life cycle costs of different design features vary widely, and the bulk of these costs tend to be concentrated in a limited number of features. (As an example, Figure 9.2 shows a fairly typical distribution of life cycle costs for an office building in which 20% of the elements account for 80% of the total cost. The details of the cost distribution will, of course, vary from project to project.) This fact can be used to identify high-cost areas and

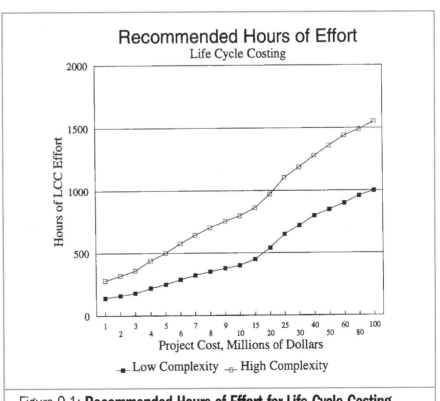

Figure 9.1: **Recommended Hours of Effort for Life Cycle Costing**

features as potential study areas. A checklist of predetermined (through experience) high-cost areas and features is useful in this regard. Figure 9.3 is a detailed list of suggested LCCA study areas according to the phase of design. It is based on the authors' experience.

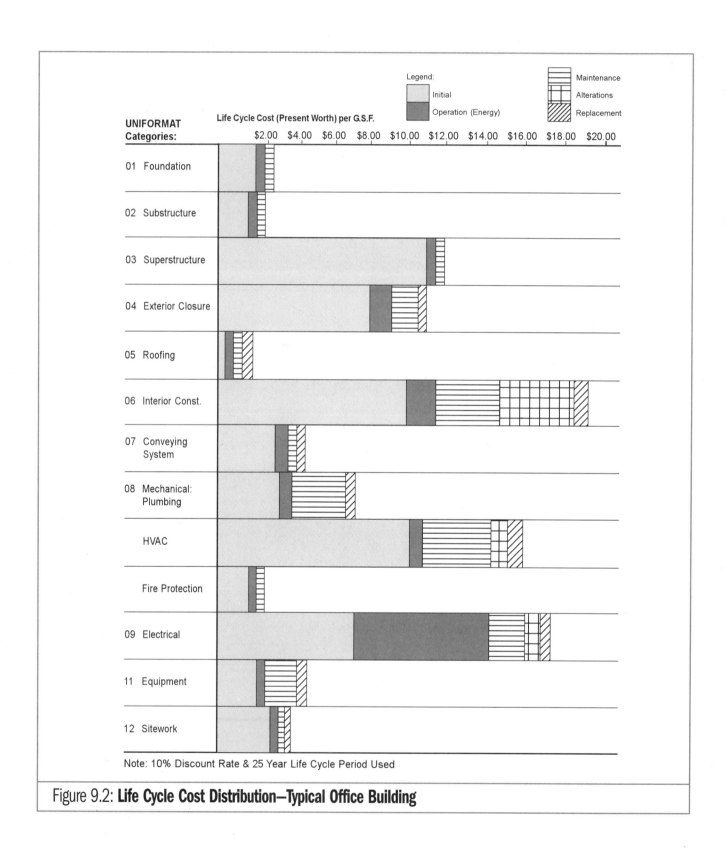

Figure 9.2: **Life Cycle Cost Distribution—Typical Office Building**

The factors most often considered in developing a list of potential study areas are:

- Staffing cost,
- Energy and maintenance costs,
- Initial cost,
- Equipment costs,

Areas of Study:	Conceptual	Schematic	Design Development
General Project Budget Layout Criteria & Standards	-Design Concepts -Program Interpretation -Site/Facility Massing -Access, Circulation -Project Budget -Design Intentions -Net to Gross Ratios	-Schematic Floor Plans -Schematic Sections -Approach to System Integration -Floor to Floor Height -Functional Space -Relationships	-Floor Plans -Sections -Typical Details -Integrated Systems -Space Circulation -Specifications
Structural Foundation Substructure Superstructure	-Performance Requirements -Structural Bay Sizing -Framing Systems Exploration -Subsurface Conditions -Underground Concepts -Initial Framing Review -Structural Load Criteria	-Schematic Basement Plan -Selection of Foundation System -Structural System Selection -Framing Plan Outline -Sizing of Elements	-Basement Floor Plan -Key Foundation Elements, Details -Floor & Roof Framing Plans -Sizing of Major Elements -Outline Specifications
Architectural Exterior Closure Roofing Interior Construction Elevators Equipment	-Approach to Elevations -Views to/from Building -Roof Type & Pitch -Interior Design -Configuration of Key Rooms -Organization of Circulation Scheme -Need & Types of Vertical Circulation -Impact of Key Equipment on Facility & Size -Passive Solar Usage	-Concept Elaboration -Selection of Wall Systems -Schematic Elevations -Selection of Roof Systems -Room Design -Selection of Partitions -Circulation Sizing -Basic Elevator & Vertial Transportation Concepts -Impact of Key Equipment on Room Design	-Elevations -Key Elevation Details -Key Roofing Details -Initial Finish Schedules -Interior Construction Elements -Integration of Structural Framing -Key Interior Elevations -Outline Specification of Equipment Items
Mechanical HVAC Plumbing Fire Protection	-Basic Energy Concepts -Impact of Mechnaical Concepts on Facility -Initial Systems Selection Space Allocation -Performance Requirements for Plumbing, HVAC, Fire Protection	-Mechanical Systems Selection -Refinement of Service & Distribution Concepts -Input to Schematic Plans -Energy Conservation	-Detailed System Selection -Initial System Drawings & Key Details -Distribution & Riser Diagrams -Outline Specifications for System Elements
Electrical Service & Distribution Lighting & Power	-Basic Power Supply -Approaches to Use of Natural & Artificial Lighting -Performance Requirements for Lighting -Need for Special Electrical Systems	-Window/Skylight Design & Sizing Selection of Lighting & Electrical Systems -General Service, Power & Distribution Concepts	-Detailed Systems Selection -Distribution Diagrams -Key Space Lighting Layouts -Outline Specification for Electrical Elements
Site Preparation Utilities Landscaping	-Site Selection -Site Development Criteria -Size Forms & Massing -Requirements for Access -Views to/from Facility -Utility Supply -Site Drainage	-Design Concept Elaboration -Initial Site Plan -Schematic Planting, Grading, Paving Plans	-Site Plan -Planting Plan -Typical Site Details -Outline Specification for Site Materials

Figure 9.3: **LCCA Study Areas**

- Small unit costs that are magnified to large total costs because of project size,
- Special user requirements,
- Repetitive features.

Staffing costs are particularly high for hospitals, correctional facilities, and manufacturing facilities. These costs can be the largest single component of the life cycle cost. High *energy and maintenance costs* are also often incurred for heating, ventilating, and air-conditioning (HVAC); lighting; operating equipment; and architectural finishes. These recurring costs can have considerable impact, and they warrant primary attention via LCCA. Features that tend to absorb the major part of the *initial cost* include the structural system, exterior closure, interior construction, plumbing, HVAC, electrical service, and lighting. Often the magnitude of a project, in either size or initial cost, warrants the study of areas that normally would be omitted from the economic analysis. In manufacturing and industrial facilities, the *equipment cost* can be quite significant.

In a large project, a small cost differential per square foot for, say, the lighting system, can result in large savings. This holds true for *repetitive features* as well. The study of *special user requirements*, relating to building configuration, orientation, interior layout, or equipment types and locations, may hold only small potential for savings. However, LCCA can create an awareness to top owner/management of the cost impact of such requirements. Finally, it may sometimes be worthwhile to study a lower-cost feature if the results of the study can be applied to other similar projects scheduled for design.

Life Cycle Cost Modeling

Another technique for identifying study areas for a project is to prepare a life cycle cost model of the entire project. This model would normally include all of the significant costs of ownership such as initial project cost, energy, maintenance and repairs, alterations and replacements, and associated costs such as staffing and other costs. This technique permits all costs to be illustrated on one page.

Figure 9.4 is an example of a life cycle cost model prepared for a microchip manufacturing facility. The pie chart of this model quickly focuses attention on the staffing cost, equipment cost, and initial construction cost. The process gases and chemicals, as well as the energy costs, are also significant and should be studied by the design team.

The life cycle cost model technique can be applied to all types of facilities, including offices, health care facilities, process plants, and research and development facilities. The data used to prepare the model can be very preliminary and marginally accurate, yet very helpful in focusing designer attention on the areas with highest costs.

Assigning Priorities to Study Areas

The potential study areas may be assigned priorities in terms of dollar resources to be expended on each, or in terms of relative importance. Any of a number of assignment schemes may be used; an example of several types follows.

Dollar priorities are the actual amounts of resources to be devoted to each study or group of studies. As an example, experience might show that the resources to be expended are most effectively utilized when they are allocated according to the following table of weights:

Study Area	Relative Weight (Total = 100)	Range of Relative Weight
Energy, maintenance costs	30	± 15
Initial cost	30	± 15
Small unit cost differentials	10	± 10
Special user requirements	10	± 10
Repetitive features	10	± 10
Type, complexity	10	± 10

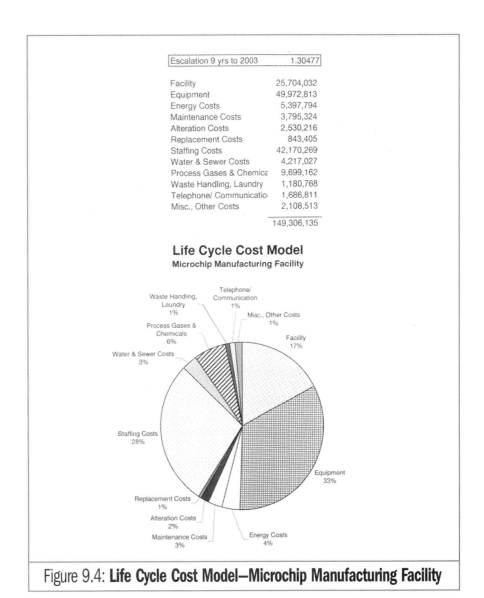

Figure 9.4: **Life Cycle Cost Model—Microchip Manufacturing Facility**

The top line of the table indicates that 15-45% of the LCC study resources available for the design phase should be expended on studies related to energy and maintenance costs, with 30% about average. In the corporate office building example, the project manager would probably assign these costs a weight of 45. Then, of the $25,000 schematic phase resource allocation, 0.45 × $25,000 = $11,250 would be spent on studies related to energy and maintenance costs. In a similar manner, such "dollar priorities" would be assigned to other groups of studies for both the schematic and design development phases. The individual studies related to, say, energy and maintenance costs, could be prioritized in the same way or in terms of relative importance.

The *priority matrix* in Figure 9.5 provides a means of assigning, to each potential study area, a priority number from 1 (highest priority) to 10 (lowest priority).

Before the matrix is used, the life cycle cost savings that will result from each study is estimated (along with the cost of the study), insofar as they are expected to be high, medium, or low. The matrix then gives a range of priorities that may be assigned to that study. Within each range, other factors may be used to choose a particular priority number. Three such factors are listed below the matrix in Figure 9.5. Consider again the corporate office building example. The project manager might estimate low life cycle cost savings and medium analysis cost for a study of lighting systems. The priority number range for this low-medium combination is 4–9. Since lighting gives rise to energy costs, which are of concern in the project, the manager might assign the lighting study a priority number from the low end of the range, say 5. Once priorities are assigned to all the

	ESTIMATED LCCA SAVINGS		
ESTIMATED LCCA STUDY COST	LOW	MED	HIGH
LOW	3–7	2–5	1–3
MED	4–9	3–7	2–5
MED	5–10	4–9	3–7

NOTE: Range shown in each block is to allow for such considerations as
- acceptability of alternatives
- ease of implementation of results
- multidiscipline or trade impact

Figure 9.5: **Matrix for Assessment of LCCA Priorities**

study areas, the study with the highest priority is performed first. Then the study with the next highest priority (lowest number) is performed, and so on, until the study resources assigned to each design phase are used up. This method of assigning and utilizing priorities ensures that the highest-priority study areas receive full attention. However, the study resources may be exhausted before all potential study areas have been treated.

Contingency Planning

The design alternatives that are lowest in life cycle cost are often not lowest in initial cost. Nothing is more frustrating, for both designers and managers, than to conduct a series of LCCA for a project, select the least life cycle cost alternatives, and then find that the funds budgeted for the project are not sufficient to cover the initial costs of the selected alternatives. Such a situation is wasteful in two ways: the facility that is eventually built will be less economical than it could be, and the resources invested in the study will produce no return. Contingency planning to avoid this type of situation is a critical part of the management of the economic study effort.

Role of Contingency Planning

The funds budgeted for a project may be insufficient to permit the implementation of the lowest life cycle cost alternatives for a number of reasons, and the situation may, in some cases, be unavoidable. However, the most common cause, a low budget estimate based on historical costs, can be avoided by contingency planning. The best (most effective) time for this planning is during the pre-schematic design phase, immediately after determining which design features will receive the bulk of the analysis effort. The primary objective of contingency planning is to develop estimates of the initial costs of the most economical design alternatives early enough so they may be considered in the final budget review. A secondary objective is to find some other means of incorporating these alternatives in the final project design, in case the primary objective cannot be met.

Planning Approaches

There are five basic approaches to contingency planning for the purpose of obtaining sufficient funds to implement the most economical design alternatives. These approaches, in descending order of normal preference, are as follows:

1. The most preferred approach is to schedule the design of all features that are to be subjected to LCCA for completion during the schematic phase. This approach normally provides sufficient time for inclusion of the initial costs of the most economical alternatives in the project budget.
2. A less effective, but sometimes necessary approach is to develop, during the schematic phase, a rough design and simplified LCCA for each design feature that is scheduled for economic analysis. The resulting data can be used to support an estimate of the probable initial costs

required to implement the lowest life cycle cost alternatives. If fully developed analyses are completed early enough in the design development phase, the rough initial cost estimate may be modified.

3. The next best approach may be used when some or all of the scheduled LCCAs cannot be completed prior to the submission of budget estimates. Then these studies may be scheduled for completion early in the design development phase. If they indicate a need to increase the project amount, a revised project estimate may be submitted in time to be included in the final budget approval process.

The three approaches discussed above have, as a common objective, the development of initial cost data early enough to be included in the final budget approval process. In some situations, this may not be possible. Often, a project has already been budgeted. Later, the design team may identify alternatives that are low in life cycle cost, but too high in initial cost. This dilemma can be solved effectively through trade-off type cost reductions that free sufficient budgeted funds to permit the inclusion of alternatives that would generate significant follow-on savings and reduce initial costs of items with little effect on follow-on costs. Such a cost reduction might be obtained by eliminating initial cost items that are either marginal or nearly marginal and do not affect the facility's basic function. An example is reducing the number of landscaping items from 100 to 50. Alternatively, a value engineering study could be used to identify initial costs that may be reduced. For example, in a high school, the original HVAC design called for 37 rooftop units. A LCCA indicated a central system would cost some $300,000 more, but it would reduce energy and maintenance costs by 33%, resulting in a 3-year break-even. As a tradeoff cost reduction, lower quality carpeting was selected that freed the required initial costs. The energy savings accrued to the owner and more than paid for a higher quality carpet for its first replacement.

If these approaches do not provide sufficient funds to implement highly desirable least-cost alternatives, additional funds may possibly be obtained either from other concurrent owner projects or from an additional funding request.

Documentation

As long as there is sufficient justification for a life cycle cost study, there is concurrent justification for the adequate documentation of that study. Adequate documentation is essential to the communication of information from the schematic through the design development phase. It provides a mechanism for the review of technical and cost data at all levels of management, and for the application of valuable information to future projects.

Level of Documentation Effort

The level of effort devoted to the documentation of LCCA is generally established and controlled on a project-by project basis. Of prime concern are the adequacy of the documentation and its cost. Normally, the documentation effort comprises 10–25% of the LCCA effort. However, it may be established at less than 10% of the LCCA effort if only an informal report (i.e., report for the record, report for the file) is required, especially with maximum use of self-documenting data sheets and checklists.

The project-related factors discussed in the previous paragraphs will influence the level of documentation effort, as will the importance of the project and its visibility at higher management levels. Generally, more documentation is needed for projects with high initial and/or energy and maintenance costs, larger projects, and highly visible projects. More than the usual documentation may be required when design decisions represent a departure from common practice, or when a high degree of uncertainty must be resolved in the analysis.

Use of Self-Documenting Forms and Microcomputer Spreadsheets

The documentation for a particular study may range from carefully organized, illustrated, and typed materials detailing most aspects of the study—to handwritten notes and computations covering only the key points. Either level may be appropriate under certain conditions, provided it is both adequate for its purpose and reasonable in cost relative to that purpose. Experience has shown that the use of standardized, self-documenting data and calculation sheets, standard report forms, and computer-generated spreadsheets can substantially reduce the effort required for documentation. This standardization and the simplification it produces can ultimately pay considerable dividends to the organization and project manager. Figure 9.6 is an example of a suitable checklist format that may be used as a guide for conducting the study, as well as documenting it. Appendix E contains sample LCCA worksheets, which may also be used to document the effort.

Summary and Conclusions

Summary

The decision-making process in building is traditionally fragmented, with major disciplines making decisions in relative isolation. This is especially true when the design team is represented by several firms, e.g., architectural design, structural consultant, mechanical, electrical, and site. As a result, each discipline's decision tends to adversely impact costs to other disciplines. In too many instances, the total cost of ownership is not adequately considered.

To date, one of the principal reasons for unnecessary costs has been the uni-disciplinary approach used by design professionals. In too many cases, the architect has dictated the design; the engineering disciplines have merely responded to those dictates. On the other hand, a multidisciplinary approach to optimizing the building as a system has produced significantly improved results for the owner. The design team concept has led the way to these results.

Of the input data required to perform LCCA, the specific project information and site data are usually available, but it is a different story for the facility components data. Where does a design professional go to get data regarding useful life, maintenance, and systems replacements? This input is needed to calculate roughly one-third of total costs. Few designers have had access to or have made an attempt to collect comprehensive data in these areas in a format facilitating LCCA. One reason for this is that there has not been a retrieval format readily available for these data. The

1. **Executive summary—short narrative**
   a. Project study area and objectives
   b. Identification of alternatives
   c. Rationale for selection of alternatives

2. **Background information**
   a. Project
      - Location/title number
      - Brief description
   b. Study member(s)
      Names/organization/telephone no./fax no.

3. **Controlling parameters for the analysis**
   a. Economic
      - Source of governing criteria
      - Approach used (e.g., PW, annualized)
      - Discount rate, analysis period, present time
   b. Facility
      - Program requirements
      - Operational mode
   c. Site
      - Climatic conditions
      - Environmental characteristics

4. **Description of alternative(s)**
   a. Brief narrative/descriptions/sketches
      - Building elements included in the analysis
      - Sources of information
   b. Initial capital investment costs
   c. Follow-on costs:
      - Energy costs
      - Maintenance, repair, replacement, alteration, and salvage costs
   d. Cost factors considered
      - Treatment of inflation
      - Sources of cost information and degree of uncertainty

5. **Life Cycle Cost Analysis**
   a. Total costs of the alternative(s)
   b. Degree of uncertainty associated with the result (including confidence index and sensitivity analyses)

6. **Recommendations**
   a. Brief description of recommended alternatives
   b. Life cycle cost summary

Figure 9.6: **LCCA Documentation Checklist**

other problem is that in order to perform LCCA (and eventually develop automated programs), the initial-cost data banks must be compatible with the follow-on cost data banks. There is an abundance of references for initial costs, including those published by RSMeans. (See estimating bibliography in the Appendix of this book.) There are few references for follow-on costs for maintenance, operations, replacements, etc., other than RSMeans' *Facilities Maintenance & Repair Cost Data*, which contains data on discrete elements.

A recognized format for LCCA is also lacking. Each owner using LCCA sets up an individual procedure. As a result, any cross-feed of data or results is difficult. It is hoped that the LCC data presented in Part 2 of this book will assist in offsetting the above difficulties.

Owners talk about increasing their initial costs for a project in return for future maintenance and energy savings. They want pay-off periods of less than 5 years, which is equivalent to an approximate 20% return on investment. To gain that return, how much are they willing to spend? They immediately will negotiate front-end design fees to the bone or, more commonly, expect these additional services within the previous fee structure. However, the authors' experience has proven there are many owners who would be willing to fund these extra services if they were confident in an improvement in their project's cost-effectiveness.

Designers talk about their interest in LCCA. Yet, they have not made this proficiency a qualification for becoming a registered design professional, nor have they made substantial efforts to meet these owner demands. Currently, there are almost no professional courses being offered on life cycle costing for design professionals. There are very few engineering and architectural schools that offer courses in life cycle costing.

Owners are becoming more concerned about the cost of ownership. A recent survey indicated that they were very interested in both life cycle costing and value engineering. Coupled with these trends is the fact that owners seek a payback of five years or less before they are willing to spend more on initial costs. The authors' experience confirms that an even earlier pay-off period is expected by industrial, profit-oriented owners, and a longer payback period is anticipated for government and non-profit-oriented owners.

Combined with this list of owner trends is the increasing number of government agencies with mandatory LCC requirements. Foremost of these is the Environmental Protection Agency (EPA), which requires a cost-effectiveness analysis of alternative processes for the early planning and design of wastewater treatment plants. The Air Force was one of the first government agencies to use LCC for housing procurement. Since then, both the Naval Facilities Engineering Command and the Corps of Engineers have published manuals on economic analysis.

Conclusions

Since the preceding edition of this book was published, private industry owners have continued to feel increased competitive business pressure to seek the most life cycle cost-effective facilities from their design professionals. Government agencies are reacting in a similar fashion, with continued reductions in tax bases and the continuing downsizing of government services. In addition, both government and private sector owners are under pressure to seek solutions that minimize facility obsolescence and that are environmentally sustainable, and contribute to the organization's reengineering of staffing effectiveness.

It is hoped that this book will set into motion renewed interest in the use of life cycle costing to improve decision-making and produce more cost-effective facilities.

Part 2 Life Cycle Cost Data

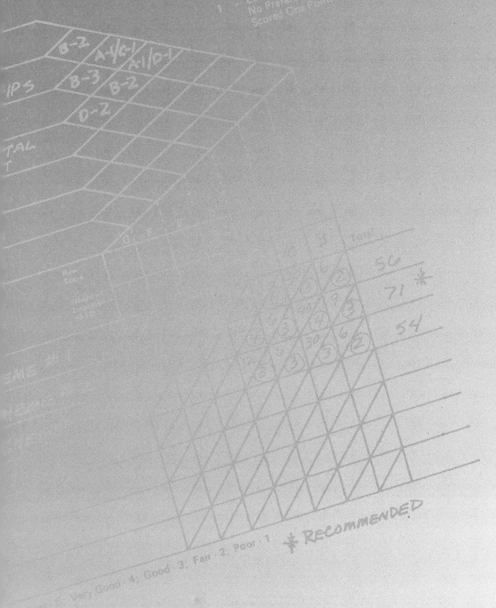

Introduction to Cost Data

This part of the book contains life cycle cost data collected by the authors over the course of life cycle cost analysis studies. Costs are provided for the following building and site elements: structural, architectural, conveying systems, mechanical, electrical, equipment, site work, and landscaping. The data is organized according to the UNIFORMAT framework, an effective system for budget estimating.

All maintenance costs are based on the assumption that the building owner will implement a preventive maintenance program that meets or exceeds the minimal standards of maintenance recommended by the manufacturers of the installed equipment or materials. Failure to provide adequate preventive maintenance will result in escalated replacement and maintenance costs, as well as premature obsolescence.

Equipment replacement costs are based on the assumption that the building owner will maintain a spare parts inventory on a repair-and-restore program, rather than on a redundant equipment standby basis.

Extreme care should be used in the application of this information. Factors that influence these costs include: labor rates, in-house versus contracted maintenance, preventive maintenance program safety, climatic conditions, functional use of the facility, and management emphasis on maintenance. The authors assume no responsibility or liability in connection with the use of this data.

Abbreviations Used in This Section:

Btu	British thermal unit
EA	Each item described
EU	Energy units
Flight	Per flight of stairs
HW	Hot water
IC	Initial cost

kW	Kilowatt (power)
LF	Lineal feet of item described
MH	Labor hours for maintenance (MH for "man-hours")
N/A	Not applicable to item described
RSF	Roof square foot of item described
SF	Square foot of item described
Stop	Per stop of elevator described
V	Volt
WSF	Wall square foot of item described

Maintenance & Replacement Estimating Data - Structural

Item Description	Unit of Measure	Maintenance Description	Maintenance Annual Cost, $ Labor	Material	Equipm't	Energy Demand (EU)	Replacement Life, yrs	Percent Replaced
Structural Data								
01 Foundations								
011 Standard Foundations								
Cast-in-place concrete foundation walls	WSF	General inspection and minor repair (1.0 min every 10 yr)	0.0296	0.00861	0.0017	N/A	75	100
Precast concrete foundation walls	WSF	General inspection and minor repair (1.0 min. every 10 yr)	0.0296	0.00861	0.0017	N/A	75	100
Masonry foundation walls	WSF	General inspection and minor repair (2.0 min every 10 yr)	0.0593	0.01722	0.0034	N/A	50	100
Concrete block foundation walls	WSF	General inspection and minor repair (2.0 min every 10 yr)	0.0593	0.01722	0.0034	N/A	50	100
Stone foundation walls	WSF	General inspection and minor repair (2.0 min every 10 yr)	0.0593	0.01722	0.0034	N/A	45	100
012 Special Foundations								
Wood pile foundations, treated	1000 LF	Underwater - no maintenance required	0.0000	0.00000	0.0000	N/A	Life	N/A
	1000 LF	Above - general inspection and repair (30.0 min every 10 yr)	0.8715	0.86100	0.0172			
Wood pile foundations, untreated	1000 LF	Underwater - no maintenance required	0.0000	0.00000	0.0000	N/A	30	100
	1000 LF	Above - general inspection and repair (1.0 h every 6 yr)	2.9108	0.25830	0.0517			
Precast concrete piles, square	1000 LF	No maintenance required	0.0000	0.00000	0.0000	N/A	Life	N/A
Prestressed concrete piles	1000 LF	No maintenance required	0.0000	0.00000	0.0000	N/A	Life	N/A
Cast-in-place concrete piles	1000 LF	No maintenance required	0.0000	0.00000	0.0000	N/A	Life	N/A
Steel pipe piles, concrete - filled	1000 LF	No maintenance required	0.0000	0.00000	0.0000	N/A	Life	N/A
Steel pipe piles, nonfilled	1000 LF	No maintenance required	0.0000	0.00000	0.0000	N/A	Life	N/A
Steel "H" piles	1000 LF	No maintenance required	0.0000	0.00000	0.0000	N/A	Life	N/A
Wood with cast-in-place concrete composite piles	1000 LF	Underwater - no maintenance required	0.0000	0.00000	0.0000	N/A	Life	N/A
	1000 LF	Above - general inspection and repair (30.0 min every 10 yr)	0.8715	0.08610	0.0172			
Wood with precast concrete composite piles	1000 LF	Underwater - no maintenance required	0.0000	0.00000	0.0000	N/A	Life	N/A
	1000 LF	Above - general inspection and repair (30.0 min every 10 yr)	0.8715	0.08610	0.0172			
Raft concrete slab foundation	SF	General inspection and minor repair (1.0 min every 10 yr)	0.0296	0.00861	0.0017	N/A	Life	N/A
02 Substructure								
021 Slab on Grade								
CONCRETE ON GRADE								
Standard 4-in slab on grade floor	SF	General inspection and minor repair (1.0 min every 10 yr) Note: additional for industrial facilities	0.0296	0.00861	0.0017	N/A	50	100
Standard 5-in slab on grade floor	SF	General inspection and minor repair (1.0 min every 10 yr) Note: additional for industrial facilities	0.0296	0.00861	0.0017	N/A	50	100
Structural 4-in slab on grade floor	SF	General inspection and minor repair (0.5 min every 10 yr)	0.0139	0.00344	0.0000	N/A	50	100

Maintenance & Replacement Estimating Data - Structural

Item Description	Unit of Measure	Maintenance Description	Maintenance Annual Cost, $			Energy Demand (EU)	Replacement Life, yrs	Percent Replaced
			Labor	Material	Equipm't			
Structural 5-in slab on grade floor	SF	Note: additional for industrial facilities. General inspection and minor repair (0.5 min every 10 yr)	0.0139	0.00344	0.0000	N/A	50	100
Concrete grade beams	1000 LF	Note: additional for industrial facilities. No maintenance required	0.0000	0.00000	0.0000	N/A	Life	N/A
Concrete steps on grade	SF	General inspection and minor repair (2.0 min every 10 yr)	0.0593	0.01722	0.0034	N/A	60	100
023 Basement Walls								
Nonreinforced concrete basement walls	WSF	General inspection and minor repair (1.0 min every 10 yr)	0.0436	0.00861	0.0017	N/A	40	100
Reinforced concrete basement walls	WSF	General inspection and minor repair (.5 min every 10 yr)	0.0227	0.00344	0.0000	N/A	75	100
Hollow concrete block basement walls	WSF	General inspection and minor repair (0.5 min every 5 yr)	0.0872	0.01722	0.0034	N/A	50	100
Solid concrete block basement walls	WSF	General inspection and minor repair (0.5 min every 5 yr)	0.0872	0.01722	0.0034	N/A	50	100
Brick masonry basement walls	WSF	General inspection and minor repair (0.5 min every 5 yr)	0.0872	0.01722	0.0034	N/A	50	100
03 Superstructure								
031 Floor Construction								
STRUCTURAL FRAME								
Steel structural frame (includes columns, beams, girders, trusses, spandrels, bracing, and fireproofing)	SF	Concrete encasement fireproofing repair (0.1 min every 10 yr)	0.0035	0.00172	0.0000	N/A	75	100
	SF	Masonry encasement fireproofing repair (0.1 min every 10 yr)	0.0035	0.00172	0.0000	N/A	75	100
	SF	Sprayed mineral fiber fireproofing repair (0.3 min every 10 yr)	0.0105	0.00517	0.0000	N/A	75	100
	SF	Lath and plaster fireproofing repair (0.2 min every 10 yr)	0.0070	0.00344	0.0000	N/A	75	100
	SF	Gypsum wallboard fireproofing repair (0.3 min every 10 yr)	0.0105	0.00517	0.0000	N/A	75	100
Reinforced concrete structural frame (includes columns, beams, and miscellaneous frame elements)	SF	General inspection and minor repair (0.2 min. every 10 yr)	0.0070	0.00344	0.0000	N/A	75	100
Precast concrete structural frame (includes columns, beams, and miscellaneous frame elements)	SF	General inspection and minor repair (0.2 min every 10 yr)	0.0070	0.00344	0.0000	N/A	75	100
Wood structural frame (includes posts, girts, plates, studs,	SF	General inspection and minor repair (1.0 min every 5 yr)	0.0070	0.00344	0.0086	N/A	50	100
Metal joist structural frame (includes metal joists and accessories)	SF	No maintenance required	0.0000	0.00000	0.0000	N/A	75	100
INTERIOR STRUCTURAL WALLS								
Interior concrete block load-bearing walls	WSF	Repointing joints (2.0 min every 15 yr)	0.0486	0.01599	0.0016	N/A	60	100
Interior brick load-bearing walls	WSF	Repointing joints (4.0 min every 15 yr)	0.0971	0.03198	0.0016	N/A	75	100
Interior concrete load-bearing walls	WSF	Minor repair (1.0 min every 15 yr)	0.0162	0.00800	0.0016	N/A	75	100
Interior wood load-bearing walls	WSF	Minor repair (2.0 min every 15 yr)	0.0486	0.03198	0.0016	N/A	50	100

Maintenance & Replacement Estimating Data - Structural

Item Description	Unit of Measure	Maintenance Description	Maintenance Annual Cost, $ Labor	Material	Equipm't	Energy Demand (EU)	Replacement Life, yrs	Percent Replaced
FLOOR SLABS & DECKS								
Reinforced concrete floor slabs (includes slab and beams)	SF	General inspection and minor repair (1.0 min every 10 yr)	0.0307	0.00892	0.0018	N/A	50	100
Post-tensioned concrete floor slabs	SF	General inspection and minor repair (1.0 min every 10 yr)	0.0307	0.00892	0.0018	N/A	50	100
Precast prestressed concrete floor slabs (may include planks, concrete, fees, floor channels, and structural concrete topping)	SF	General inspection and minor repair (1.0 min every 10 yr)	0.0307	0.00892	0.0018	N/A	50	100
Noncellular open metal decking with structural concrete topping	SF	General inspection and minor repair (1.0 min every 10 yr)	0.0307	0.00892	0.0018	N/A	50	100
Cellular metal decking with structural concrete topping	SF	General inspection and repair (1.0 min every 10 yr)	0.0307	0.00892	0.0018	N/A	50	100
032 Roof Construction								
Reinforced concrete roof slabs (includes slab and beams)	RSF	General inspection and minor repair (1.0 min every 10 yr)	0.0361	0.01784	0.0000	N/A	50	100
Post-tensioned concrete roof slabs	RSF	General inspection and minor repair (1.0 min every 10 yr)	0.0361	0.01784	0.0000	N/A	50	100
Precast prestressed concrete roof slabs (may include planks, concrete tees, and structural concrete topping)	RSF	General inspection and minor repair (1.0 min every 10 yr)	0.0361	0.01784	0.0000	N/A	50	100
Structural wood framing (includes sheating, joists, beams, etc.)	RSF	General inspection and minor repair (2.0 min every 10 yr)	0.0722	0.01784	0.0000	N/A	40	100
Corrugated metal deck with light weight concrete topping	RSF	General inspection and repair (1.0 min every 10 yr)	0.0722	0.01784	0.0000	N/A	50	100
Corrugated metal deck only	RSF	General inspection and repair (1.0 min every 10 yr)	0.0722	0.01784	0.0000	N/A	30	100
Precast concrete (hollow core) roof slab	RSF	General inspection and repair (1.0 min every 10 yr)	0.0722	0.01784	0.0000	N/A	30	100
Poured-in-place gypsum concrete over formboards	RSF	General inspection and repair (1.0 min every 10 yr)	0.0722	0.01784	0.0000	N/A	50	100
Metal-edged gypsum plank	RSF	General inspection and repair (1.0 min every 10 yr)	0.0722	0.01784	0.0000	N/A	40	100
Cement fiber planks	RSF	General inspection and repair (1.0 min every 10 yr)	0.0722	0.01784	0.0000	N/A	40	100
033 Stair Construction								
Precast concrete	Flight	Nonfinish: sweep stairs and landings, pick up trash (8.0 min/day)	361.0500	53.50500	3.5670	N/A	50	100
	Flight	Minor repair per year	6.3184	3.56700	0.0000			
Steel pan type, filled with concrete	Flight	Nonfinish: sweep stairs and landings, pick up trash (8.0 min/day)	361.0500	53.50500	3.5670	N/A	40	100
	Flight	Minor repair per year	12.6368	7.13400	0.0000			
Steel tread and riser	Flight	Non-finish: sweep stairs and landings, pick-up trash (10.0 min/day)	406.1813	66.88125	4.4588	N/A	40	100
	Flight	Minor repair per year	36.1050	8.91750	1.7835			
Prefabricated steel form filled with concrete	Flight	Nonfinish: sweep stairs and landings, pick up trash (8.0 min/day	361.0500	53.50500	3.5670	N/A	50	100
	Flight	Minor repair per year	6.3184	3.56700	0.0000			

Maintenance & Replacement Estimating Data - Structural/Architectural

Item Description	Unit of Measure	Maintenance Description	Maintenance Annual Cost, $			Energy Demand (EU)	Replacement Life, yrs	Percent Replaced
			Labor	Material	Equipm't			
Steel frame, precast concrete treads, and risers	Flight	Nonfinish: sweep stairs and landings, pick up trash (8.0 min/day	361.0500	53.50500	3.5670	N/A	40	100
	Flight	Minor repair per year	12.6368	7.13400	0.0000			

Architectural Data

04 Exterior Closure
041 Exterior Walls
0411 Exterior Wall Construction

Item Description	Unit of Measure	Maintenance Description	Labor	Material	Equipm't	Energy Demand (EU)	Replacement Life, yrs	Percent Replaced
Masonry veneer: 4-in brick and 4-in block, insulation and vapor barrier	WSF	Repointing joints (4.0 min every 15 yr)	0.1114	0.03660	0.00183	N/A	75	100
Precast concrete veneer insulation and vapor barrier	WSF	Sand blast cleaning (6.0 min every 12 yr)	0.1486	0.09150	0.00183	N/A	75	100
Stucco on metal studs: insulation and vapor barrier	WSF	Minor repair (6.0 min every 8 yr)	0.3529	0.07320	0.00183	N/A	35	100
Stone veneer: block backup insulation, and vapor barrier	WSF	Repointing joints (6.0 min every 15 yr)	0.1672	0.05490	0.00183	N/A	75	100
Aluminum panel: insulation and vapor barrier	WSF	Minor repair, cleaning (2.0 min every 8 yr)	0.1486	0.01830	0.00183	N/A	50	100
Metal panel: insulation and vapor barrier	WSF	Minor repair, cleaning (4.0 min every 6 yr)	0.2972	0.01830	0.00183	N/A	40	100
Cast-in-place 8-in concrete wall: insulation and vapor barrier	WSF	Sand-blast cleaning (6.0 min every 12 yr)	0.1486	0.07320	0.00183	N/A	Life	100
Concrete block (standard) 8-in wall: insulation and vapor barrier	WSF	Reporting joints (2.0 min every 15 yr)	0.0743	0.03660	0.00183	N/A	Life	100
Split-face concrete block 8-in wall: insulation and vapor barrier	WSF	Repointing joints (2.0 min every 15 yr)	0.0743	0.03660	0.00183	N/A	Life	100
Plywood siding, texture 1-11 with wood studs: insulation and vapor barrier	WSF	Clean and retain (2.0 min every 5 yr)	0.2972	0.07320	0.00183	N/A	30	100
Cedar siding, rough-sawn with wood studs: insulation and vapor barrier	WSF	Minor repair (0.5 min every 5 yr)	0.0371	0.01830	0.00183	N/A	40	100
Redwood siding, board, and batten: insulation and vapor barrier	WSF	Minor repair (0.5 min every 5 yr)	0.0371	0.01830	0.00183	N/A	40	100

0412 Exterior Louvers and Screens

Item Description	Unit of Measure	Maintenance Description	Labor	Material	Equipm't	Energy Demand (EU)	Replacement Life, yrs	Percent Replaced
Screen louvers, galvanized steel	WSF	Repair, paint, and miscellaneous (1.0 min every 8 yr)	0.0557	0.01830	0.00183	N/A	15	100
Screen louvers, copper	WSF	Repair, paint, and miscellaneous (1.0 min every 8 yr)	0.0557	0.01830	0.00183	N/A	25	100
Storm proof louvers, galvanized steel	WSF	Repair, paint, and miscellaneous (1.0 min every 8 yr)	0.0557	0.01830	0.00183	N/A	15	100
Storm proof louvers, copper	WSF	Repair, paint, and miscellaneous (1.0 min every 8 yr)	0.0173	0.00568	0.00183	N/A	15	100

0413 Sun Control Devices

Item Description	Unit of Measure	Maintenance Description	Labor	Material	Equipm't	Energy Demand (EU)	Replacement Life, yrs	Percent Replaced
Glass screen and metal frame	WSF	Repair and paint frame (1.0 min every 8 yr)	0.0557	0.01830	0.00183	N/A	15	25
Preformed metal screen and metal frame	WSF	Repair and paint frame (1.0 min every 8 yr)	0.0557	0.01830	0.00183	N/A	15	25
Fabric screen and metal frame		Repair and paint frame (1.0 min every 8 yr)	0.0557	0.01830	0.00183	N/A	15	25

0414 Balcony Walls and Handrails

Item Description	Unit of Measure	Maintenance Description	Labor	Material	Equipm't	Energy Demand (EU)	Replacement Life, yrs	Percent Replaced
Cast-in-place concrete	WSF	Sand-blast cleaning (6.0 min every 12 yr)	0.1486	0.07320	0.00183	N/A	75	100
Precast concrete	WSF	Sand-blast cleaning (6.0 min every 12 yr)	0.1486	0.07320	0.00183	N/A	75	100
Brick masonry	WSF	Repointing joints (4.0 min every 15 yr)	0.1486	0.03660	0.00183	N/A	75	100
Concrete unit masonry	WSF	Reporting joints (2.0 min every 15 yr)						
Stone	WSF	Repointing joints (6.0 min every 15 yr)	0.1672	0.05490	0.00183	N/A	75	100

Maintenance & Replacement Estimating Data - Architectural

Item Description	Unit of Measure	Maintenance Description	Maintenance Annual Cost, $ Labor	Material	Equipm't	Energy Demand (EU)	Replacement Life, yrs	Percent Replaced
Metal panels	WSF	Clean and paint (2.0 min every 6 yr)	0.1486	0.03660	0.00183	N/A	30	100
Glass panels	WSF	Minor repair, cleaning (2.0 min every 6 yr)	0.1486	0.01830	0.00183	N/A	40	100
	WSF	Window washing (0.18 min every 6 months)	0.1300	0.01830	0.00915	N/A	40	100
	WSF	Repair glazing	0.0037	0.00915	0.00183			
0415 Exterior Soffits								
Exterior gypsum board including metal hangers	SF	Minor repair (2.0 min every 6 yr)	0.1486	0.05490	0.00183	N/A	12	100
	SF	Painting - 2 coats (1.0 min every 6 yr)	0.0743	0.01830	0.00183			
Metal panels including metal hangers	SF	Minor repair (0.6 min every 15 yr)	0.0371	0.03660	0.00183	N/A	40	100
042 Exterior Doors and Windows								
0421 Windows								
Fixed glazing, frame, hardware	WSF	Lobby, storefront: wash and squeegee dry both sides of glass (0.18 min/wk)	3.2263	3.27240	0.0654	N/A	40	100
	WSF	Office, other areas: wash and squeegee dry (0.18 min/quarter)	0.2415	0.03272	0.0164			
	WSF	Repair glazing, frame, and hardware	0.0173	0.01636	0.0016			
Operable glazing, frame, hardware	WSF	Lobby, storefront: wash and squeegee dry both sides of glass (0.18 min/wk)	3.2263	0.32724	0.0654	N/A	35	100
	WSF	Office, other areas: wash and squeegee dry (0.18 min/quarter)	0.2415	0.03272	0.0164			
	WSF	Repair glazing, frame, and hardware	0.0690	0.04909	0.0164			
Single glazing, fixed frame, hardware	WSF	Lobby, storefront: wash and squeegee dry both sides of glass (0.18 min/wk)	3.2263	0.32724	0.0654	N/A	40	100
	WSF	Office, other areas: wash and squeegee dry (0.18 min/quarter)	0.2415	0.03272	0.0164			
	WSF	Repair glazing, frame, and hardware	0.0173	0.01636	0.0016			
Double glazing, fixed frame, hardware	WSF	Lobby, storefront: wash and squeegee dry both sides of glass (0.18 min/week)	3.2263	0.32724	0.0654	N/A	40	100
	WSF	Office, other areas: wash and squeegee dry (0.18 min/quarter)	0.2415	0.03272	0.0164			
	WSF	Repair glazing, frame, and hardware	0.0173	0.03272	0.0016			
Reflective single glazing, fixed frame, hardware	WSF	Lobby, storefront: wash and squeegee dry both sides of glass (0.18 min/wk)	3.2263	0.32724	0.0654	N/A	40	100
	WSF	Office, other areas: wash and squeegee dry (0.18 min/quarter)	0.2415	0.03272	0.0164			
	WSF	Repair glazing, frame, and hardware	0.0173	0.03272	0.0016			
Tinted single glazing, fixed frame, hardware	WSF	Lobby, storefront: wash and squeegee dry both sides of glass (0.18 min/wk)	3.2263	0.32724	0.0654	N/A	40	100
	WSF	Office, other areas: wash and squeegee dry (0.18 min/quarter)	0.2415	0.03272	0.0164			
	WSF	Repair glazing, frame, and hardware	0.0173	0.03272	0.0016			

Maintenance & Replacement Estimating Data - Architectural

Item Description	Unit of Measure	Maintenance Description	Maintenance Annual Cost, $			Energy Demand (EU)	Replacement Life, yrs	Percent Replaced
			Labor	Material	Equipm't			
0422 Curtain Walls								
Aluminum spandrel panel	WSF	Minor repair, cleaning (2.0 min every 6 yr)	0.1380	0.01636	0.0016	N/A	50	100
Stainless steel panel	WSF	Minor repair, cleaning (2.0 min every 6 yr)	0.1380	0.01636	0.0016	N/A	50	100
Porcelain enamel panel	WSF	Minor repair, cleaning (2.0 min every 6 yr)	0.1380	0.01636	0.0016	N/A	50	100
Weathering steel panel	WSF	No maintenance required	0.0000	0.00000	0.0000	N/A	50	100
Opaque colored-glass panel	WSF	Window washing (0.18 min every 6 months)	0.1208	0.01636	0.0082	N/A	40	100
	WSF	Repair glazing	0.0035	0.00818	0.0016			
Ceramic tile facing on panel	WSF	Minor repair, cleaning (2.0 min every 6 yr)	0.1380	0.01636	0.0016	N/A	50	100
Stone facing on panel	WSF	Repointing joints (6.0 min every 15 yr)	0.1380	0.04909	0.0016	N/A	75	100
0423 Exterior Doors								
Hollow metal door, frame, hardware	WSF	Damp-clean both sides (0.12 min/quarter)	0.1380	0.01636	0.0164	N/A	40	100
	WSF	Repair door, frame, hardware	0.1725	0.08181	0.0327			
	WSF	Paint, 2 coats every 4 yr	0.1208	0.03272	0.0327			
Solid-core wood door	WSF	Damp-clean both sides (0.12 min/quarter)	0.1380	0.01636	0.0164	N/A	40	100
	WSF	Repair door, frame, hardware	0.2070	0.09817	0.0327			
	WSF	Paint, 2 coats every 4 yr	0.1208	0.03272	0.0327			
Overhead metal service door, frame, hardware	WSF	Repair door, frame, hardware	0.1380	0.06545	0.0164	N/A	30	100
Rolling metal service door, frame hardware	WSF	Repair door, frame, hardware	17.2530	0.08181	0.0327	N/A	30	100
Telescoping metal service door, frame, hardware	WSF	Repair door, frame, hardware	17.2530	0.08181	0.0327	N/A	25	100
Revolving door, frame, hardware	EA	Damp-clean all sides (0.1 MH/quarter)	6.9012	65.44800	16.3620	N/A	25	100
	EA	Remove obstructions and clean track, clean pivot points, inspect (1 MH/6 months)	51.7590	8.18100	81.8100			
Automatic sliding door, mechanism, frame, hardware (2 horsepower)	WSF	Damp-clean both sides (0.12 min/quarter)	0.1380	0.01636	6.5448	1.5kW	15	50
	WSF	Check and adjust operating functions adjust and service (3.0 min/6 months)	2.5880	0.40905	0.0818			
0424 Storefronts								
Aluminum panel, framing, insulation	WSF	Minor repair, cleaning (2.0 min every 6 yr)	0.1380	0.01636	0.0016	N/A	50	100
Hollow metal panel, framing	WSF	Minor repair, cleaning (2.0 min every 6 yr)	0.1380	0.01636	0.0016	N/A	40	100
05 Roofing								
0501 Roof Coverings								
4 Ply Organic Felt w/ Coal Tar Pitch & Gravel	RSF	Preventative inspection (0.01 min/yr)	0.0138		N/A	N/A	20	100
Polyisocyanurate + Fiberglass R=12.04	RSF	Minor repair (0.03 min/yr)	0.0413	0.00702	N/A			
4 Ply Fiberglass Type IV w/ Asphalt	RSF	Preventative inspection (0.01 min/yr)	0.0138		N/A	N/A	20	100
Polyisocyanurate + Fiberglass R= 12.04	RSF	Minor repair (0.03 min/yr)	0.0413	0.00564	N/A			
3 Ply Fiberglass Type IV Felt w/ Asphalt	RSF	Preventative inspection (0.01 min/yr)	0.0138		N/A	N/A	15	100
Polyisocyanurate + Fiberglass R= 12.04	RSF	Minor repair (0.03 min/yr)	0.0413	0.00432	N/A			
IRMA 45 mil EPDM Membrane Polystyrene 2.5" Thick R=12.5	RSF	Preventative inspection (0.01 min/yr)	0.0138		N/A	N/A	20	100
		Minor repair (0.01 min/yr)	0.0138	0.00115	N/A			

Maintenance & Replacement Estimating Data - Architectural

Item Description	Unit of Measure	Maintenance Description	Maintenance Annual Cost, $			Energy Demand (EU)	Replacement Life, yrs	Percent Replaced
			Labor	Material	Equipm't			
EPDM 60 mil Fully Adhered Polyisocuanurate 2.3" R=12.79	RSF	Preventative inspection (0.01 min/yr)	0.0138		N/A	N/A	20	100
		Minor repair (0.02 min/yr)	0.0275	0.00370	N/A			
PVC Reinforced 80 mil Polyisocuanurate 2.3" R=12.79	RSF	Preventative inspection (0.01 min/yr)	0.0138		N/A	N/A	20	100
		Minor repair (0.02 min/yr)	0.0275	0.00326	N/A			
2 Ply Modified Bitumen w/ Min. Surfacing Polyisocyanurate + Fiberglass R= 12.04	RSF	Preventative inspection (0.01 min/yr)	0.0138		N/A	N/A	15	100
		Minor repair (0.02 min/yr)	0.0275	0.00408	N/A			
Prepared roll roofing, 15# felt	RSF	Preventative inspection (0.01 min/yr)	0.0138	0.00000	0.00000	N/A	12	100
	RSF	Minor repair (0.03 min/yr)	0.0413	0.00338	0.00000			
Copper roofing: flat, standing, or batten seam	RSF	Preventative inspection (0.01 min/yr)	0.0138	0.00000	0.00000	N/A	50	100
	RSF	Minor repair (0.01 min/yr)	0.0138	0.00116	0.00000			
Galvanized steel sheet metal roofing	RSF	Preventative inspection (0.01 min/yr)	0.0138	0.00000	0.00000	N/A	30	100
	RSF	Minor repair (0.02 min/yr)	0.0275	0.00169	0.00000			
Teflon film one-piece loose membrane	RSF	Preventative inspection (0.01 min/yr)	0.0138	0.00000	0.00000	N/A	25	100
	RSF	Minor repair (0.01 min/yr)	0.0138	0.00169	0.00000			
0502 Traffic Toppings and Paving Membrane								
Fluid applied membrane over structural slab, covered with ¼-in protection boards, and 2½- to 3-in reinforced concrete	RSF	Preventative inspection (0.01 min/yr)	0.0018	0.0000	0.0000	N/A	40	100
	RSF	Minor repair (0.01 min/yr)	0.0181	0.0169	0.0000			
Teflon film membrane over structural slab, covered with ¼-in protection boards, and 2½- to 3-in reinforced concrete	RSF	Preventative inspection (0.01 min/yr)	0.0018	0.0000	0.0000	N/A	40	100
	RSF	Minor repair (0.01 min/yr)	0.0181	0.0169	0.0000			
0503 Roof Insulation and Fill								
Expanded perlite board	RSF	Minor repair (0.01 min/yr)	0.0018	0.0017	0.0000	N/A	40	100
Perlite polyurethane	RSF	Minor repair (0.01 min/yr)	0.0018	0.0017	0.0000	N/A	40	100
Poured-in-place insulation	RSF	Minor repair (0.01 min/yr)	0.0018	0.0017	0.0000	N/A	75	100
Fiber board	RSF	Minor repair (0.01 min/yr)	0.0018	0.0017	0.0000	N/A	40	100
0504 Flashing and Trim								
Copper flashing	LF	Preventative maintenance (0.01 min every 10 yr)	0.0018	0.0017	0.0000	N/A	50	100
Aluminum flashing	LF	Preventative maintenance (0.1 min every 10 yr)	0.0018	0.0017	0.0000	N/A	40	100
Stainless steel flashing	LF	Preventative maintenance (0.1 min every 10 yr)	0.0018	0.0017	0.0000	N/A	50	100
Lead flashing	LF	Preventative maintenance (0.1 min every 10 yr)	0.0018	0.0017	0.0000	N/A	25	100
Galvanized steel	LF	Preventative maintenance (0.1 min every 10 yr)	0.0018	0.0017	0.0000	N/A	30	100
0505 Roof Openings								
Skylights, double-glazed	SF	Exterior and interior cleaning (0.5 min/quarter)	67.2300	0.0786	0.0314	N/A	40	100
	SF	Repair glazing, frame, and hardware (0.5 min/yr)	16.8075	31.4280	0.0157			

Maintenance & Replacement Estimating Data - Architectural

Item Description	Unit of Measure	Maintenance Description	Maintenance Annual Cost, $ Labor	Material	Equipm't	Energy Demand (EU)	Replacement Life, yrs	Percent Replaced
Roof hatches, painted steel	SF	Minor repair and painting (1.0 min every 8 yr)	0.0504	0.0157	0.0016	N/A	24	100
Roof hatches, galvanized	SF	Minor repair (0.5 min every 12 yr)	0.0168	0.0079	0.0016	N/A	30	100
Roof hatches, stainless	SF	Minor repair (0.5 min every 12 yr)	0.0084	0.0079	0.0016	N/A	40	100

06 Interior Construction
061 Partitions
0611 Fixed Partitions

Item Description	Unit of Measure	Maintenance Description	Labor	Material	Equipm't	Energy Demand (EU)	Replacement Life, yrs	Percent Replaced
Brick partitions, exposed	WSF	Repointing joints (2.0 min every 25 yr)	0.0371	0.01830	0.0018	N/A	75	100
Concrete block partitions lightweight, exposed	WSF	Repointing joints (1.0 min every 25 yr)	0.0186	0.00915	0.0000	N/A	60	100
Structural clay facing tile partition, exposed	WSF	Repointing joints (1.0 min every 25 yr)	0.0186	0.00915	0.0000	N/A	60	100
Drywall partitions, metal or wood studs	WSF	Minor repair (2.0 min every 10 yr)	0.1114	0.05490	0.0037	N/A	25	100
Lath and plaster partitions, metal or wood studs	WSF	Minor repair (1.0 min every 10 yr)	0.0557	0.01830	0.0018	N/A	35	100
Glazed partitions, bank-height metal or wood framing	WSF	Window washing (0.18 min/quarter)	0.2600	0.05490	0.0183	N/A	30	100
	WSF	Repair glazing, frame & hardware (1.0 min. every 10 years)	0.0557	0.01830	0.0018			

0612 Demountable Partitions

Item Description	Unit of Measure	Maintenance Description	Labor	Material	Equipm't	Energy Demand (EU)	Replacement Life, yrs	Percent Replaced
Baked-enamel steel partitions, demountable, full or bank height	WSF	Damp-clean both sides (0.12 min/quarter)	0.1435	0.01767	0.0000	N/A	25	100
	WSF	Minor repair (1.0 min every 10 yr)	0.0359	0.01767	0.0018			
Vinyl-covered steel partitions, demountable, full or bank height	WSF	Damp-clean both sides (0.12 min/quarter)	0.1435	0.01767	0.0000	N/A	25	100
Gypsum plain-finish partitions, movable, full or bank height	WSF	Repainting, 2 coats (1.0 min every 6 yr)	0.0538	0.01767	0.0018	N/A	20	100
	WSF	Minor repair (1.0 min every 6 yr)	0.0538	0.01767	0.0018			
Gypsum prefinished painted partitions, movable, full or bank height	WSF	Repainting, 2 coats (1.0 min every 6 yr)	0.0538	0.01767	0.0018	N/A	20	100
	WSF	Minor repair (1.0 min every 6 yr)	0.0538	0.01767	0.0018			
Gypsum vinyl-covered partitions, movable, full or bank height	WSF	Damp-clean both sides (0.12 min/quarter)	0.1435	0.01767	0.0000	N/A	20	100
	WSF	Minor repair (1.0 min every 6 yr)	0.0538	0.01767	0.0018			
Gypsum plastic-laminated partitions, movable, full or bank height	WSF	Damp-clean both sides (0.12 min/quarter)	0.1435	0.01767	0.0000	N/A	20	100
	WSF	Minor repair (1.0 min every 6 yr)	0.0538	0.01767	0.0018			

0613 Retractable Partitions

Item Description	Unit of Measure	Maintenance Description	Labor	Material	Equipm't	Energy Demand (EU)	Replacement Life, yrs	Percent Replaced
Steel-lined vinyl folding partitions, manual	WSF	Damp-clean both sides (0.12 min/quarter)	0.1435	0.01767	0.0000	N/A	20	100
	WSF	Minor repair (1.0 min every 10 yr)	0.0359	0.01767	0.0018			
Vinyl-clad steel folding partitions, manual	WSF	Damp-clean both sides (0.12 min/quarter)	0.1435	0.01767	0.0000	N/A	25	100
	WSF	Minor repair (1.0 min every 10 yr)	0.0359	0.01767	0.0018			
Aluminum-faced folding partitions, manual	WSF	Damp-clean both sides (0.12 min/quarter)	0.1435	0.01767	0.0000	N/A	20	100
	WSF	Minor repair (1.0 min every 10 yr)	0.0359	0.01767	0.0018			
Enameled-steel folding partitions, manual	WSF	Damp-clean both sides (0.12 min/quarter)	0.1435	0.01767	0.0000	N/A	25	100
	WSF	Minor repair (1.0 min every 10 yr)	0.0359	0.01767	0.0018			
Hardwood veneer folding partitions, manual	WSF	Damp-clean both sides (0.12 min/quarter)	0.1435	0.01767	0.0000	N/A	25	100
	WSF	Minor repair (1.0 min every 10 yr)	0.0359	0.01767	0.0018			

Maintenance & Replacement Estimating Data - Architectural

Item Description	Unit of Measure	Maintenance Description	Maintenance Annual Cost, $			Energy Demand (EU)	Replacement Life, yrs	Percent Replaced
			Labor	Material	Equipm't			
Plastic-laminated folding partitions, manual	WSF	Damp-clean both sides (0.12 min/quarter)	0.1435	0.01767	0.0000	N/A	25	100
	WSF	Minor repair (1.0 min every 10 yr)	0.0359	0.01767	0.0018			
0614 Compartments and Cubicles								
Metal baked-enamel toilet partition: frame, door, and hardware	EA	Damp-wipe toilet partition and door; dust partition top (1.0 min/day)	77.6542	7.06720	0.8834	N/A	25	100
	EA	Minor repair (10.0 min every 5 yr)	0.5918	0.35336	0.1767			
Laminated-plastic toilet partition: frame, door, and hardware	EA	Damp-wipe toilet partition and door; dust partition tops (1.0 min/day)	77.6542	7.06720	0.8834	N/A	25	100
	EA	Minor repair (10.0 min every 5 yr)	0.5918	0.35336	0.0177			
Stainless steel toilet partition: frame, door, and hardware	EA	Damp-wipe toilet partition and door; dust partitions and top (1.0 min/day)	77.6542	7.06720	0.8834	N/A	35	100
	EA	Minor repair (10.0 min every 5 yr)	0.5918	7.06720	0.8834	N/A	35	100
Procelain enamel toilet partition: frame, door, and hardware	EA	Damp-wipe toilet partition and door: dust partition top (1.0 min/day)	77.6542	7.06720	0.8834	N/A	25	100
	EA	Minor repair (10.0 min every 5 yr)	0.5918	0.35336	0.1767			
Painted plywood toilet partition: frame, door, and hardware	EA	Damp-wipe toilet partition and door; dust partition top (1.0 min/day)	77.6542	7.06720	0.8834	N/A	20	100
	EA	Minor repair (20.0 min every 6 yr)	1.1836	0.88340	0.3534			
	EA	Repainting 2 coats (1.0 M every 6 yr)	3.5868	1.76680	0.3534			
Marble toilet partition: frame, door, and hardware	EA	Damp-wipe toilet partition and door: dust partition top (1.0 min/day)	77.6542	7.06720	0.8834	N/A	75	100
	EA	Minor repair (5.0 min every 10 yr)	0.1435	0.03534	0.0018			
0615 Interior Balustrades and Screens								
BALUSTRADES								
Steel railing and handrail, pipe or bar	100LF	Damp-clean (15.0 min/quarter)	17.9340	1.76680	0.1767	N/A	30	100
	100LF	Repainting (1.0 MH every 6 yr)	4.4835	1.76680	0.1767			
Aluminum railing and handrail	100LF	Damp-clean (15.0 min/quarter)	17.9340	1.76680	0.1767	N/A	40	100
						N/A		
Stainless steel railing and handrail	100LF	Damp-clean (15.0 min/quarter)	17.9340	1.76680	0.1767	N/A	50	100
Bronze railing and handrail	100LF	Damp-clean (15.0 min/quarter)	17.9340	1.76680	0.1767	N/A	50	100
Wood railing and handrail	100LF	Damp-clean (15.0 min/quarter)	17.9340	1.76680	0.1767	N/A	25	100
	100LF	Repainting (1.0 MH every 6 yr)	4.4835	1.76680	0.1767			
SCREENS								
Wood screen	WSF	Dust and clean both sides (0.12 min/quarter)	0.1435	0.01767	0.0000	N/A	25	100
	WSF	Repainting (1.0 min every 6 yr)	0.0538	0.01767	0.0018			
Concrete block	WSF	Dust and clean both sides (0.12 min/quarter)	0.1435	0.01767	0.0000	N/A	40	100

Maintenance & Replacement Estimating Data - Architectural

Item Description	Unit of Measure	Maintenance Description	Maintenance Annual Cost, $			Energy Demand (EU)	Replacement Life, yrs	Percent Replaced
			Labor	Material	Equipm't			
0616 Interior Doors and Frames								
Hollow metal door and frame, hardware	WSF	Damp-clean both sides (0.12 min/quarter)	0.1431	0.01697	0.0000	N/A	30	100
	WSF	Repair door, frame, hardware	0.1431	0.06787	0.0170			
	WSF	Painting - 2 coats (1.0 min every 6 yr)	0.0537	0.01697	0.0017			
Hollow-core wood door with metal frame, hardware	WSF	Damp-clean both sides (0.12 min/quarter)	0.1431	0.01697	0.0000	N/A	20	100
	WSF	Repair door, frame, hardware	17.8920	0.08484	0.0339			
	WSF	Painting - 2 coats (1.0 min every 6 yr)	0.0537	0.01697	0.0017			
Solid-core wood door with metal frame, hardware	WSF	Damp-clean both sides (0.12 min/quarter)	0.1431	0.01697	0.0000	N/A	30	100
	WSF	Repair door, frame, hardware	0.1431	0.06787	0.0170			
	WSF	Painting - 2 coats (1.0 min every 6 yr)	0.0537	0.01697	0.0017			
Hollow-core wood door with wood frame, hardware	WSF	Damp-clean both sides (0.12 min/quarter)	0.1431	0.01697	0.0000	N/A	20	100
	WSF	Repair door, frame, hardware	0.2147	0.10181	0.0339			
	WSF	Painting - 2 coats (1.0 min every 6 yr)	0.0537	0.01697	0.0017			
Solid-core wood door with wood frame, hardware	WSF	Damp-clean both sides (0.12 min/quarter)	0.1431	0.01697	0.0000	N/A	30	100
	WSF	Repair door, frame, hardware	17.8920	0.08484	0.0339			
	WSF	Painting - 2 coats (1.0 min every 6 yr)	0.0537	0.01697	0.0017			
Special (security) metal door	WSF	Damp-clean both sides (0.12 min/quarter)	0.1431	0.01697	0.0000	N/A	40	100
	WSF	Repair Door, frame, hardware	0.1431	0.01697	0.0170			
	WSF	Painting - 2 coats (1.0 min every 6 yr)	0.0537	0.01697	0.0017			
Plastic laminate wood door	WSF	Damp-clean both sides (0.12 min/quarter)	0.1431	0.01697	0.0000	N/A	25	100
	WSF	Repair door, frame, hardware	0.1431	0.06787	0.0339			
062 Interior Finishes								
0621 Wall Finishes								
Interior paint on masonry	WSF	High-use areas: paint - 2 coats (1.0 min every 2 yr)	0.2386	0.04885	0.0326	N/A	N/A	N/A
	WSF	Low-use areas: paint - 2 coats (1.0 min every 7 yr)	0.0734	0.01628	0.0163	N/A	N/A	N/A
Interior paint on plaster	WSF	High-use areas: paint - 2 coats (1.0 min every 2 yr)	0.2386	0.04885	0.0326	N/A	N/A	N/A
	WSF	Low-use areas: paint - 2 coats (1.0 min every 5 yr)	0.0918	0.03257	0.0163	N/A	N/A	N/A
Interior paint on drywall	WSF	Change of color: paint - 2 costs (1.0 min every 2 yr)	0.2386	0.04885	0.0326	N/A	N/A	N/A
	WSF	Otherwise: paint - 2 coats (1.0 min every 7 yr)	0.0734	0.01628	0.0163	N/A	N/A	N/A
Wall paper, light to medium weight	WSF	High-use areas: complete replacement (every 4 yr)	0.1101	0.09770	0.0163	N/A	N/A	N/A
	WSF	Low-use areas: complete replacement (every 7 yr)	0.0734	0.06513	0.0163	N/A	N/A	N/A
Vinyl, light to medium weight	WSF	High-use areas: damp-cleaning (0.06 min/day)	4.7728	0.16284	0.0326	N/A	12	100
	WSF	Low-use areas: Replacement only (12 yr)	0.0000	0.00000	0.0000			
Ceramic tile, glazed with organic adhesive	WSF	High-use areas: damp-cleaning daily (0.06 min/day)	4.7728	16.28350	0.0326	N/A	25	100

Maintenance & Replacement Estimating Data - Architectural

Item Description	Unit of Measure	Maintenance Description	Maintenance Annual Cost, $ Labor	Material	Equipm't	Energy Demand (EU)	Replacement Life, yrs	Percent Replaced
	WSF	Minor repair (yearly)	0.0184	0.01628	0.0016			
	WSF	Low-use areas: minor cleaning (0.06 min once a month)	18.3570	0.03257	0.0000			
	WSF	Minor repairs (yearly)	0.0184	0.01628	0.0000			
Ceramic mosaics, unglazed with organic adhesive	WSF	Average usage: damp-cleaning (0.06 min/wk)	0.2937	0.03257	0.0000	N/A	25	100
	WSF	Minor repairs (yearly)	0.0184	0.01628	0.0000			
Stone veneer	WSF	Reporting joints and minor repair (6.0 min every 20 yr)	0.1285	0.03257	0.0016	N/A	75	100
Wood veneer, stain or varnish	WSF	Paint - 2 coats (1.0 min every 7 yr)	0.0734	0.01628	0.0163	N/A	40	100
	WSF	Minor repairs (yearly)	0.0184	0.01628	0.0000	N/A	N/A	N/A
0622 Floor Finishes								
Oak parquet and block flooring, solid	SF	Sweep daily (0.01 min/day)	0.7343	0.00000	0.0163	N/A	35	100
	SF	Wax (0.2 min every 2 months)	0.3671	0.13027	0.0163			
	SF	Minor repairs (yearly)	0.0367	0.01628	0.0000			
Maple gym flooring	SF	Sweep daily (0.01 min/day)	0.7343	0.00000	0.0163	N/A	35	100
	SF	Wax (0.2 min every 2 months)	0.4589	0.13027	0.0163			
	SF	Minor repairs (yearly)	0.0367	0.01628	0.0000			
Resilient asphalt tile, 1/8 in thick	SF	Cleaning, waxing (0.2 min every month)	0.7343	17.91185	0.0326	N/A	15	100
Resilient vinyl tile, 1/8 in thick	SF	Cleaning, waxing (0.2 min every month)	0.7343	17.91185	0.0326	N/A	20	100
Vinyl asbestos tile, 1/8 in thick	SF	Cleaning, waxing (0.2 min every month)	0.7343	17.91185	0.0326	N/A	18	100
Carpeting, standard arylic or nylon	SF	Vacuuming, shampoo (0.02 min/ wk)	0.3121	0.03257	0.0163	N/A	12	100
Ceramic tile, glazed with trim, organic	SF	Mop weekly (0.02 min/wk)	0.2937	0.01628	0.0163	N/A	25	100
	SF	Minor repairs (yearly)	0.0367	0.03257	0.0000			
Ceramic mosaics, unglazed with organic adhesive	SF	Mop weekly (0.02 min/wk)	0.2937	0.01628	0.0163	N/A	25	100
	SF	Minor repairs (yearly)	0.0367	0.03257	0.0000			
Quarry tile 3/4-in portland cement bed	SF	Mop weekly (0.02 min/wk)	0.2937	0.01628	0.0163	N/A	30	100
	SF	Minor repairs (yearly)	0.0367	0.03257	0.0000			
Terrazzo, 2½-3 in thick	SF	Cleaning, waxing (0.2 min every month)	0.7343	0.17912	0.0326	N/A	50	100
Brick, unglazed pavers	SF	Mop weekly (0.02 min/wk)	0.2937	0.01628	0.0163	N/A	35	100
	SF	Minor repairs (yearly)	0.0367	0.17912	0.0326			
Raised access floor - plastic laminate 30 x 30-in panels	SF	Cleaning, waxing (0.2 min every month)	0.7343	0.17912	0.0326	N/A	25	100
Raised access floor - carpeted 24 x 24-in panels	SF	Vacuuming, shampoo (0.02 min/every wk)	0.3121	0.03257	0.0163	N/A	10	20
Raised access floor - carpeted 30 x 30-in panels	SF	Vacuuming, shampoo (0.02 min/wk)	0.3121	0.03257	0.0163	N/A	10	20
0623 Ceiling Finishes								
Acousical tile, concealed zee splines	SF	Cleaning and repair (0.1 min/yr)	0.0184	0.01628	0.0000	N/A	15	100
Acoustical tile, exposed 2 x 4-ft grid with hangers	SF	Cleaning and repair (0.05 min/yr)	0.0092	0.00814	0.0000	N/A	15	100
Acoustical tile, mineral fiber, 12 x 12-in	SF	Cleaning and repair (0.01 min/yr)	0.0018	0.00163	0.0000	N/A	17.5	100
Acoustical tile, mineral fiber lay-in panels with painted face, 24 x 24 -in	SF	Painting (1.0 min every 5 yr)	0.1101	0.01628	0.0163	N/A	20	100
	SF	Cleaning and repair (0.05 min/yr)	0.0092	0.00814	0.0000			
Acoustical tile, exposed 2 x 2-ft grid with hangers	SF	Cleaning and repair (0.05 min/yr)	0.0092	0.00814	0.0000	N/A	15	100
Gypsum board, painted	SF	Change of color: Painted - 2 coats (1.0 min every 2 yr)	0.2386	0.04885	0.0326	N/A	N/A	N/A
	SF	Useful life: Painting - 2 coats (1.0 min every 7 yr)	0.0734	0.01628	0.0163			

Maintenance & Replacement Estimating Data - Architectural/Mechanical

Item Description	Unit of Measure	Maintenance Description	Maintenance Annual Cost, $			Energy Demand (EU)	Replacement Life, yrs	Percent Replaced
			Labor	Material	Equipm't			
07 Conveying System (Also see Appendix C, "Elevator Selection: Rules of Thumb")								
0701 Elevators								
Passengers elevators - high speed, automatic (25 hp; 75% efficiency)	EA	Remove gum, sweep and damp-mop or vacuum carpet; damp-wipe walls, trim, and doors; wax or shampoo as necessary (10.0 min twice weekly)	641.28	27.99	5.60	25.0 kW	30	100
	Stop	Inspection and repair (35.0 MH/yr)	971.36	933.08	93.31			
Passenger elevators - hydraulic (25 hp; 75% efficiency)	EA	Remove gum, sweep, and damp-mop or vacuum carpet; damp-wipe walls, trim, and doors: wax or shampoo as necessary (20.0 min twice weekly)	641.28	27.99	5.60	25.0 kW	30	100
	Stop	Inspection and repair (14.0 MH/yr)	388.54	373.23	74.65			
Freight elevators - hydraulic (35 hp; 75% efficiency)	EA	Sweep floor, dust walls and doors (10.0 min twice weekly)	320.64	14.00	2.80	35.0 kW	30	100
	Stop	Inspection and repair (14.0 MH/yr)	388.54	373.23	74.65			
0702 Moving Stairs and Walks								
Single-width (32-in) escalator (7½ hp; 75% efficiency)	EA	Cleaning and damp-wipe steps, walls, and trim (30.0 min twice weekly)	962.10	65.32	9.33	7.5 kW	15	50
	EA	Preventative Inspection and general repair (148.0 MH/yr)	4,107.44	933.08	46.65			
Moving walk (4 ft 0 in wide) (4 hp; 75% efficiency)	100LF	Cleaning and damp-wipe walk, walls, and trim (0.1 min twice weekly)	320.64	18.66	1.87	3.0 kW	5	25
	EA	Preventative Inspection and general repair (75.0 MH/yr)	2,081.48	373.23	27.99			
0703 Dumbwaiters								
Hand-operated dumbwaiter, 1000 lb	EA	Damp-wipe walls, trim, and doors (5.0 min twice weekly)	160.41	18.66	3.73	N/A	20	50
	Stop	Inspection and repair (8.0 h/yr)	185.02	111.97	37.32			
Electric-operated dumbwaiter, 5000 lb (5 hp; 75% efficiency)	EA	Damp-wipe walls, trim, and doors (5.0 min twice weekly)	160.41	18.66	3.73	5.0 kW	20	50
	Stop	Inspection and repair (10.0 h/yr)	277.53	186.62	37.32			

Mechanical data

08 Mechanical
081 Plumbing
0811 Domestic Water Supply System

PIPE AND PIPE FITTING

Item Description	Unit of Measure	Maintenance Description	Labor	Material	Equipm't	Energy Demand (EU)	Replacement Life, yrs	Percent Replaced
Black steel pipe, schedule 40, ½-in to 8-in sizes	1000LF	Maintenance and repair as required (0.01 MH/yr)	0.2788	0.01734	0.0173	N/A	30	20
Copper, type K, including fittings and supports, 3/8-in to 3-in sizes	1000LF	Maintenance and repair as required (0.01 MH/yr)	0.2788	0.01734	0.0173	N/A	35	30
Copper, type L, including fittings and supports, 3/8-in to 3-in sizes	1000LF	Maintenance and repair as required (0.01 MH/yr)	0.2788	0.01734	0.0173	N/A	35	30

Maintenance & Replacement Estimating Data - Mechanical

Item Description	Unit of Measure	Maintenance Description	Maintenance Annual Cost, $			Energy Demand (EU)	Replacement Life, yrs	Percent Replaced
			Labor	Material	Equipm't			
PVC pipe, schedule 40 & 80, all sizes	1000LF	Maintenance and repair as required (0.01 MH/yr)	0.2788	0.01734	0.0173	N/A	35	30
VALVES								
Bronze gate valves, 3/8-in to 1-in sizes	EA	Maintenance and repair as required (0.1 MH/yr)	2.7884	0.86710	0.1734	N/A	15	50
Iron body, bronze mounted gate valves, 6-in size	EA	Maintenance and repair as required (2.0 MH/yr)	55.7670	6.93680	1.7342	N/A	15	50
Brass tee and lever handle type, ½ to 3/4-in sizes	EA	Maintenance and repair as required (0.1 MH/yr)	2.7884	0.43355	0.0867	N/A	15	100
Hose gate drain valves, bronze, 2-in size	EA	Maintenance and repair as required (0.5 MH/yr)	13.9418	1.73420	17.3420	N/A	15	100
SHOCK ABSORBERS								
Shock absorbers, 3/4 x 4 in long	EA	Maintenance and repair as required (0.01 MH/yr)	0.2788	0.01734	0.0000	N/A	25	100
WATER METERS								
Disk type water meters, 3/4-in to 2-in diameter sizes	EA	Maintenance and repair as required (0.1 MH/yr)	2.7884	86.71000	0.0867	N/A	25	100
INSULATION								
Piping insulation, ½-in to 2½-in sizes	1000LF	Maintenance and repair as required (0.1 MH/yr)	2.7884	17.34200	0.0000	N/A	15	75
CIRCULATING PUMPS (IN-LINE)								
Iron body circulation pump, 1/12 hp	EA	Preventative maintenance and repair (0.5 MH/yr)	13.9418	4.33550	0.8671	0.06 kW	15	50
Iron body circulation pump, 1/8 hp	EA	Preventative maintenance and repair (0.8 MH/yr)	18.5890	6.93680	1.3874	0.12 kW	15	50
Iron body circulating pump, ½ hp	EA	Preventative maintenance and repair (1.0 MH/yr)	22.3068	8.67100	1.7342	0.37 kW	15	50
DOMESTIC HOT WATER GENERATORS								
Gas-fired hot water generator, commercial, cement lined, 70% efficient, 500-gal/hr recovery rate	EA	Preventative maintenance and repair (20 MH/yr) Note: Water usage should be included in LCC	557.67	173.42	26.01	1070 Btu/gal (50 degree F incoming water)	20	100
Gas-fired hot water generator, commercial, cement lined, 75% efficient, 1000-gal/hr recovery rate	EA	Preventative maintenance and repair (30 MH/yr) Note: Water usage should be included in LCC	807.66	251.16	33.49	1070 Btu/gal (50 degree F incoming water)	20	100
Electric-heated hot water generator, residential, glass lined, 100% efficient, 8-120 gal/hr recovery rate	EA	Preventative maintenance and repair (0.5 MH/yr) Note: Water usage should be included in LCC	13.46	1.67	0.17	0.22 kW/gal (50 degree F incoming water)	20	100
0812 Sanitary Waste and Vent System								
PIPE & PIPE FITTINGS								
Cast iron soil pipe, extra heavy (bell), 2-in to 6-in sizes	1000LF	Maintenance and repair as required (0.1 MH/yr)	2.7884	1.67440	0.8372	N/A	40	100
Cast iron soil pipe, no hub, 1½-in to 2-in sizes	1000LF	Maintenance and repair as required (0.1 MH/yr)	2.7884	1.67440	0.8372	N/A	40	100
PVC pipe, schedule 40 & 80, all sizes	1000LF	Maintenance and repair as required (0.1 MH/yr)	2.7884	1.67440	0.8372	N/A	35	30

Maintenance & Replacement Estimating Data - Mechanical

Item Description	Unit of Measure	Maintenance Description	Maintenance Annual Cost, $			Energy Demand (EU)	Replacement Life, yrs	Percent Replaced
			Labor	Material	Equipm't			
FLOOR DRAINS								
Cast iron flat round-top floor drains, 3-in to 5-in outlet sizes	EA	Maintenance and repair as required (0.5 MH/yr)	13.4610	1.67440	0.8372	N/A	40	100
Cast iron flat square-top floor drains, 3-in to 5-in outlet sizes	EA	Maintenance and repair as required (0.5 MH/yr)	13.4610	1.67440	0.8372	N/A	40	100
Rough brass top funnel-type floor drains, 3-in to 4-in outlet sizes	EA	Maintenance and repair as required (0.5 MH/yr)	13.4610	1.67440	0.8372	N/A	40	100
Cast iron top floor drain with bucket, 3-in to 6-in outlet sizes	EA	Preventative maintenance and repair (2 MH/yr)	53.8440	3.34880	0.8372	N/A	40	100
0813 Rainwater Drainage System								
ROOF DRAINS								
Cast iron roof drains, domed top, 3-in to 6-in outlet sizes	EA	Preventative maintenance and repair (0.5 MH/yr)	13.4610	1.67440	0.8372	N/A	40	100
Cast iron roof drains, promenade top, 3-in to 6-in outlet sizes	EA	Preventative maintenance and repair (0.5 MH/yr)	13.4610	1.67440	0.8372	N/A	40	100
ROOF SCUPPERS								
Cast iron roof scuppers, grate, 2-in to 4-in outlet sizes	EA	Preventative maintenance and repair (0.5 MH/yr)	10.5765	1.31560	0.6578	N/A	40	100
AREA DRAINS								
Cast iron area drains, grate, 3-in throat size	EA	Preventative maintenance and repair (0.5 MH/yr)	10.5765	1.31560	0.6578	N/A	40	100
TRENCH DRAINS								
Trench drain, light duty, 2-in to 4-in outlet - 2 ft overall	EA	Preventative maintenance and repair (0.5 MH/yr)	10.5765	1.31560	0.6578	N/A	40	100
0814 Plumbing Fixtures								
WATER CLOSETS								
Floor-mounted water closets, washdown and siphon jet types	EA	Clean interior and exterior surfaces (2.5 min/day)	58.5233	6.57800	1.3156	N/A	35	100
	EA	Maintenance and repair as required (1.0 MH/yr) Note: Water usage should be included in LCC	21.1530	2.63120	0.6578			
Walll-mounted water closets, washdown and siphon jet types	EA	Clean interior and exterior surfaces (2.25 min/day)	58.9464	6.57800	1.3156	N/A	35	100
	EA	Maintenance and repair as required (1.0 MH/yr) Note: Water usage should be included in LCC	21.1530	2.63120	0.6578			
URINALS								
Pedestal-type urinals, washdown and siphon jet types	EA	Clean interior and exterior surfaces (2.25 min/day)	53.5876	6.57800	1.3156	N/A	35	100
	EA	Maintenance and repair as required (1.0 MH/yr) Note: Water usage should be included in LCC	21.1530	2.63120	0.6578			

Maintenance & Replacement Estimating Data - Mechanical

Item Description	Unit of Measure	Maintenance Description	Maintenance Annual Cost, $			Energy Demand (EU)	Replacement Life, yrs	Percent Replaced
			Labor	Material	Equipm't			
Wall-hung urinals, washdown, blowout, and siphon jet types	EA	Clean interior and exterior surfaces (2.0 min/day)	47.2417	5.92020	1.3156	N/A	35	100
	EA	Maintenance and repair as required (1.0 MH/yr) Note: Water usage should be included in LCC	21.1530	2.63120	0.6578			
Floor-mounted urinals, washdown and women's type	EA	Clean interior and exterior surfaces (2.5 min/day)	58.5233	6.57800	1.3156	N/A	35	100
	EA	Maintenance and repair as required (1.0 MH/yr) Note: Water usage should be included in LCC	21.1530	2.63120	0.6578			
LAVATORIES								
Vitreous china, wall-hung lavatory, 20 x 18-in size	EA	Clean interior and exterior surfaces (2.25 min/day)	53.5876	6.57800	1.3156	N/A	35	100
	EA	Maintenance and repair as required (1.0 MH/yr) Note: Water usage should be included in LCC	21.1530	2.63120	0.6578			
Iron enamel, wall-hung lavatory, 20 x 18-in size	EA	Clean interior and exterior surfaces (2.25 min/day)	53.5876	6.57800	1.3156	N/A	40	100
	EA	Maintenance and repair as required (1.0 MH/yr) Note: Water usage should be included in LCC	21.1530	2.63120	0.6578			
Enameled steel, wall-hung lavatory, 20 x 8-in size	EA	Clean interior and exterior surfaces (2.25 min/day)	53.5876	6.57800	1.3156	N/A	35	100
	EA	Maintenance and repair as required (1.0 MH/yr) Note: Water usage should be included in LCC	21.1530	2.63120	0.6578			
BATHTUBS								
Cast iron enamel bathtub, 5 ft 0 in recessed	EA	Clean interior and exterior surfaces (5.0 min/wk)	24.6785	2.63120	0.6578	N/A	40	100
	EA	Maintenance and repair as required (1.0 MH/yr) Note: Water usage should be included in LCC	21.1530	2.63120	0.6578			
Enameled steel bathtub, 5 ft 0 in recessed	EA	Clean interior and exterior surfaces (5.0 min/yr)	24.6785	2.63120	0.6578	N/A	40	100
	EA	Maintenance and repair as required (1.0 MH/yr) Note: Water usage should be included in LCC	21.1530	2.63120	0.6578			
SHOWERS								
Terrazzo shower receptor, 32 x 48 in	EA	Clean interior and exterior surfaces (2.25 min/day)	12.6918	1.31560	0.3947	N/A	50	100
	EA	Maintenance and repair as required (1.0 MH/yr) Note: Water usage should be included in LCC	21.1530	2.63120	0.6578			
Enameled steel shower receptor 32 x 48 in	EA	Clean interior and exterior surfaces (2.25 min/day)	12.6918	1.31560	0.3947	N/A	35	100

Maintenance & Replacement Estimating Data - Mechanical

Item Description	Unit of Measure	Maintenance Description	Maintenance Annual Cost, $			Energy Demand (EU)	Replacement Life, yrs	Percent Replaced
			Labor	Material	Equipm't			
	EA	Maintenance and repair as required (1.0 MH/yr) Note: Water usage should be included in LCC	21.1530	2.63120	0.6578			
Plastic shower receptor, 32 x 48 in	EA	Clean interior and exterior surfaces (2.25 min/day)	12.6918	1.31560	0.3947	N/A	20	100
	EA	Maintenance and repair as required (1.0 MH/yr) Note: Water usage should be included in LCC	21.1530	2.63120	0.6578			
Aluminum and glass shower, commercial grade	EA	Clean interior and exterior surfaces (2.25 min/day)	12.6918	1.31560	0.3947	N/A	25	100
	EA	Maintenance and repair as required (1.0 MH/yr) Note: Water usage should be included in LCC	21.1530	2.63120	0.6578			
WASH SINKS								
Iron enamel, highback, single sink, 24 x 48 in	EA	Clean interior and exterior surfaces (2.5 min/week)	12.6918	1.31560	0.6578	N/A	25	100
	EA	Maintenance and repair as required (1.0 MH/yr) Note: Water usage should be included in LCC	21.1530	2.63120	0.6578			
Enameled steel, highback, single sink, 24 x 48 in	EA	Clean interior and exterior surfaces (2.5 min/wk)	12.6918	1.19600	0.3588	N/A	35	100
	EA	Maintenance and repair as required (1.0 MH/yr) Note: Water usage should be included in LCC	21.1530	2.39200	0.5980			
Stainless steel, highback, single sink	EA	Clean interior and exterior surfaces (2.5 min/wk)	12.6918	1.31560	0.3947	N/A	40	100
	EA	Maintenance and repair as required (1.0 MH/yr) Note: Water usage should be included in LCC	21.1530	2.63120	0.6578			
Plastic, highback, single sink	EA	Clean interior and exterior surfaces (2.5 min/wk)	12.6918	1.31560	0.3947	N/A	15	100
	EA	Maintenance and repair as required (1.0 MH/yr) Note: Water usage should be included in LCC	21.1530	2.63120	0.6578			
DRINKING FOUNTAINS								
Stainless steel electric drinking fountain	EA	Clean all surfaces (0.5 min/day)	17.6275	1.31560	0.1316	0.01 kW	20	100
	EA	Maintenance and repair as required (2.0 MH/yr) Note: Water usage should be included in LCC	42.3060	6.57800	0.6578			

082 HVAC
0821 Energy Supply System

PIPE AND FITTINGS (up to 3 in)

Item Description	Unit of Measure	Maintenance Description	Labor	Material	Equipm't	Energy Demand (EU)	Replacement Life, yrs	Percent Replaced
Gas and oil	1000 LF	Preventative inspection and general maintenance (0.1 MH/yr)	0.2692	0.01674	0.0167	N/A	20	40

Maintenance & Replacement Estimating Data - Mechanical

Item Description	Unit of Measure	Maintenance Description	Maintenance Annual Cost, $			Energy Demand (EU)	Replacement Life, yrs	Percent Replaced
			Labor	Material	Equipm't			
VALVES AND COCKS (up to 3 in)								
Gas and oil	EA	Preventative inspection and general maintenance - repack glands, etc. (0.1 MH/yr)	2.6922	0.83720	0.1674	N/A	20	50
PIPING SPECIAL TEES AND ACCESSORIES (up to 3 in)								
Gas and oil	EA	Preventative inspection and general maintenance (0.1 MH/yr)	1.9230	0.59800	0.1196	N/A	20	50
METERS (up to 3 in)								
Gas	EA	Preventative inspection and general maintenance (0.1 MH/yr)	1.9230	0.59800	0.1196	N/A	20	20
Oil	EA	Preventative inspection and general maintenance (0.2 MH/yr)	3.8460	1.19600	0.1196	N/A	20	20
TANKS								
Oil storage	EA	Preventative inspection and general maintenance (0.2 MH/yr)	3.8460	1.19600	0.1196	N/A	20	100
PUMPS (up to 3 in)								
Oil	EA	Preventative inspection and general maintenance - repack glands, seals, etc. (1.0 MH/yr)	19.2300	5.98000	1.1960	3.7 kW	15	50
EQUIPMENT								
Gas compressor	EA	Preventative inspection and general maintenance (20 MH/yr)	384.6000	119.60000	17.9400	Per size	15	20
Oil preheater (steam)	EA	Preventative inspection and general maintenance (20 MH/yr)	384.6000	119.60000	17.9400	Per size	15	25
Preheater (electric)	EA	Preventative inspection and general maintenance (20 MH/yr)	192.3000	59.80000	8.9700	Per size	10	50
INSULATION	1000 LF	Preventative inspection and general maintenance - patch and replace (0.1 MH/yr)	1.9230	0.11960	0.0120	N/A	15	75
0822 Heat Generating System								
BOILERS (steam)								
Pakaged marine type (No. 2 oil)								
40 hp	EA	Preventative and insurance inspection and general maintenance (350 MH/yr)	6,410.0000	418.60000	119.6000	1.25 kW	35	100
50 hp	EA	Preventative and insurance inspection and general maintenance (350 MH/yr)	6,410.0000	478.40000	143.5200	1.6 kW	35	100

Maintenance & Replacement Estimating Data - Mechanical

Item Description	Unit of Measure	Maintenance Description	Maintenance Annual Cost, $			Energy Demand (EU)	Replacement Life, yrs	Percent Replaced
			Labor	Material	Equipm't			
150 hp	EA	Preventative and, insurance inspection and general maintenance (400 MH/yr)	7,692.0000	598.00000	179.4000	6.4 kW	35	100
700 hp	EA	Preventative and insurance inspection maintenance (500 MH/yr)	9,615.0000	1,794.00000	598.0000	28 kW	35	100
Packaged marine type (No. 6 oil)								
80 hp	EA	Preventative and insurance inspection and general maintenance (500 MH/yr)	9,615.0000	598.00000	179.4000	8.7 kW	35	100
250 hp	EA	Preventative and insurance inspection maintenance (650 MH/yr)	12,499.5000	1,196.00000	418.6000	14 kW	35	100
700 hp	EA	Preventative and insurance inspection and general maintenance (800 MH/yr)	15,384.0000	2,392.00000	837.2000	37 kW	35	100
Steel-fire box, base, jacket (light oil)								
25 hp	EA	Preventative and insurance inspection and general maintenance (350 MH/yr)	6,410.0000	239.20000	83.7200	0.25 kW	30	100
160 hp	EA	Preventative and insurance inspection and general maintenance (400 MH/yr)	7,692.0000	598.00000	179.4000	5.5 kW	30	100
540 hp	EA	Preventative and insurance inspection and general maintenance (500 MH/yr)	9,615.0000	598.00000	179.4000	19 kW	30	100
Steel-fire box, base, etc. packaged								
50 hp	EA	Preventative and insurance inspection and general maintenance (350 MH/yr)	6,410.0000	358.80000	119.6000	0.55 kW	30	100
170 hp	EA	Preventative and insurance inspection and general maintenance (500 MH/yr)	9,615.0000	598.00000	179.4000	5.5 kW	30	100
460 hp	EA	Preventative and insurance inspection and general maintenance (750 MH/yr)	14,422.5000	1,794.00000	598.0000	15 kW	30	100
Cast iron sectional with jacket, trim, and burner (No. 6 oil)								
60 hp	EA	Preventative and insurance inspection and general maintenance (350 MH/yr)	6,410.0000	239.20000	83.7200	Per size	30	100
120 hp	EA	Preventative and insurance inspection and general maintenance (3400 MH/yr)	7,692.0000	598.00000	179.4000	Per size	30	100
170 hp	EA	Preventative and insurance inspection and general maintenance (500 MH/yr)	9,615.0000	598.00000	239.2000	Per size	30	100
Cast iron sectional with jacket, trim, and burner (steam and gas)								
40 hp	EA	Preventative and insurance inspection and general maintenance (166 MH/yr)	9,615.0000	598.00000	239.2000	0.5 kW	35	100

Maintenance & Replacement Estimating Data - Mechanical

Item Description	Unit of Measure	Maintenance Description	Maintenance Annual Cost, $			Energy Demand (EU)	Replacement Life, yrs	Percent Replaced
			Labor	Material	Equipm't			
100 hp	EA	Preventative and insurance inspection and general maintenance (250 MH/yr)	4,807.5000	358.80000	119.6000	3 kW	35	100
170 hp	EA	Preventative and insurance inspection and general maintenance (300 MH/yr)	5,769.0000	478.40000	179.4000	5 kW	35	100
Hot water (gas)	EA	Preventative and insurance inspection and general maintenance (300 MH/yr)	5,769.0000	358.80000	119.6000	Per size	30	100
Hot water (electric)	EA	Preventative and insurance inspection and general maintenance (200 MH/yr)	3,846.0000	478.40000	179.4000	Per size	20	100
FURNACES								
Upflow, gas-fired								
80,000 Btu	EA	Preventative and insurance inspection and general maintenance (20 MH/yr)	384.6000	11.96000	4.7840	0.25 kW	20	100
105,000 Btu	EA	Preventative and insurance inspection and general maintenance (20 MH/yr)	384.6000	11.96000	4.7840	0.35 kW	20	100
Upflow, oil-fired								
85,000 Btu	EA	Preventative and insurance inspection and general maintenance (30 MH/yr)	576.9000	29.90000	9.5680	0.15 kW	20	100
100,000 Btu	EA	Preventative and insurance inspection and general maintenance (30 MH/yr)	576.9000	35.88000	11.9600	0.25 kW	20	100
125,000 Btu	EA	Preventative and insurance inspection and general maintenance (30 MH/yr)	576.9000	29.90000	14.3520	0.37 kW	20	100
Upflow, electric heat								
77,400 Btu, 22 kW, 240 V	EA	Preventative and insurance inspection and general maintenance (10 MH/yr)	192.3000	23.92000	8.3720	0.25 kW	15	100
96,200 Btu, 27.5 kW, 240 V	EA	Preventative and insurance inspection and general maintenance (10 MH/yr)	192.3000	23.92000	8.3720	0.37 kW	15	100
Counterflow, gas-fired								
120,000 Btu	EA	Preventative and insurance inspection and general maintenance (30 MH/yr)	673.0500	35.88000	11.9600	0.37 kW	20	100
140,000 Btu	EA	Preventative and insurance inspection and general maintenance (35 MH/yr)						
Horizontal flow, gas-fired								
105,000 Btu	EA	Preventative and insurance inspection and general maintenance (35 MH/yr)	673.0500	29.90000	9.5680	0.25 kW	20	100
125,000 Btu	EA	Preventative and insurance inspection and general maintenance (35 MH/yr)	673.0500	35.88000	11.9600	0.37 kW	20	100

Maintenance & Replacement Estimating Data - Mechanical

Item Description	Unit of Measure	Maintenance Description	Maintenance Annual Cost, $			Energy Demand (EU)	Replacement Life, yrs	Percent Replaced
			Labor	Material	Equipm't			
PUMPS								
Horizontal split case type								
3 x 2½ in, 1½ hp	EA	Preventative and insurance inspection and general maintenance - rings, bearing, etc. (20 MH/yr)	384.6000	19.13600	5.9800	1.1 kW	20	100
5 x 4 in, 20 hp	EA	Preventative and insurance inspection and general maintenance - rings, bearings, etc. (40 MH/yr)	769.2000	179.40000	59.8000	15 kW	20	100
10 x 8 in, 125 hp	EA	Preventative and insurance inspection and general maintenance - rings, bearing, etc. (100 MH/yr)	1,923.0000	358.80000	119.6000	93 kW	20	100
End suction type								
1½ x 3¼ in, 3/4 hp	EA	Preventative and insurance inspection and general maintenance - rings, bearings, etc. (10 MH/yr)	153.8400	59.80000	17.9400	0.56 kW	15	100
3 x 2½ in, 1½ hp	EA	Preventative and insurance inspection and general maintenance - rings, bearings etc. (15 MH/yr)	205.1200	89.70000	29.9000	1.1 kW	15	100
5 x 4 in, 7½ hp	EA	Preventative and insurance inspection and general maintenance - rings, bearings, etc. (30 MH/yr)	576.9000	119.60000	59.8000	5.6 kW	15	100
AIR CONTROL								
Boiler fittings	EA	Preventative and insurance inspection and general maintenance (30 MH/yr)	1,153.8000	29.90000	9.5680	Per size	20	100
Air separators								
56 gal/min (2 in screwed)	EA	Preventative and insurance inspection and general maintenance (40 MH/yr)	1,115.3400	86.71000	26.0130	Per size	20	100
300 gal/min (4 in flanged)	EA	Preventative and insurance inspection and general maintenance (50 MH/yr)	1,394.1750	104.05200	34.6840	Per size	20	100
2,950 gal/min (12 in flanged)	EA	Preventative and insurance inspection and general maintenance (75 MH/yr)	2,091.2625	416.20800	138.7360	Per size	20	100
Compression tank and fittings								
15 gal (13-in diameter x 34½ in)	Set	Preventative and insurance inspection and general maintenance (50 MH/yr)	1,394.1750	8.67100	3.4684	Per size	20	100
180 gal (30-in diameter x 70 in)	Set	Preventative and insurance inspection and general maintenance (60 MH/yr)	1,673.0100	43.35500	13.8736	Per size	20	100
280 gal (36-in diameter x 74 in)	Set	Preventative and insurance inspection and general maintenance (70 MH/yr)	1,951.8450	52.02600	17.3420	Per size	20	100

Maintenance & Replacement Estimating Data - Mechanical

Item Description	Unit of Measure	Maintenance Description	Maintenance Annual Cost, $			Energy Demand (EU)	Replacement Life, yrs	Percent Replaced
			Labor	Material	Equipm't			
AUXILIARY EQUIPMENT (up to 500 hp)								
Feed water treatment	Set	Preventative and insurance inspection and general maintenance (300 MH/yr)	8,365.0500	520.26000	173.4200	Per size	15	100
Deaerators	EA	Preventative and insurance inspection and general maintenance (300 MH/yr)	8,365.0500	346.84000	121.3940	Per size	20	100
Breeching	Set	Preventative and insurance inspection and general maintenance (50 MH/yr)	1,394.1750	173.42000	60.6970	Per size	20	100
Flues	Set	Preventative and insurance inspection and general maintenance (50 MH/yr)	1,394.1750	173.42000	60.6970	Per size	20	100
Draft control	Set	Preventative and insurance inspection and general maintenance (100 MH/yr)	2,788.3500	173.42000	60.6970	Per size	15	100
HEAT EXCHANGERS								
DX	EA	Preventative and insurance inspection and general maintenance (50 MH/yr)	961.5000	119.60000	41.8600	Per size	20	100
Water or steam	EA	Preventative and insurance inspection and general maintenance (50 MH/yr)	961.5000	119.60000	41.8600	Per size	20	100
HEAT RECOVERY								
All types	EA	Preventative and insurance inspection and general maintenance (50 MH/year)	961.5000	119.60000	41.8600	Per size	15	100
EQUIPMENT INSULATION								
All types	SF	Preventative and insurance inspection and general maintenance - patch and repair (0.01 MH/yr)	0.1923	0.01196	0.0120	Per size	15	100
0823 Cooling Generating System								
WATER CHILLERS								
Chiller-Air Cool, Reciprocating								
5-ton capacity	EA	Preventative and insurance inspection and general maintenance (15 MH/yr)	288.45	80.73	26.91	Per size	20	100
10-ton capacity	EA	Preventative and insurance inspection and general maintenance (20 MH/yr)	384.60	107.64	35.88	Per size	20	100
15-ton capacity	EA	Preventative and insurance inspection and general maintenance (25 MH/yr)	480.75	134.55	44.85	Per size	20	100
20-ton capacity	EA	Preventative and insurance inspection and general maintenance (30 MH/yr)	576.90	161.46	53.82	Per size	20	100
50-ton capacity	EA	Preventative and insurance inspection and general maintenance (50 MH/yr)	961.50	269.10	89.70	Per size	20	100

Maintenance & Replacement Estimating Data - Mechanical

Item Description	Unit of Measure	Maintenance Description	Maintenance Annual Cost, $			Energy Demand (EU)	Replacement Life, yrs	Percent Replaced
			Labor	Material	Equipm't			
100-ton capacity	EA	Preventative and insurance inspection and general maintenance (75 MH/yr)	1,442.25	403.65	134.55	Per size	20	100
Chiller-Water Cool, Reciprocating								
10-ton capacity	EA	Preventative and insurance inspection and general maintenance (25 MH/yr)	480.75	134.55	44.85	Per size	25	100
20-ton capacity	EA	Preventative and insurance inspection and general maintenance (30 MH/yr)	576.90	161.46	53.82	Per size	25	100
50-ton capacity	EA	Preventative and insurance inspection and general maintenance (45 MH/yr)	865.35	242.19	80.73	Per size	25	100
100-ton capacity	EA	Preventative and insurance inspection and general maintenance (60 MH/yr)	1,153.80	322.92	107.64	Per size	25	100
200-ton capacity	EA	Preventative and insurance inspection and general maintenance (75 MH/yr)	1,442.25	403.65	134.55	Per size	25	100
Chiller-Hermetic, Centrifugal								
100-ton capacity	EA	Preventative and insurance inspection and general maintenance (100 MH/yr)	1,923.00	538.20	179.40	Per size	25	100
300-ton capacity	EA	Preventative and insurance inspection and general maintenance (120 MH/yr)	2,307.60	645.84	215.28	Per size	25	100
900-ton capacity	EA	Preventative and insurance inspection and general maintenance (150 MH/yr)	2,884.50	807.30	269.10	Per size	25	100
Chiller-Open, Centrifugal								
300-ton capacity	EA	Preventative and insurance inspection and general maintenance (120 MH/yr)	2,307.60	645.84	215.28	Per size	25	100
900-ton capacity	EA	Preventative and insurance inspection and general maintenance (200 MH/yr)	3,846.00	1,076.40	358.80	Per size	25	100
Chiller-Double Bundle, Hermetic								
100-ton capacity	EA	Preventative and insurance inspection and general maintenance (110 MH/yr)	2,115.30	592.02	197.34	Per size	25	100
300-ton capacity	EA	Preventative and insurance inspection and general maintenance (120 MH/yr)	2,307.60	645.84	215.28	Per size	25	100
900-ton capacity	EA	Preventative and insurance inspection and general maintenance (160 MH/yr)	3,076.80	861.12	287.04	Per size	25	100
Chiller-One Stage Absorption								
100-ton capacity	EA	Preventative and insurance inspection and general maintenance (50 MH/yr)	961.50	269.10	89.70	Per size	25	100
300-ton capacity	EA	Preventative and insurance inspection and general maintenance (55 MH/yr)	1,057.65	296.01	98.67	Per size	25	100

Maintenance & Replacement Estimating Data - Mechanical

Item Description	Unit of Measure	Maintenance Description	Maintenance Annual Cost, $			Energy Demand (EU)	Replacement Life, yrs	Percent Replaced
			Labor	Material	Equipm't			
900-ton capacity	EA	Preventative and insurance inspection and general maintenance (65 MH/yr)	1,249.95	349.83	116.61	Per size	25	100
Chiller-Two Stage Absorption								
300-ton capacity	EA	Preventative and insurance inspection and general maintenance (55 MH/yr)	1,057.65	296.01	98.67	Per size	25	100
900-ton capacity	EA	Preventative and insurance inspection and general maintenance (65 MH/yr)	1,249.95	349.83	116.61	Per size	25	100
HEAT PUMPS								
Single package, air to air								
24,000 Btu	EA	Preventative and insurance inspection and general maintenance (4 MH/yr)	76.92	4.78	2.39	Per size	10	100
48,000 Btu	EA	Preventative and insurance inspection and general maintenance (10 MH/yr)	192.30	23.92	8.37	Per size	15	100
60,000 Btu	EA	Preventative and insurance inspection and general maintenance (10 MH/yr)	192.30	23.92	8.37	per size	15	100
Split system, air to air								
36,000 Btu outdoor	EA	Preventative and insurance inspection and general maintenance (10 MH/yr)	115.38	7.18	2.39	Per size	15	100
36,000 Btu indoor	EA	Preventative and insurance inspection and general maintenance (5 MH/yr)	96.15	11.96	4.78	Per size	15	100
COOLING TOWERS								
Packaged centrifugal blowthrough								
200-ton capacity	EA	Preventative and insurance inspection and general maintenance (70 MH/yr)	1,346.10	526.24	167.44	Per size	15	100
Packaged axial flow								
200-ton capacity	EA	Preventative and insurance inspection and general maintenance (70 MH/yr)	1,346.10	717.60	179.40	Per size	15	100
Packaged draw through								
150 ton capacity	EA	Preventative and insurance inspection and general Maintenance (70 MH/yr)	1,346.10	669.76	179.40	Per size	15	100
Ejector type								
250-ton capacity	EA	Preventative and insurance inspection and general maintenance (50 MH/yr)	384.60	92.09	47.84	Per size	20	100
CONDENSERS								
Water-cooled condenser								
30-ton capacity	EA	Preventative and insurance inspection and general maintenance (40 MH/yr)	769.20	184.18	95.68	Per size	20	100

Maintenance & Replacement Estimating Data - Mechanical

Item Description	Unit of Measure	Maintenance Description	Maintenance Annual Cost, $			Energy Demand (EU)	Replacement Life, yrs	Percent Replaced
			Labor	Material	Equipm't			
Air-cooled condenser								
100-ton capacity	EA	Preventative and insurance inspection and general maintenance (20 MH/yr)	384.60	239.20	71.76	Per size	20	100
Evaporative condenser								
10-ton capacity	EA	Preventative and insurance inspection and general maintenance (10 MH/yr)	192.30	26.31	11.96	Per size	20	100
PIPE AND FITTINGS	1000 LF	Preventative and insurance inspection and general maintenance (0.01 MH/yr)	0.1923	0.01196	0.0120	N/A	20	40
DIVERTING VALVES	EA	Preventative and insurance inspection and general maintenance (0.1 MH/yr)	1.9230	0.59800	0.1196	N/A	15	100
FREEZE PROTECTION	LS	Preventative and insurance inspection and general maintenance (10 MH/yr)	192.30	119.60	59.80	N/A	15	100
PUMPS								
Horizontal split case type								
3 x 2½ in, 1½ hp	EA	Preventative and insurance inspection and general maintenance (20 MH/yr)	384.60	19.14	5.98	1.1 kW	20	100
5 x 4 in, 20 hp	EA	Preventative and insurance inspection and general maintenance (40 MH/yr)	769.20	179.40	59.80	15 kW	20	100
10 x 8 in, 125 hp	EA	Preventative and insurance inspection and general maintenance (100 MH/yr)	1,923.00	358.80	119.60	93 kW	20	100
End suction								
1½ x 1¼ in 3/4 hp	EA	Preventative and insurance inspection and general maintenance (10 MH/yr)	153.84	59.80	17.94	0.56 kW	15	100
3 x 2½ in, 1½ hp	EA	Preventative and insurance inspection and general maintenance (15 MH/yr)	205.12	89.70	29.90	1.1 kW	15	100
5 x 4 in, 7½ hp	EA	Preventative and insurance inspection and general maintenance (30 MH/yr)	576.90	119.60	59.80	5.6 kW	15	100
DIRECT EXPANSION SYSTEM								
Refrigerant circulation system	LS	Preventative and insurance inspection and general maintenance (25 MH/yr)	480.75	119.60	119.60	N/A	30	100
Pipe and fittings	LS	Preventative and insurance inspection and general maintenance (50 MH/yr)	961.50	239.20	59.80	N/A	30	100
Accessories	LS	Preventative and insurance inspection and general maintenance (25 MH/yr)	480.75	239.20	59.80	N/A	25	100

Maintenance & Replacement Estimating Data - Mechanical

Item Description	Unit of Measure	Maintenance Description	Maintenance Annual Cost, $			Energy Demand (EU)	Replacement Life, yrs	Percent Replaced
			Labor	Material	Equipm't			
INSULATION (piping and equipment)	1000 LF	Preventative and insurance inspection and general maintenance (0.1 MH/yr)	1.9230	0.11960	0.0120	N/A	15	75
0824 Distribution Systems								
PIPE AND FITTINGS	1000 LF	Preventative and insurance inspection and general maintenance (0.01 MH/yr)	0.1923	0.01196	0.0120	N/A	20	40
VALVES								
Gate	EA	Preventative and insurance inspection and general maintenance (1.0 MH/yr)	19.2300	2.39200	2.3920	N/A	15	50
Butterfly	EA	Preventative and insurance inspection and general maintenance (0.1 MH/yr)	1.9230	0.11960	0.0120	N/A	15	50
Plug	EA	Preventative and insurance inspection and general maintenance (0.1 MH/yr)	1.9230	0.11960	0.0120	N/A	15	50
OS&Y (outside screw and yoke)	EA	Preventative and insurance inspection and general maintenance (0.2 MH/yr)	1.9230	0.11960	0.0239	N/A	15	50
Check	EA	Preventative and insurance inspection and general maintenance (0.2 MH/yr)	3.8460	0.23920	0.0239	N/A	15	50
Piping Special Tees and accessories	EA	Preventative and insurance inspection and general maintenance (0.2 MH/yr)	3.8460	0.23920	0.0239	N/A	10	50
PUMPS								
Horizontal split case type								
2-in to 3-in size up to 1½ hp	EA	Preventative and insurance inspection and general maintenance (20 MH/yr)	384.60	19.14	5.98	1.1 kW	20	100
4-in to 5-in size up to 20 hp	EA	Preventative and insurance inspection and general maintenance (40 MH/yr)	769.20	179.40	59.80	15 kW	20	100
8-in to 10-in size up to 125 hp	EA	Preventative and insurance inspection and general maintenance (100 MH/yr)	1,923.00	358.80	119.60	93 kW	20	100
End suction								
1¼-in to 1½-in size up to 3/4 hp	EA	Preventative and insurance inspection and general maintenance (10 MH/yr)	153.84	59.80	17.94	0.56 kW	15	100
2-in to 3-in size up to 1½ hp	EA	Preventative and insurance inspection and general maintenance (15 MH/yr)	205.12	89.70	29.90	1.1 kW	15	100
4-in to 5-in size up to 7½ hp	EA	Preventative and insurance inspection and general maintenance (30 MH/yr)	576.90	119.60	59.80	5.6 kW	15	100

Maintenance & Replacement Estimating Data - Mechanical

Item Description	Unit of Measure	Maintenance Description	Maintenance Annual Cost, $			Energy Demand (EU)	Replacement Life, yrs	Percent Replaced
			Labor	Material	Equipm't			
DISTRIBUTION SYSTEMS								
Steam	1000 LF	Preventative and insurance inspection and general maintenance (0.02 MH/yr)	0.3846	0.02392	0.0239	N/A	20	50
Glycol	1000 LF	Preventative and insurance inspection and general maintenance (0.02 MH/yr)	0.3846	0.02392	0.0239	N/A	20	50
Other liquid	1000 LF	Preventative and insurance inspection and general maintenance (0.02 MH/yr)	0.1923	0.01196	0.0120	N/A	20	50
AIR-HANDLING EQUIPMENT								
Single zone with mixing box HW coil, CW coil, flat filter								
1750-2750 cfm	EA	Preventative and insurance inspection and general maintenance (20 MH/yr)	384.6000	107.64000	35.8800	0.75 kW	20	100
Single zone with mixing box HW coil, CW coil, manual roll filter								
1750-2750 cfm	EA	Preventative and insurance inspection and general maintenance (25 MH/yr)	480.7500	197.34000	47.8400	0.75 kW	20	100
Single zone with mixing box, HW coil, CW coil, auto roll filter								
1750-2750 cfm	EA	Preventative and insurance inspection and general maintenance (40 MH/yr)	769.2000	239.20000	59.8000	Per size	20	100
Single zone with mixing box HW coil, DX coil, flat filter								
1750-2750 cfm	EA	Preventative and insurance inspection and general maintenance (25 MH/yr)	480.7500	215.28000	41.8600	0.75 kW	20	100
Roof top unit								
1750-2750 cfm	EA	Preventative and insurance inspection and general maintenance (30 MH/yr)	576.9000	263.12000	59.8000	0.75 kW	15	100
Single zone with mixing box, HW coil, DX coil, roll filter								
1750-2750 cfm	EA	Preventative and insurance inspection and general maintenance (32 MN/yr)	615.3600	167.44000	35.8800	0.75 kW	20	100
Four zone with mixing box, dampers, HW coil, roll filter								
1750-2750 cfm	EA	Preventative and insurance inspection and general maintenane (37 MH/yr)	711.5100	221.26000	41.8600	0.75 kW	20	100
Four zone with mixing box dampers, HW coil, CW coil, auto roll filter								
1750-2750 cfm	EA	Preventative and insurance inspection and general maintenance (42 MH/yr)	807.6600	263.12000	59.8000	0.75 kW	20	100

Maintenance & Replacement Estimating Data - Mechanical

Item Description	Unit of Measure	Maintenance Description	Maintenance Annual Cost, $			Energy Demand (EU)	Replacement Life, yrs	Percent Replaced
			Labor	Material	Equipm't			
Air tempering (packaged)								
24,000 Btu	EA	Preventative and insurance inspection and general maintenance (40 MH/yr)	769.2000	29.90000	59.8000	Per size	15	100
Air tempering (split system) outdoor section								
24,000 Btu	EA	Preventative and insurance inspection and general maintenance (40 MH/yr)	769.2000	35.88000	59.8000	Per size	15	100
60,000 Btu	EA	Preventative and insurance inspection and general maintenance (60 MH/yr)	1,153.8000	83.72000	71.7600	Per size	15	100
Air tempering (split system) indoor section								
24,000 Btu	EA	Preventative and insurance inspection and general maintenance (30 MH/yr)	576.9000	23.92000	41.8600	Per size	20	100
60,000 Btu	EA	Preventative and insurance inspection and general maintenance (40 MH/yr)	769.2000	23.92000	59.8000	Per size	20	100
AIR TEMPERING (incremental)								
Through wall type, fixed								
11,700 Btu cool 13,300 Btu Heat (HW)	EA	Preventative and insurance inspection and general maintenance (8 MH/yr)	153.8400	5.98000	5.9800	Per size	10	100
Through wall type								
11,700 Btu Cool 15,300 Btu Heat (elect)	EA	Preventative and insurance inspection and general maintenance (8 MH/yr)	153.8400	4.78400	5.9800	Per size	10	100
Through wall type, remove chassis								
14,600 Btu	EA	Preventative and insurance inspection and general maintenance (2 MH/yr)	38.4600	2.39200	3.5880	Per size	10	100
DUCTWORK								
Round								
Low pressure	1000 LF	Preventative and insurance inspection and general maintenance (1 MH/yr)	19.2300	239.20000	0.0000	N/A	Life	100
Medium pressure	1000 LF	Preventative and insurance inspection and general maintenance (1 MH/yr)	19.2300	269.10000	0.0000	N/A	Life	100
High pressure	1000 LF	Preventative and insurance inspection and general maintenance (1 MH/yr)	19.2300	358.80000	0.0000	N/A	Life	100
Rectangular								
Low pressure	1000 LF	Preventative and insurance inspection and general maintenance (1.5 MH/yr)	28.8450	239.20000	0.0000	N/A	Life	100

Maintenance & Replacement Estimating Data - Mechanical

Item Description	Unit of Measure	Maintenance Description	Maintenance Annual Cost, $			Energy Demand (EU)	Replacement Life, yrs	Percent Replaced
			Labor	Material	Equipm't			
Medium pressure	1000 LF	Preventative and insurance inspection and general maintenance (1.5 MH/yr)	28.8450	269.10000	0.0000	N/A	Life	100
High pressure	1000 LF	Preventative and insurance inspection and general maintenance (1.5 MH/yr)	28.8450	358.80000	0.0000	N/A	Life	100
Plenums	SF	Preventative and insurance inspection and general maintenance (0.01 MH/yr)	0.1923	0.01196	0.0120	N/A	Life	100
REGISTERS AND GRILLS								
12 x 8 in	EA	Preventative and insurance inspection and general maintenance (0.1 MH/yr)	1.9230	0.00000	0.0000	N/A	25	100
DIFFUSERS								
8-in neck	EA	Preventative and insurance inspection and general maintenance (0.1 MH/yr)	1.9230	0.00000	0.0000	N/A	25	100
DAMPERS								
8-in round	EA	Preventative and insurance inspection and general maintenance (0.1 MH/yr)	1.9230	0.00000	0.0000	N/A	25	100
Troffers (Light fixture type)	EA	Preventative and insurance inspection and general maintenance (0.1 MH/yr)	1.9230	1.79400	0.5980	N/A	20	100
Air Treatment Equipment	EA	Preventative and insurance inspection and general maintenance (50 MH/yr)	961.5000	119.60000	11.9600	N/A	20	100
Heat Recovery Equipment	EA	Preventative and insurance inspection and general maintenance (50 MH/yr)	961.5000	119.60000	11.9600	N/A	15	100
Anti-Vibration Equipment	Set	Preventative and insurance inspection and general maintenance (5 MH/yr)	96.1500	59.80000	17.9400	N/A	15	100
Insulation								
Cooling (vapor barrier)	SF	Preventative and insurance inspection and general maintenance (0.01 MH/yr)	0.1923	0.01196	0.0120	N/A	15	75
Heating	SF	Preventative and insurance inspection and general maintenance (0.1 MH/yr)	0.1923	0.01196	0.0120	N/A	15	75
EXHAUST FANS								
Direct drive, ¼ hp	EA	Preventative and insurance inspection and general maintenance (10 MH/yr)	192.3000	17.94000	23.9200	Per size	20	100
Belt drive, ½ hp and over	EA	Preventative and insurance inspection and general maintenance (15 MH/yr)	365.3700	23.92000	35.8800	Per size	20	100

Maintenance & Replacement Estimating Data - Mechanical

Item Description	Unit of Measure	Maintenance Description	Maintenance Annual Cost, $			Energy Demand (EU)	Replacement Life, yrs	Percent Replaced
			Labor	Material	Equipm't			
ventilators	EA	Preventative and insurance inspection and general maintenance (4 MH/yr)	76.9200	11.96000	11.9600	Per size	15	100
makeup air unit	EA	Preventative and insurance inspection and general maintenance (20 MH/yr)	384.6000	59.80000	35.8800	Per size	20	100
Inside/Outside	EA	Preventative and insurance inspection and general maintenance (30 MH/yr)	576.9000	119.60000	47.8400	Per size	15	100

0825 Terminal and Package Units

BASEBOARD HEATING UNITS (hot water)

Radiant, cast iron panel

Item Description	Unit of Measure	Maintenance Description	Labor	Material	Equipm't	Energy Demand (EU)	Replacement Life, yrs	Percent Replaced
7 ¼ in high	EA	Preventative and insurance inspection and general maintenance; assume 10 PSI (1 MH/yr)	19.2300	1.19600	1.1960	Per size	30	100

Nonferrous element

Item Description	Unit of Measure	Maintenance Description	Labor	Material	Equipm't	Energy Demand (EU)	Replacement Life, yrs	Percent Replaced
4 in deep x 36 in long	EA	Preventative and insurance inspection and general maintenance (6 MH/yr)	115.3800	4.78400	3.5880	Per size	25	100

CONVECTOR HEATING UNITS

Baseboard panel with 9 1/16-in -high enclosure

Item Description	Unit of Measure	Maintenance Description	Labor	Material	Equipm't	Energy Demand (EU)	Replacement Life, yrs	Percent Replaced
1-in tube	EA	Preventative and insurance inspection and general maintenance (2 MH/yr)	38.4600	2.39200	2.3920	Per size	20	100

Free standing or semi-recessed

Item Description	Unit of Measure	Maintenance Description	Labor	Material	Equipm't	Energy Demand (EU)	Replacement Life, yrs	Percent Replaced
24 in high x 36 in long	EA	Preventative and insurance inspection and general maintenance (1 MH/yr)	19.2300	1.19600	1.1960	Per size	25	100

INDUCTION UNIT WITH CABINET

Item Description	Unit of Measure	Maintenance Description	Labor	Material	Equipm't	Energy Demand (EU)	Replacement Life, yrs	Percent Replaced
90 to 510 cfm	EA	Preventative and insurance inspection and general maintenance (10 MH/yr)	192.3000	14.35200	2.3920	Per size	20	100

FAN COIL UNITS WITH CABINETS

Item Description	Unit of Measure	Maintenance Description	Labor	Material	Equipm't	Energy Demand (EU)	Replacement Life, yrs	Percent Replaced
155 to 215 cfm	EA	Preventative and insurance inspection and general maintenance (10 MH/yr)	192.3000	87.30800	8.3720	Per size	20	100

RADIATORS

Cast iron, free standing

Item Description	Unit of Measure	Maintenance Description	Labor	Material	Equipm't	Energy Demand (EU)	Replacement Life, yrs	Percent Replaced
Six tube, 32 in high	EA	Preventative and insurance inspection and general maintenance (1½ MH/yr)	29.4860	1.19600	1.1960	Per size	40	100
Five tube, 22 in high	EA	Preventative and insurance inspection and general maintenance (1½ MH/yr)	29.4860	1.19600	1.1960	Per size	40	100

Maintenance & Replacement Estimating Data - Mechanical

Item Description	Unit of Measure	Maintenance Description	Maintenance Annual Cost, $ Labor	Material	Equipm't	Energy Demand (EU)	Replacement Life, yrs	Percent Replaced
Three tube, 25 in high	EA	Preventative and insurance inspection and general maintenance (1 MH/yr)	19.2300	1.19600	1.1960	Per size	40	100
FINNED TUBE ELEMENTS								
Copper fin-tube								
48 fins/ft	100 LF	Preventative and insurance inspection and general maintenance (2 MH/yr)	38.4600	1.19600	1.1960	Per size	35	100
Steel Fin-Tube								
40 fins/ft	100 LF	Preventative and insurance inspection and general maintenance (2 MH/yr)	38.4600	1.19600	1.1960	Per size	35	100
DUCT-MOUNTED COIL SECTIONS								
Duct-mounted coil sections, steam	EA	Preventative and insurance inspection and general maintenance (10 MH/yr)	192.3000	29.90000	5.9800	Per size	20	100
Duct-mounted coil sections, hot water	EA	Preventative and insurance inspection and general maintenance (10 MH/yr)	192.3000	29.90000	5.9800	Per size	20	100
Duct-mounted coil sections, electric	EA	Preventative and insurance inspection and general maintenance (5 MH/yr)	96.1500	11.96000	2.3920	Per size	15	100
RADIANT HEATING UNITS								
Radiant heating units, electric	EA	Preventative and insurance inspection and general maintenance (3 MH/yr)	57.6900	11.96000	2.3920	Per size	15	100
Radiant heating units, hot water	EA	Preventative and insurance inspection and general maintenance (5 MH/yr)	96.1500	11.96000	2.3920	Per size	25	100
UNIT HEATERS								
Unit heaters, gas	EA	Preventative and insurance inspection and general maintenance (5 MH/yr)	96.1500	17.94000	3.5880	Per size	15	100
Unit heaters, electric	EA	Preventative and insurance inspection and general maintenance (3 MH/yr)	57.6900	11.96000	2.3920	Per size	15	100
Unit heaters, hot water	EA	Preventative and insurance inspection and general maintenance (5 MH/yr)	96.1500	11.96000	2.3920	Per size	20	100
Unit heaters, steam	EA	Preventative and insurance inspection and general maintenance (10 MH/yr)	192.3000	23.92000	4.7840	Per size	20	100
Space heater, steam/hot water	EA	Preventative and insurance inspection and general maintenance (10 MH/yr)	192.3000	17.94000	2.3920	Per size	20	100
Air curtains, steam/hot water	EA	Preventative and insurance inspection and general maintenance (10 MH/yr)	192.3000	11.96000	1.1960	Per size	20	100
Unit air conditioners with heating	EA	Preventative and insurance inspection and general maintenance (30 MH/yr)	576.9000	23.92000	4.7840	Per size	15	100

Maintenance & Replacement Estimating Data - Mechanical

Item Description	Unit of Measure	Maintenance Description	Maintenance Annual Cost, $			Energy Demand (EU)	Replacement Life, yrs	Percent Replaced
			Labor	Material	Equipm't			
Package humidifiers	EA	Preventative and insurance inspection and general maintenance (10 MH/yr)	192.3000	35.88000	3.5880	Per size	10	100
Package dehumidifiers	EA	Preventative and insurance inspection and general maintenance (25 MH/yr)	480.7500	59.80000	5.9800	Per size	15	100

0826 Controls and Instrumentation

ROOM THERMOSTATS

Item Description	Unit of Measure	Maintenance Description	Labor	Material	Equipm't	Energy Demand (EU)	Replacement Life, yrs	Percent Replaced
Low-voltage heating	EA	Preventative inspection and general maintenance (1 MH/yr)	19.2300	1.19600	0.1196	N/A	25	100
Low-voltage cooling	EA	Preventative inspection and general maintenance (1 MH/yr)	19.2300	1.19600	0.1196	N/A	25	100
Line voltage heating	EA	Preventative inspection and general maintenance (1 MH/yr)	19.2300	2.39200	0.1196	N/A	25	100
Low-voltage heating and cooling	EA	Preventative inspection and general maintenance (1 MH/yr)	19.2300	1.19600	0.1196	N/A	25	100
Line voltage heating, heavy duty	EA	Preventative inspection and general maintenance (1 MH/yr)	19.2300	1.19600	0.1196	N/A	25	100

POSITIONAL DAMPER-
Motor-actuated

Item Description	Unit of Measure	Maintenance Description	Labor	Material	Equipm't	Energy Demand (EU)	Replacement Life, yrs	Percent Replaced
Modulating type with external return spring-transformer, 115 V AC	EA	Preventative inspection and general maintenance (1 MH/yr)	19.2300	5.98000	1.1960	Per size	20	100
Modulating type with internal return spring-transformer, 115 V AC	EA	Preventative inspection and general maintenance (1 MH/yr)	19.2300	5.98000	1.1960	Per size	20	100
1-in modulating motorized valves	EA	Preventative inspection and general maintenance (2 MH/yr)	38.4600	5.98000	1.1960	Per size	15	100
1¼-in modulating motorized valves	EA	Preventative inspection and general maintenance (2 MH/yr)	38.4600	5.98000	1.1960	Per size	15	100
1½-in modulating motorized valves	EA	Preventative inspection and general maintenance (2 MH/yr)	38.4600	8.37200	1.1960	Per size	15	100
2-in modulating motorized valves	EA	Preventative inspection and general maintenance (2 MH/yr)	38.4600	9.56800	1.1960	Per size	15	100

UNIVERSAL RELAYS

Item Description	Unit of Measure	Maintenance Description	Labor	Material	Equipm't	Energy Demand (EU)	Replacement Life, yrs	Percent Replaced
SPST, use with low-voltage controls, heat only	EA	Preventative inspection and general maintenance (1 MH/yr)	20.5120	0.00000	0.0000	Per size	25	100
SPDT, use with low-voltage controls, heat or cool only	EA	Preventative inspection and general maintenance (1 MH/yr)	19.2300	0.00000	0.0000	Per size	25	100
DPDT, use with low-voltage controls, heat and cool	EA	Preventative inspection and general maintenance (1 MH/yr)	19.2300	0.00000	0.0000	Per size	25	100

Maintenance & Replacement Estimating Data - Mechanical

Item Description	Unit of Measure	Maintenance Description	Maintenance Annual Cost, $			Energy Demand (EU)	Replacement Life, yrs	Percent Replaced
			Labor	Material	Equipm't			
AQUASTATS								
External bellows type, close on pressure drop, 2 to 50-lb range	EA	Preventative inspection and general maintenance (1 MH/yr)	19.2300	0.00000	0.0000	Per size	25	100
Remote bulb type, mercury tube thermostat	EA	Preventative inspection and general maintenance (1 MH/yr)	19.2300	0.00000	0.0000	Per size	25	100
Make circuit on drop-line voltage	EA	Preventative inspection and general maintenance (1 MH/yr)	19.2300	0.00000	0.0000	Per size	25	100
Make circuit on rise-line voltage	EA	Preventative inspection and general maintenance (1 MH/yr)	19.2300	0.00000	0.0000	Per size	25	100
OTHER CONTROLS								
All parts, components, devices, tubing wiring, and accessories necessary to control air and liquid distribution systems, components, and equipment	1000 LF	Preventative inspection and general maintenance (0.2 MH/yr)	3.8460	0.35880	0.0598	N/A	25	100
All parts, components, devices, tubing wiring, and accessories necessary to monitor, record, or otherwise indicate status of any of the components of the distribution systems or equipment	1000 LF	Preventative inspection and general maintenance (0.2 MH/yr)	3.8460	0.35880	0.0598	N/A	25	100
0827 System Testing and Balancing								
All parts, components, devices, accessories, and equipment necessary to detect and repair leaks and to make alignments, inspections, and sampling and trial and final startup of HVAC systems and equipment	Per HVAC zone	Testing and balancing (80 MH/occurrence)	153.8400	11.96000	11.9600	N/A	N/A	N/A
0828 Special HVAC Systems								
All parts, components, devices, piping or duct systems, accessories, and equipment for special cooling or heating systems, storage cells, dust and fume collectors, deodorizing system, carbon monoxide equipment, special sound attenuating equipment, air curtains, paint spray booth, and ventilation system	LS	Preventative inspection and general maintenance (per item MH/yr)	Per item	Per item	Per item	Per item	Per item	100
0829 Process Mechanical Systems								
All parts, components, devices, piping or duct systems, accessories, and equipment	LS	Preventative inspection and general maintenance (per item MH/yr)	Per item	Per item	Per item	Per item	Per item	100
083 Fire Protection								
0831 Water Supply (Fire Protection)								
(See 0811 - Domestic Water Supply System)								
0832 Sprinklers								
Automatic sprinkler system, wet type, concealed piping	Head	Preventative maintenance and repair (0.05 MH/yr)	0.9615	11.9600	0.0598	N/A	25	100

Maintenance & Replacement Estimating Data - Mechanical/Electrical

Item Description	Unit of Measure	Maintenance Description	Maintenance Annual Cost, $			Energy Demand (EU)	Replacement Life, yrs	Percent Replaced
			Labor	Material	Equipm't			
0833 Standpipe Systems								
Simplex-type fire pumps, 20 hp 500 gpm, 1750 rpm	EA	Preventative maintenance and repair (10.0 MH/yr)	192.3000	47.8400	11.9600	20 kW	20	100
Fire hose cabinets, primed steel, 125-ft hose, recessed	EA	Preventative maintenance and repair (2.0 MH/yr)	38.4600	5.9800	1.1960	N/A	Life	100
Fire hose rack, 125-ft hose	EA	Preventative maintenance and repair (0.5 MH/yr)	9.6150	1.1960	0.5980	N/A	Life	100
Siamese connection, brass, 2½ x 2½ x 4 in	EA	Preventative maintenance and repair (0.5 MH/yr)	9.6150	1.1960	0.5980	N/A	Life	100
Roof manifold, brass (vertical), 2½ x 2½ x 4 in	EA	Preventative maintenance and repair (0.5 MH/yr)	9.6150	1.1960	0.5980	N/A	Life	100
0834 Fire Extinguishers								
Dry chemical enameled steel extinguisher	EA	Chemical (replaced every three years)	Per size	Per size	Per size	N/A	Life	100
084 Special Mechanical Systems								
0841 Special Plumbing Systems								
Simplex air compressor, 1 hp with 30-gal recovery	EA	Preventative maintenance and repair (5 MH/yr)	96.1500	11.9600	2.3920	1 kW	25	100
Vacuum pumps, controls, & accessories, 1 hp with 30-gal recovery	EA	Preventative maintenance and repair (5 MH/yr)	96.1500	11.9600	2.3920	1 kW	25	100
0842 Special Fire Protection Systems								
Carbon dioxide cylinders, simplex and duplex	EA	Preventative maintenance and repair (2 MH/yr)	38.4600	5.9800	1.1960	N/A	25	100

Electrical data

09 Electrical
091 Service and Distribution
0911 High-Tension Service and Distribution

Item Description	Unit of Measure	Maintenance Description	Labor	Material	Equipm't	Energy Demand (EU)	Replacement Life, yrs	Percent Replaced
Circuit breakers, metal clad drawout, 0-599 V, all sizes	EA	Preventative inspection & general maintenance (2.0 MH/yr)	40.1100	1.1960	0.1076	N/A	Life	N/A
	EA	Repair failed component (0.0027 failures/yr x 47.2 MH/failure)	2.4868	0.0027 x I.C.	N/A			
	EA	Replacement of component with spare (0.0027 failures/yr x 2.9 MH/failure)	0.0134	0.0027 x I.C.	N/A			
Circuit breakers, metal clad drawout, 600 V and over, all sizes	EA	Preventative inspection & general maintenance (4 MH/yr)	80.2200	2.3920	0.2152	N/A	Life	N/A
	EA	Repair failed component (0.0036 failure/yr x 62.4 MH/failure)	4.3854	0.0036 x I.C.	N/A			
	EA	Replacement of component with spare (0.0036 failures/yr x 5.2 MH/failure)	0.3610	0.0036 x I.C.	N/A			
Circuit breakers, fixed type, 0-599 V, all sizes	EA	Preventative inspection & general maintenance (1.0 MH/yr)	19.5202	1.1960	0.1076	N/A	Life	N/A

Maintenance & Replacement Estimating Data - Electrical

Item Description	Unit of Measure	Maintenance Description	Maintenance Annual Cost, $			Energy Demand (EU)	Replacement Life, yrs	Percent Replaced
			Labor	Material	Equipm't			
	EA	Repair failed component (0.004 failures/yr x 6.0 MH/failure)	0.4680	0.0004 x I.C.	N/A			
	EA	Replacement of component with spare (0.0044 failures/yr x 2.0 MH/failure)	0.1738	0.0044 x I.C.	N/A			
Circuit breaker, fixed type, 600 V and over, all sizes	EA	Preventative inspection & general maintenance (4.0 MH/yr)	80.2200	2.3920	0.2152	N/A	Life	N/A
	EA	Repair failed component (0.0176 failures/yr x 44.5 MH/failure)	15.3087	0.0176 x I.C.	N/A			
	EA	Replacement of component with spare (0.0176 failures/yr x 12.0 MH/failure)	4.1313	0.0210	0.0000			
Disconnect switches, enclosed, all sizes	EA	Preventative inspection & general maintenance (1.0 MH/yr)	19.5202	1.1960	0.1076	N/A	Life	N/A
	EA	Repair failed component (0.0061 failures/yr x 50.1 MH/failure)	0.4145	0.0061 x I.C.	N/A			
	EA	Replacement of component with spare (0.0061 failures/yr x 13.7 MH/failure)						
Transformers, liquid-filled, 0-750 kVA, below 600 V	EA	Preventative inspection & general maintenance (2.0 MH/yr)	40.1100	1.1960	0.1076	1 ½-3% kW loss	Life	100
	EA	Repair failed component (0.0037 failures/yr x 49.0 MH/failure)	3.5431	0.0037 x I.C.	N/A			
	EA	Replacement of component with spare (0.0037 failures/yr x 3.7 MH/failure)	0.2674	0.0037 x I.C.	N/A			
Transformers, liquid-filled, 500-2499 kVA, above 600 V	EA	Preventative inspection & general maintenance (2.0 MH/yr)	40.1100	2.3920	0.2152	1 ½-3% kW loss	Life	100
	EA	Repair failed component (0.0025 failures/yr x 297.0 MH/failure)	14.5198	0.0025 x I.C.	N/A			
	EA	Replacement of component with spare (0.0032 failures/yr x 150.0 MH/failure	9.3590	0.0025 x I.C.	N/A			
Transformers, dry type, 0-750 kVA, below 15,000 V	EA	Preventative inspection & general maintenance (2.0 MH/yr)	40.1100	1.1960	0.1076	1 ½-3% kW loss	30	100
	EA	Repair failed component (0.0036 failures/yr x 67.0 MH/failure)	4.6795	0.0036 x I.C.	N/A			
	EA	Replacement of component with spare (0.0036 failure/yr x 39.9 MH/failure)	2.8077	0.0036 x I.C.	N/A			
Transformers, dry type, 500-2499 kVA, above 15,000 V	EA	Preventative inspection & general maintenance (2.0 MH/yr)	40.1100	2.3920	0.2152	1 ½-3% kW loss	30	100
	EA	Repair failed component (0.0130 failures/yr X 367.0 MH/failure)	93.2558	0.013 x I.C.	N/A			

Maintenance & Replacement Estimating Data - Electrical

Item Description	Unit of Measure	Maintenance Description	Maintenance Annual Cost, $			Energy Demand (EU)	Replacement Life, yrs	Percent Replaced
			Labor	Material	Equipm't			
	EA	Replacement of component with spare (0.0130 failure/yr x 71.5 MH/failure)	18.1698	0.013 x I.C.	N/A			
Rectifier, 600 V and over	EA	Preventative inspection & general maintenance (10.0 MH/yr)	200.5500	35.8800	2.1520	N/A	Life	N/A
	EA	Repair failed component with (0.0298 failures/yr failures/yr x 300	134.7696	0.0356	0.0000			
Switchgear bus, indoor- and outdoor-insulated, 601-15,000 V	Per circuit breaker	Preventative inspection & general maintenance (0.1 MH/yr)	1.8718	0.1196	0.0108	N/A	20	100
	Per circuit breaker	Repair failed component (0.00063 failures/yr x 41.0 MH/failure)	0.5081	0.0006 x I.C.	N/A			
	Per circuit breaker	Replacement of component with spare (0.00063 failures/yr x 66.0 MH/failure)	0.8156	0.0006 x I.C.	N/A			
Switchgear bus, indoor- and outdoor-bare, 0-600 V	Per circuit breaker	Preventative inspection & general maintenance (0.05 MH/yr)	0.9760	0.0598	0.0108	N/A	20	100
	Per circuit breaker	Repair failed component (0.00034 failures/yr x 41.5 MH/failure)	0.2808	0.0004	0.0000			
	Per circuit breaker	Replacement of component with spare (0.00034 failures/yr x MH/failure)	0.1604	0.0003 x I.C.	N/A			
Switchgear bus, indoor-and outdoor-bare, 601 V and over	Per circuit breaker	Preventative inspection & general maintenance (0.1 MH/yr)	1.8718	0.1196	0.0108	N/A	20	100
	Per circuit breaker	Repair failed component (0.0017 failures/yr x 20.6 MH/failure)	0.6819	0.0017 x I.C.	N/A			
	Per circuit breaker	Replacement of component with spare (0.0017 failures/yr x 7.3 MH/failure)	0.2407	0.0017 x I.C.	N/A			
Bus duct, indoor and outdoor, all voltages	LF	Preventative inspection & general maintenance (0.01 MH/yr)	0.1872	0.0120	0.0011	N/A	20	100
	LF	Repair failed component (0.000125 failures/yr x 12.9 MH/failure)	0.0267	0.0001 x I.C.	N/A			
	LF	Replacement of component with spare (0.000125 failures/yr x 6.0 MH/failure)	0.0134	0.0001 x I.C.	N/A			
Inverters, all sizes	EA	Preventative inspection & general maintenance (15.0 MH/yr)	293.2041	23.9200	1.0760	N/A	20	100
	EA	Repair failed component (1.254 failures/yr x 5.0 MH/failure)	122.6029	1.2540 x I.C.	N/A			
	EA	Replacement of component with spare (1.254 failures/yr x 8.0 MH/failure)	196.1379	1.2450 x I.C.	N/A			
Rectifiers, all sizes under 600 V	EA	Preventative inspection & general maintenance (5.0 MH/yr)	97.7347	23.9200	2.1520	N/A	20	100
	EA	Repair failed component (0.038 failures/yr x 41.5 MH/failure)	30.7510	0.0380 x I.C.	N/A			

Maintenance & Replacement Estimating Data - Electrical

Item Description	Unit of Measure	Maintenance Description	Maintenance Annual Cost, $			Energy Demand (EU)	Replacement Life, yrs	Percent Replaced
			Labor	Material	Equipm't			
	EA	Replacement of component with spare (0.038 failures/yr x 12.0 MH/failure)	8.9178	0.0380 x I.C.	N/A			
Cable, thermoplastic, 601-15,000 V	1000 LF or circuit	Preventative inspection & general maintenance (1.0 MH/yr)	20.0550	1.1960	1.0760	N/A	Life	N/A
	1000 LF	Repair failed component (0.00387 failures/yr x 22.5 MH/failure)	1.6980	0.0038 x I.C.	N/A			
	1000 LF	Replacement of component with spare (0.00387 failures/yr x 29.3 MH/failure)	2.2194	0.0038 x I.C.	N/A			
Cable, thermosetting, 601-15,000 V	1000 LF or circuit	Preventative inspection & general maintenance (1.0 MH/yr)	20.0550	1.1960	0.1076	N/A	Life	N/A
	1000 LF	Repair failed component (0.00889 failures/yr x 27.2 MH/failure)	4.7330	0.0089 x I.C.	N/A			
	1000 LF	Replacement of component with spare (0.00889 failure/yr x 55.2 MH/failure)	9.5863	0.0089 x I.C.	N/A			
Cable, paper-insulated, lead-covered, 601-15,000 V	1000 LF	Preventative inspection & general maintenance (1.0 MH/yr)	9.7601	1.1960	0.1076	N/A	Life	N/A
	1000 LF	Repair failed component (0.00912 failures/yr x 17.3 MH/failure)	3.0885	0.0109	0.0000			
	1000 LF	Replacement of component with spare (0.00912 failures/yr x 18.3 MH/failure)	3.2623	0.0091 x I.C.	N/A			
Cable, other, 601-15,000 V	1000 LF	Preventative inspection & general maintenance (1.0 MH/yr)	20.0550	1.1960	0.1076	N/A	Life	N/A
	1000 LF	Repair failed component (0.01832 failures/yr x 23.2 MH/failure)	8.3028	0.0183 x I.C.	N/A			
	1000 LF	Replacement of component with spare (0.01832 failures/yr x 44.8 MH/failure)	16.0440	0.0183 x I.C.	N/A			
Cable joints, all types of insulation in duct or conduit below ground, 601-15,000 V	EA	Repair failed component (0.000864 failures/yr x 14.7 MH/failure)	0.2540	0.0010	0.0000	N/A	Life	N/A
	EA	Replacement of component with spare (0.00084 failures/yr x 5.5 MH/failure)	0.0936	0.0008 x I.C.	N/A			
Cable joints, thermoplastic insulation, 601-15,000 V	EA	Repair failed component (0.000754 failures/yr x 12.6 MH/failure)	0.1872	0.0008 x I.C.	N/A	N/A	Life	N/A
	EA	Replacement of component of spare (0.000754 failures/yr x 12.6 MH/failure)	0.1872	0.0008 x I.C.	N/A			
Cable joints, paper-insulated, lead-covered, 601-15,000 V	EA	Repair failed component (0.001037 failures/yr x 30.0 MH/failure)	0.6017	0.0010 x I.C.	N/A	N/A	Life	N/A
Cable terminations, all types of insulation, above ground and aerial, 0-600 V	EA	Repair failed component (0.000127 failures/yr x 8.0 MH/failure)	0.0134	0.0001 x I.C.	N/A	N/A	Life	N/A

Maintenance & Replacement Estimating Data - Electrical

Item Description	Unit of Measure	Maintenance Description	Maintenance Annual Cost, $ Labor	Material	Equipm't	Energy Demand (EU)	Replacement Life, yrs	Percent Replaced
	EA	Replacement of component with spare (0.000127 failures/yr x 8.0)	0.0134	0.0001 x I.C.	N/A			
Cable termination, all types of insulation, above ground and aerial, 601-15,000 V	EA	Repair failed component (0.000879 failures/yr x 34.6 MH/failure)	0.5883	0.0009 x I.C.	N/A	N/A	Life	N/A
	EA	Replacement of component spare (0.000879 failures/yr x 40.6 MH/failure)	0.6952	0.0009 x I.C.	N/A			
Cable terminations, all types of insulation, aerial, 601-15,000 V	EA	Repair failed component (0.001848 failures/yr x 15.3 MH/failure)	0.5482	0.0019 x I.C.	N/A	N/A	Life	100
	EA	Replacement of component with spare (0.001848 failures/yr x 18.0 MH/failure)	0.6551	0.0019 x I.C.	N/A			
Cable terminations, all types of insulation, in trays above ground 601-15,000 V	EA	Repair failed component (0.000333 failures/yr x 48.8 MH/failure)	0.3209	0.0003 x I.C.	N/A	N/A	Life	N/A
	EA	Replacement of component with spare (0.000333 failures/yr x 58.3 MH/failure)	0.3744	0.00003 x I.C.	N/A			
Cable terminations, all types of insulation, in duct or conduit below ground, 601-15,000 V	EA	Repair failed component (0.000303 failures/yr x 28.8 MH/failure)	0.1738	0.0003 x I.C.	N/A	N/A	Life	N/A
	EA	Replacement of component with spare (0.000303 failures/yr x 30.0 MH/failure)	0.1738	0.0003 x I.C.	N/A			
Cable terminations, thermoplastic insulation, all applications, 601-15,000 V	EA	Repair failed component (0.004192 failures/yr x 12.0 MH/failure)	0.9894	0.0042 x I.C.	N/A	N/A	Life	N/A
	EA	Replacement of component with spare (0.004192 failures/yr x 12.0 MH/failure)	0.9894	0.0042 x I.C.	N/A			
Cable terminations, thermosetting insulation, all applications, 601-15,000 V	EA	Repair failed component (0.000307 failures/yr x 30.2 MH/failure)	0.1872	0.0003 x I.C.	N/A	N/A	Life	N/A
	EA	Replacement of component spare (0.000307 failures/yr x 42.8 MH/failure)	0.2540	0.0003 x I.C.	N/A			
Cable terminations, paper-insulated, lead-covered, all applications, 601-15,000 V	EA	Repair failed component (0.000781 failures/yr x 39.0 MH/failure)	0.6017	0.0008 x I.C.	N/A	N/A	Life	N/A
	EA	Replacement of component spare (0.000781 failures/yr x 30.0 MH/failure)	0.4546	0.0008 x I.C.	N/A			
0912 Low-Tension Service and Distribution								
Motor starters, contact type, 0-600 V	EA	Preventative inspection and general maintenance (0.5 MH/yr)	9.7601	1.7940	0.1076	N/A	18	100
	EA	Repair failed component (0.0139 failures/yr x 8.0 MH/failure)	2.1392	0.0139 x I.C.	N/A			
	EA	Replacement of component with spare (0.0139 failures/yr x 4.6 MH/failure)	1.2434	0.0139 x I.C.	N/A			
Motor starters, contact type, 601-15,000 V	EA	Preventative inspection and general maintenance (1.0 MH/yr)	19.5202	3.5880	0.2690	N/A	18	100
	EA	Repair failed component (0.0153 failures/yr x 23.6 MH/failure)	7.0594	0.0015 x I.C.	N/A			
	EA	Replacement of component with spare (0.0153 failures/yr x 13.8 MH/failure)	4.1313	0.0015 x I.C.	N/A			
Motors, synchronous, 0-600 V	EA	Preventative inspection and general maintenance (2.0 MH/yr)	39.0404	5.9800	0.5380	Per size	15	100

Maintenance & Replacement Estimating Data - Electrical

Item Description	Unit of Measure	Maintenance Description	Maintenance Annual Cost, $ Labor	Material	Equipm't	Energy Demand (EU)	Replacement Life, yrs	Percent Replaced
	EA	Repair failed component (0.0109 failures/yr x 32.0 MH/failure)	6.8187	0.0109 x I.C.	N/A			
	EA	Replacement of component with spare (0.0109 failures /yr x 10.0 MH/failure)	2.1258	0.0109 x I.C.	N/A			
Motors, synchronous, 601-15,000 V	EA	Preventative inspection and general maintenance (4.0 MH/yr)	78.2145	11.9600	1.0760	Per size	15	100
	EA	Repair failed component (0.0318 failures/yr x 146.0 MH/failure)	90.7823	0.0318 x I.C.	N/A			
	EA	Replacement of component with spare (0.0318 failures/yr x 18.7 MH/failure	11.6319	0.0318 x I.C.	N/A			
Motors, direct current, all sizes	EA	Preventative inspection and general maintenance (2.0 MH/yr)	32.4089	4.7840	0.4304	Per size	15	100
	EA	Repair failed component (0.0556 failures/yr x 69.0 MH/failure)	75.0057	0.0556 x I.C.	N/A			
	EA	Replacement of component with spare (0.0556 failures/yr x 5.3 MH/failure)	5.7491	0.0556 x I.C.	N/A			
Motors, induction, 0-600 V	EA	Preventative inspection and general maintenance (2.0 MH/yr)	39.0404	5.9800	0.5380	Per size	15	100
	EA	Repair failed component (0.0109 failures/yr x 50.2 MH/failures)	10.6960	0.0109 x I.C.	N/A			
	EA	Replacement of component with spare (0.0109 failures/yr x 1.3 MH/failure)	2.8077	0.0109 x I.C.	N/A			
Motors, induction, 601-15,000 V	EA	Preventative inspection and general maintenance (4.0 MH/yr)	78.2145	11.9600	1.0760	Per size	15	100
	EA	Repair failed component (0.0404 failures/yr x 71.4 MH/failures)	56.4214	0.0404 x I.C.	N/A			
	EA	Replacement of component with spare (0.0404 failures/yr x 19.7 MH/failure)	15.3755	0.0404 x I.C.	N/A			

092 Lighting and Power
0921 Branch Wiring

Item Description	Unit of Measure	Maintenance Description	Labor	Material	Equipm't	Energy Demand (EU)	Replacement Life, yrs	Percent Replaced
Wire, 0-600 V	1000 LF per circuit	Preventative inspection and general maintenance (0.1 MH/yr)	2.00550	0.5380	0.05380	N/A	25	100
	1000 LF	Repair failed component (0.0189 failure/yr x 4.6 MH/failure)	1.73810	0.0189 x I.C.	N/A			
	1000 LF	Replacement of component with spare (0.0189 failures/yr x 8.0 MH/failure)	2.94140	0.0189 x I.C.	N/A			
Wire, 601 V and over	1000 LF per circuit	Preventative inspection and general maintenance (0.3 MH/yr)	5.88280	1.0760	0.10760	N/A	35	100
	1000 LF	Repair failed component (0.0075 failures/yr x 8.0 MH/failure)	1.20330	0.0075 x I.C.	N/A			

0922 Lighting Equipment

Item Description	Unit of Measure	Maintenance Description	Labor	Material	Equipm't	Energy Demand (EU)	Replacement Life, yrs	Percent Replaced
Fluorescent interior lighting fixtures, 2 each, 40 W tubes (20,000 burning hours)	EA	Dusting - feather duster (0.25 min/quarter)	0.25403	0.0108	0.00000	0.1kW	20	100
	EA	Washing fixture lens, etc. low-use areas (10.0 min/year)	2.56704	0.0108	0.03228			
	EA	Relamp fixture (5.0 min every replacement)	0.50806	0.6886	0.00000			
	EA	Repair fixture (0.01 failures/yr x 1.67 MH/failure)	0.13370	0.1076	0.00000			

Maintenance & Replacement Estimating Data - Electrical

Item Description	Unit of Measure	Maintenance Description	Maintenance Annual Cost, $			Energy Demand (EU)	Replacement Life, yrs	Percent Replaced
			Labor	Material	Equipm't			
Incandescent interior lighting fixtures, 1 each, 200 W (1,000 burning hours)	EA	Dusting - feather duster (0.25 min/quarter)	0.25403	0.0108	0.00000	0.2kW	20	100
	EA	Washing fixture lens, etc. high-use areas (5.0 min/quarter)	5.13408	0.2152	0.05380			
	EA	Washing fixture lens, etc. low-use areas (5.0 min/yr)	1.28352	0.0538	0.01076			
	EA	Relamp fixture (4.0 min every replacement)	4.11796	8.0700	0.04304			
	EA	Repair fixture (0.01 failures /yr x 1.67 MH/failure)	0.13370	0.1076	0.00000			
High-intensity mercury vapor lighting fixtures, 250 W (24,000 burning hours)	EA	Cleaning (5.0 min/yr)	1.28352	0.1076	0.00000	0.285 kW	20	100
	EA	Relamp fixture (20 min every replacement)	1.69799	4.8420	0.00000			
	EA	Repair fixture (0.01 failures/yr x 1.67 MH/failure)	0.13370	0.1076	0.00000			
High-intensity metal-halide (multivapor) lighting fixtures, 250 W (10,000 burning hours)	EA	Cleaning (5.0 min/yr)	1.28352	0.1076	0.00000	0.285 kW	20	100
	EA	Relamp fixture (20.0 min every replacement)	3.59653	19.3680	0.00000			
	EA	Repair fixture (0.01 failures /yr x 1.67 MH/failures)	0.13370	0.1076	0.00000			
High-pressure sodium vapor lighting fixtures, 250 W (20,000 burning hours)	EA	Cleaning (5.0 min/yr)	1.28352	0.1076	0.00000	0.285 kW	20	100
	EA	Relamp fixture (20.0 min every replacement	2.07235	19.3680	0.00000			
	EA	Repair fixture (0.01 failures /yr x 1.67 MH/failure)	0.13370	0.1076	0.00000			
Low-pressure sodium vapor lighting fixtures, 100 W (18,000 burning hours)	EA	Cleaning (5.0 min/yr)	1.28352	0.1076	0.00000	0.1 kW	20	100
	EA	Relamp fixture (20.0 min every replacement)	2.13920	16.1400	0.00000			
	EA	Repair fixture (0.01 failures /year x 1.67 MH/failure)	0.13370	0.1076	0.00000			

093 Special Electrical Systems
0931 Communications and Alarm

Item Description	Unit of Measure	Maintenance Description	Labor	Material	Equipm't	Energy Demand (EU)	Replacement Life, yrs	Percent Replaced
Public address systems	SF	Preventative inspection and general maintenance and repair (assume 10% of initial cost/yr)	0.02674	0.0108	0.00000	Per size	15	100
Central music systems	SF	Preventative inspection and general maintenance and repair (assume 10% of initial cost/yr)	0.00669	0.0022	0.00000	Per size	15	100
Intercommunication systems	SF	Preventative inspection and general maintenance and repair (assume 10% of initial cost/yr)	0.02674	0.0108	0.00000	Per size	15	100
Paging systems	SF	Preventative inspection and general maintenance and repair (assume 10% of initial cost/yr)	0.02674	0.0108	0.00000	Per size	15	100
Utility telephone systems	SF	Preventative inspection and general maintenance and repair (assume 10% of initial cost/yr)	0.13370	0.0215	0.00000	Per size	15	100
Nurses' call systems	SF	Preventative inspection and general maintenance and repair (assume 10% of initial cost/yr)	0.05348	0.0108	0.00000	Per size	15	100
Television systems	SF	Preventative inspection and general maintenance and repair (assume 5% of initial cost/yr)	0.00669	0.0011	0.00000	Per size	15	100

Maintenance & Replacement Estimating Data - Electrical

Item Description	Unit of Measure	Maintenance Description	Maintenance Annual Cost, $			Energy Demand (EU)	Replacement Life, yrs	Percent Replaced
			Labor	Material	Equipm't			
Clock and program systems	SF	Preventative inspection and general maintenance and repair (assume 5% of initial cost/yr)	0.06685	0.0108	0.00000	Per size	15	100
Fire alarm systems	SF	Preventative inspection and general maintenance and repair (assume 10% of initial cost/yr)	0.02674	0.0108	0.00000	Per size	15	100
Burglar alarm systems	SF	Preventative inspection and general maintenance and repair (assume 10% of initial cost/yr)	0.01337	0.0054	0.00000	Per size	15	100
0933 Emergency Light and Power								
Generators, steam turbine driven, 1000 kW (600 psi @ 750°F with 4-in mercury back pressure)	EA	Preventative inspection and general maintenance (50 MH/yr)	1,022.81	322.8000	80.70000	14,000 lb/h	25	100
	EA	Repair failed component (0.350 failures/yr x 234.0 MH/failure)	1,675.26	645.6000	80.70000			
	EA	Replacement of component with spare (0.350 failures/yr x 201.0 MH/failure)	1,438.61	860.8000	80.70000			
Generators, gas turbine driven, 1000 kW	EA	Preventative inspection and general maintenance (72 MH/yr)	1,470.70	430.4000	107.60000	106 gal/h (JP fuel)	25	100
	EA	Repair failed component (0.550 failures/yr x 190.0 MH/failure)	2,137.86	860.8000	107.60000			
	EA	Replacement of component with spare (0.550 failures/yr x 400 MH/failure)	4,500.34	1291.2000	107.60000			
Generators, reciprocating diesel, 1000 kW	EA	Preventative inspection and general maintenance (1 failure/yr x 72 MH/failure)	1,470.70	430.4000	107.60000	74 gal/h (diesel fuel)	25	100
	EA	Repair failed component (1 failure/yr x 133 MH/failure)	2,720.80	1076.0000	134.50000			
	EA	Replacement of component with spare (1 failure/yr x 280 MH/failure)	5,727.71	1291.2000	134.50000			
0934 Electric Heating								
Baseboard heating units, prewired, including accessories.	EA	Preventative and insurance inspection and general maintenance (2.0 MH/yr)	39.04	2.1520	1.07600	Per size	20	100
Wall-mounted heating units, recessed heaters with fan	EA	Preventative and insurance inspection and general maintenance (2.0 MH/yr)	39.04	2.1520	1.07600	Per size	20	100
Radiant suspended commercial indoor/outdoor heating units	EA	Preventative and insurance inspection and general maintenance (3.0 MH/yr)	58.69	3.2280	1.07600	Per size	15	100
Infrared suspended commercial indoor/outdoor heating units	EA	Preventative and insurance inspection and general maintenance (3.0 MH/yr)	58.69	3.2280	1.07600	Per size	15	100
Standard suspended industrial heating units	EA	Preventative and insurance inspection and general maintenance (3.0 MH/yr)	58.69	3.2280	1.07600	Per size	15	100
Explosion proof industrial heating units	EA	Preventative and insurance inspection and general maintenance (3.0 MH/yr)	58.69	3.2280	1.07600	Per size	15	100

Maintenance & Replacement Estimating Data - Equipment/Sitework

Item Description	Unit of Measure	Maintenance Description	Maintenance Annual Cost, $ Labor	Material	Equipm't	Energy Demand (EU)	Replacement Life, yrs	Percent Replaced

Equipment data

10 General Items

WASTE REMOVAL

Item Description	Unit of Measure	Maintenance Description	Labor	Material	Equipm't	Energy Demand (EU)	Replacement Life, yrs	Percent Replaced
Trash collection, office space	SF	Collect and remove trash from space (0.01 min/day)	0.71328	0.08366	0.00000	N/A	N/A	N/A
Litter, site	Acre	Pick up and sack, 50% of area covered (2.0 MH/qtr)	132.70	0.00000	0.00000	N/A	N/A	N/A
Trash and light debris removal	Per can	Empty refuse in loadpacker walking distance 25 ft/can (2.0 min/day)	143.82	0.00000	0.00000	N/A	N/A	N/A

SOIL ENTRAPMENT, ENTRANCEWAYS

Item Description	Unit of Measure	Maintenance Description	Labor	Material	Equipm't	Energy Demand (EU)	Replacement Life, yrs	Percent Replaced
Throw rugs, carpet (Note: Controls soil entry to facility; decreases all floor maintenance 5-10%)	SF	Vacuum and clean daily (5.0 min/day)	359.46	41.83	8.37	N/A	2	100
Carpet and drain (Note: Controls soil entry to facility; decreases all floor maintenance 8-12%)	SF	Vacuum and clean daily (5.0 min/day)	359.46	41.83	8.37	N/A	5	100

11 Equipment
112 Furnishings
1122 Window Treatment

Item Description	Unit of Measure	Maintenance Description	Labor	Material	Equipm't	Energy Demand (EU)	Replacement Life, yrs	Percent Replaced
Venetian blinds	EA	Dust both sides of slats with treated dust cloth (5.0 min/month)	6.63520	0.50193	0.08366	N/A	7	100
	EA	Wash venetian blinds on both sides, repair as necessary (30.0 min/year)	3.31760	0.83655	0.16731			

Sitework data

12 Sitework
122 Site Improvements
1221 Parking Lots

Item Description	Unit of Measure	Maintenance Description	Labor	Material	Equipm't	Energy Demand (EU)	Replacement Life, yrs	Percent Replaced
1" Asphalt on 1 1/2" Asphalt on 6" Base	Stall	Preventative maintenance and repair	22.00	1.02 0.5%	0.051 per year	N/A	15	100
1" Asphalt on 1 1/2" Asphalt on 8" Base	Stall	Preventative maintenance and repair	16.00	1.02 0.5%	0.051 per year	N/A	25	100
1" Asphalt on 3" Asphalt on 8" Base	Stall	Preventative maintenance and repair	12.00	1.53 0.5%	0.077 per year	N/A	30	100
1" Asphalt on 2" Asphalt on 10" Base	Stall	Preventative maintenance and repair	14.00	1.20 0.5%	0.060 per year	N/A	30	100

Maintenance & Replacement Estimating Data - Sitework

Item Description	Unit of Measure	Maintenance Description	Maintenance Annual Cost, $			Energy Demand (EU)	Replacement Life, yrs	Percent Replaced
			Labor	Material	Equipm't			
1222 Roads, Walks, and Terraces								
ROADS								
Light Use Roads:								
7" Unreinforced Concrete on 7" Base	1000 SF	Preventative maintenance and repair of joints	167.70	17.30 1.0% per year	0.86500	N/A	25	100
8" Reinforced Concrete on 8" Base	1000 SF	Preventative maintenance and repair of joints	112.20	19.80 1.0% per year	0.99000	N/A	40	100
1" Asphalt on 1-1/2" Asphalt on 6" Base	1000 SF	Preventative maintenance and repair	66.20	6.80 1.0% per year	0.34000	N/A	8	100
1" Asphalt on 1-1/2" Asphalt on 8" Base	1000 SF	Preventative maintenance and repair	47.20	6.80 1.0% per year	0.34000	N/A	12	100
1" Asphalt on 3" Asphalt on 8" Base	1000 SF	Preventative maintenance and repair	32.80	10.20 1.0% per year	0.51000	N/A	20	100
8" Gravel of 3/4" Size, 22A Natural Road	1000 SF	Preventative maintenance and repair	51.50	34.50 5.0% per year	1.72500	N/A	5	100
Heavy Use Roads								
7" Unreinforced Concrete on 7" Base	1000 SF	Preventative maintenance and repair of joints	185.70	17.30 1.0% per year	0.86500	N/A	15	100
8" Reinforced Concrete on 8" Base	1000 SF	Preventative maintenance and repair of joints	133.20	19.80 1.0% per year	0.99000	N/A	40	100
1" Asphalt on 3" Asphalt on 8" Base	1000 SF	Preventative maintenance and repair	75.90	5.10 1.0% per year	0.25500	N/A	15	100
1" Asphalt on 2" Asphalt on 10" Base	1000 SF	Preventative maintenance and repair	88.00	4.00 1.0% per year	0.20000	N/A	15	100
WALKS								
Bituminous paving	1000 SF	Sweep walks (0.19 MH/wk)	197.60 52 sweepings per year	0.00	0.198	N/A	10	100
	1000 SF	Preventative Maintenance and repair (0.1 MH/yr)	2.00	0.26000	0.02600			
Cast-in-place concrete paving	1000 SF	Sweep walks (0.19 MH/wk)	197.60 52 sweepings per year	0.00	0.198	N/A	35	100
	1000 SF	Preventative maintenance and repair (0.05 MH/yr)	1.00	0.13000	0.02600			
Precast concrete paving	1000 SF	Sweep walks (0.19 MH/wk)	197.60 52 sweepings per year	0.00	0.198	N/A	35	100
	1000 SF	Preventative maintenance and repair (0.05 MH/yr)	1.00	0.13000	0.02600			

Maintenance & Replacement Estimating Data - Sitework

Item Description	Unit of Measure	Maintenance Description	Maintenance Annual Cost, $ Labor	Material	Equipm't	Energy Demand (EU)	Replacement Life, yrs	Percent Replaced
Brick paving	1000 SF	Sweep walks (0.19 MH/wk) 52 sweepings per year	197.60	0.00	0.198	N/A	35	100
	1000 SF	Preventative maintenance and repair (0.20 MH/yr)	4.00	0.52000	0.02600			
Limestone paving	1000 SF	Sweep walks (0.19 MH/wk) 52 sweepings per year	197.60	0.00	0.198	N/A	35	100
	1000 SF	Preventative maintenance and repair (0.10 MH/yr)	2.00	0.52000	0.02600			
Flagstone paving	1000 SF	Sweep walks (0.19 MH/wk) 52 sweepings per year	197.60	0.00	0.198	N/A	35	100
	1000 SF	Preventative maintenance and repair (0.10 MH/yr)	2.00	0.52000	0.02600			
Granite paving	1000 SF	Sweep walks (0.19 MH/wk) 52 sweepings per year	197.60	0.00	0.198	N/A	35	100
	1000 SF	Preventative maintenance and repair (0.01 MH/yr)	0.20	0.52000	0.02600			

1223 Site Development

PLAYING FIELD AND SPORTS FACILITIES

Item Description	Unit of Measure	Maintenance Description	Labor	Material	Equipm't	Energy Demand (EU)	Replacement Life, yrs	Percent Replaced
Baseball field	Field	Drag baseball field (0.32 MH/occurrence) 50 events per year	320.00	0.00	0.320	N/A	N/A	N/A
	Field	Rake baseball field (0.62 MH/occurence) 50 events per year	620.00	0.00	0.620	N/A	N/A	N/A
	Field	Water baseball field (0.82 MH/occurrence) 12 waterings per year Grass cutting (See 1224 Landscaping)	196.80	19.68	0.197	N/A	N/A	N/A
Soccer field	Field	Mark soccer field (3.7 MH/occurrence) 20 events per year	1,480.00	74.00	1.480	N/A	N/A	N/A
	Field	Water soccer field (0.50 MH/occurrence) 12 waterings per year	120.00	12.00	0.120	N/A	N/A	N/A
Tennis court, asphalt	10,000 SF	Wash tennis court (2.6 MH/occurrence) 25 washings per year	1,300.00	130.00	1.300	N/A	N/A	N/A
Tennis court, concrete	10,000 SF	Wash tennis court (1.8 MH/occurrence) 25 washings per year	900.00	90.00	0.900	N/A	N/A	N/A

1224 Landscaping

PLANTING

Item Description	Unit of Measure	Maintenance Description	Labor	Material	Equipm't	Energy Demand (EU)	Replacement Life, yrs	Percent Replaced
Lawns	Acre	Aerate, tractor-drawn (0.57 MH/occurrence) 1 aeration per year	11.40	0.00	1.14	N/A	N/A	N/A
	Acre	Fertilize vehicle-drawn (0.24 MH/occurrence) 2 fertilizer events per year	9.60	4.80	0.48	N/A	N/A	N/A
	Acre	Fertilize, hand spreader (0.35 MH/occurrence) 2 fertilizer events per year	14.00	4.80	0.14	N/A	N/A	N/A
	Acre	Mowing, 3-gang mower, tractor-drawn, 84-in net (0.39 MH/occurrence) 12 cuttings per year	93.60	0.00	1.87	N/A	N/A	N/A
	Acre	Mowing, 5-gang mower, tractor-drawn, 123-in net (0.27 MH/occurrence) 12 cuttings per year	64.80	0.00	2.59	N/A	N/A	N/A
	Acre	Mowing, 7-gang mower, tractor-drawn, 30-in reels (0.10 MH/occurrence) 12 cuttings per year	24.00	0.00	1.44	N/A	N/A	N/A
	Acre	Mowing, triplex mower 76-in (0.95 MH/occur.) 12 cuttings per year	228.00	0.00	4.56	N/A	N/A	N/A
	Acre	Mowing, 18-in power reel (2.8 MH/occurrence) 12 cuttings per year	672.00	0.00	33.60	N/A	N/A	N/A

Maintenance & Replacement Estimating Data - Sitework

Item Description	Unit of Measure	Maintenance Description	Maintenance Annual Cost, $ Labor	Maintenance Annual Cost, $ Material	Maintenance Annual Cost, $ Equipm't	Energy Demand (EU)	Replacement Life, yrs	Percent Replaced
	Acre	Mowing, 24-in power reel (2.2 MH/occurrence) 12 cuttings per year	528.00	0.00	26.40	N/A	N/A	N/A
	Acre	Sweeping, power sweeper, 18-in (2.8 MH/occurrence) 12 sweeping per year	228.00	0.00	11.40	N/A	N/A	N/A
	100 LF	Edgings, hand sweep, and pickup (0.09 MH/occurrence) 12 edgings per year	21.60	0.00	1.08	N/A	N/A	N/A
		Automatic irrigation (See 1231 - Water Supply and Distribution)						
Deciduous trees	100 SF	Water by hand with hose (0.2 MH/occurrence) 4 waterings per year	16.00	1.60	0.80	N/A	35	100
	100 SF	Leaf raking (0.06 MH/occurrence) 2 raking events per year	8.00	0.00	0.40	N/A	N/A	N/A
	EA	Stake and tie tree (0.05 MH/occurrence) 2 events per year	2.00	0.00	0.10	N/A	N/A	N/A
	Per limb	Remove/trim limb (0.08 MH/occurrence) 1 event per year	1.60	0.00	0.08	N/A	N/A	N/A
Evergreen trees	EA	Stake an tie tree (0.05 MH/occurrence) 2 events per year	2.00	0.00	0.10	N/A	25	100
	EA	Trim by hand, disposal (0.29 MH/occurrence) 2 events per year	11.60	0.00	0.58	N/A	N/A	N/A
	EA	Disease control, spray (0.04 MH/occurrence) 2 events per year	1.60	0.00	0.08	N/A	N/A	N/A
Shrubs	EA	Pruning, up to 6 ft high, rake into pile (1.4 MH/occurrence) 2 events per year	56.00	0.00	2.80	N/A	15	100
	EA	Disease control, spray (0.04 MH/occurrence) 2 events per year	1.60	0.16	0.08			
Hedges	100 LF	Trim hedge by hand, top and both sides, including disposal (2.9 MH/occurrence) 2 events per year	116.00	0.00	5.80	N/A	15	100
	100 LF	Trim hedge with electric shears, top and both sides, including disposal (1.6 MH/occurrence) 2 events per year	64.00	0.00	3.20	N/A	15	100
	100 LF	Disease control, spray (0.30 MH/occurrence) 2 events per year	12.00	1.20	0.60			
Rosebush	EA	Pruning by hand (0.07 MH/occurrence)	5.60	0.00	0.28	N/A	15	100
	EA	Disease control, spray (0.04 MH/occurrence) 2 events per year	1.60	0.16	0.08			

123 Site Utilities
1231 Water Supply and Distribution

Item Description	Unit of Measure	Maintenance Description	Labor	Material	Equipm't	Energy Demand (EU)	Replacement Life, yrs	Percent Replaced
Irrigation (sprinkler) system	Acre	Sprinkler head, remove and replace (0.29 MH/occurrence)	58.00	100.00	5.00	N/A	20	100
	Acre	Periodic inspection (0.4 MH/occurrence) 2 events per year	16.00	0.00	0.00			
	Acre	Watering, manual quick-coupler valve system (0.3 MH/occur.) 20 waterings per year	120.00	240.00	6.00			
	Acre	Watering, automatic system (0.0 MH/occur.) 20 waterings per year	0.00	240.00	12.00			

1234 Electric Distribution and Lighting Systems

Item Description	Unit of Measure	Maintenance Description	Labor	Material	Equipm't	Energy Demand (EU)	Replacement Life, yrs	Percent Replaced
Cable, all types insulation, above ground and aerial, 0-600 V	1000 LF	Preventative inspection and general maintenance (0.5 MH/yr)	15.00	2.25	0.11250	N/A	Life	N/A
	1000 LF	Repair failed component (0.00141 failures/yr x 20.8 MH/failure)	0.88	0.00141 x I.C.	N/A			

Maintenance & Replacement Estimating Data - Sitework

Item Description	Unit of Measure	Maintenance Description	Labor	Material	Equipm't	Energy Demand (EU)	Replacement Life, yrs	Percent Replaced
	1000 LF	Replacement of component with spare (0.00141 failures/yr x 39.7 MH/failure)	1.68	0.00141 x I.C.	N/A			
Cable, all types insulation, in trays above ground 601-15,000 V	1000 LF	Preventative inspection and general maintenance (0.5 MH/yr)	15.00	2.25	0.11250	N/A	Life	N/A
	1000 LF	Repair failed component (0.00923 failures/yr x 49.4 MH/failure)	13.68	0.00923 x I.C.	N/A			
	1000 LF	Replacement of component with spare (0.00923 failure/yr x 119.0 MH/failure)	32.95	0.00923 x I.C.	N/A			
Cable, all types insulation, in conduit above ground 601-15,000 V	1000 LF	Preventative inspection and general maintenance (0.5 MH/yr)	15.00	2.25	0.11250	N/A	N/A	N/A
	1000 LF	Replacement of component with spare (0.04918 failures/yr x 19.8 MH/failure)	29.21	0.04918 x I.C.	N/A			
Cable, all types insulation, aerial cable, 601-15,000 V	1000 LF	Preventative inspection and general maintenance (0.5 MH/yr)	15.00	2.25	0.11250	N/A	N/A	N/A
	1000 LF	Repair failed component (0.01437 failures/yr x 10.6 MH/failure)	4.57	0.01437 x I.C.	N/A			
	1000 LF	Replacement of component with spare (0.01437 failures/yr x 28.0 MH/failure)	5.88	0.01437 x I.C.	N/A			
Cable, all types insulation, above ground 15,000 V and over	1000 LF	Preventative inspection and general maintenance (0.5 MH/yr)	15.00	2.25	0.11250	N/A	Life	N/A
	1000 LF	Repair failed component (0.00336 failures/yr x 16.0 MH/failure	1.07	0.00336 x I.C.	N/A			
Cable, all types insulation, below ground and direct burial, 0-600 V	1000 LF	Preventative inspection and general maintenance (0.5 MH/yr)	15.00	2.25	0.11250	N/A	Life	N/A
	1000 LF	Replacement of component with spare (0.00388 failures/yr x 26.8 MH/failure)	1.23	0.00388 x I.C.	N/A			
Cable, all types insulation, below ground and direct burial, 601-15,000 V, in duct or conduit	1000 LF	Preventative inspection and general maintenance (0.5 MH/yr)	15.00	2.25	0.11250	N/A	Life	N/A
	1000 LF	Repair failed component (0.00613 failures/yr x 20.9 MH/failure)	3.84	0.00613 x I.C.	N/A	N/A	Life	N/A
	1000 LF	Replacement of component with spare (0.00613 failures/yr x 26.8 MH/failure)	4.93	0.00613 x I.C.	N/A			
Mercury vapor exterior light fixtures, 250 W (24,000 burning hours)	EA	Cleaning and relamping fixture (25.0 min every 3.0 yr)	2.50260	5.78888	N/A	0.285 kW	20	100
	EA	Repair fixture (0.01 failures/yr x 1.67 MH/failure)	0.17876	0.11578	N/A			
Incandescent exterior lighting fixture, 500 W (1000 burning hours)	EA	Cleaning and relamping fixture (6.0 min every quarter)	7.32903	4.63110	N/A	0.5 kW	20	100
	EA	Repair fixture (0.01 failures/yr x 1.67 MH/failure)	0.17876	0.11578	N/A			
Fluorescent exterior lighting fixtures, two 40 W tubes (20,000 burning hours)	EA	Cleaning and relamping fixture (10.0 min every 2.5 yr)	1.35855	0.86833	N/A	0.1 kW	20	100
	EA	Repair fixture (0.01 failures/yr x 1.67 MH/failure)	0.17876	0.11578	N/A			

Part 3 Appendix

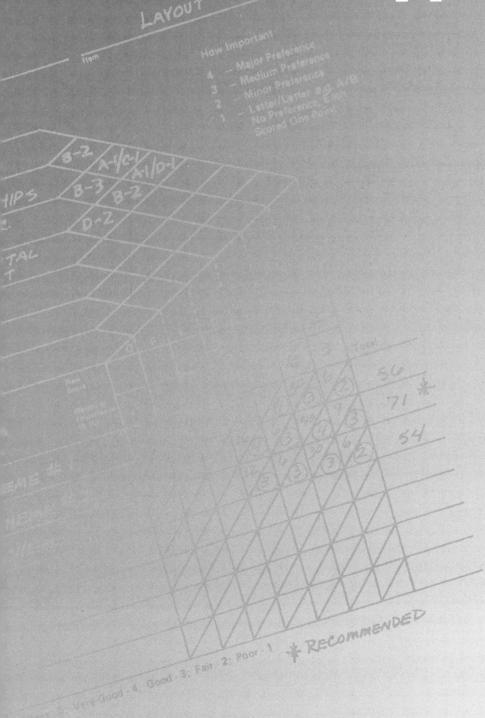

Appendix A: Economic Tables

Economic Tables

This appendix provides the reader with quick reference tables to deal with the time value of money. Several chapters in the text refer to these tables for conversion factors. This appendix provides material for use in present worth in annualized analysis for LCC. Of special significance are the years to payback tables which incorporate various escalation rates. For a discussion in the use of these tables, refer to Chapter 2, "Life Cycle Costing Fundamentals."

Factor:	SCA Single Compound Amount	PW Single Present Worth	USF Uniform Sinking Fund	PP Periodic Payment	UCA Uniform Compound Amount	PWA Present Worth Annuity	
Given:	P	F	F	P	A	A	
Find:	F	P	A	A	F	P	
n							n
1	1.0600	0.9434	1.00000	1.06000	1.000	0.943	1
2	1.1236	0.8900	0.48544	0.54544	2.060	1.833	2
3	1.1910	0.8396	0.31411	0.37411	3.184	2.673	3
4	1.2625	0.7921	0.22859	0.28859	4.375	3.465	4
5	1.3382	0.7473	0.17740	0.23740	5.637	4.212	5
6	1.4185	0.7050	0.14336	0.20336	6.975	4.917	6
7	1.5036	0.6651	0.11914	0.17914	8.394	5.582	7
8	1.5938	0.6274	0.10104	0.16104	9.897	6.210	8
9	1.6895	0.5919	0.08702	0.14702	11.491	6.802	9
10	1.7908	0.5584	0.07587	0.13587	13.181	7.360	10
11	1.8983	0.5268	0.06679	0.12679	14.972	7.887	11
12	2.0122	0.4970	0.05928	0.11928	16.870	8.384	12
13	2.1329	0.4688	0.05296	0.11296	18.882	8.853	13
14	2.2609	0.4423	0.04758	0.10758	21.015	9.295	14
15	2.3966	0.4173	0.04296	0.10296	23.276	9.712	15
16	2.5404	0.3936	0.03895	0.09895	25.673	10.106	16
17	2.6928	0.3714	0.03544	0.09544	28.213	10.477	17
18	2.8543	0.3503	0.03236	0.09236	30.906	10.828	18
19	3.0256	0.3305	0.02962	0.08962	33.760	11.158	19
20	3.2071	0.3118	0.02718	0.08718	36.786	11.470	20
21	3.3996	0.2942	0.02500	0.08500	39.993	11.764	21
22	3.6035	0.2775	0.02305	0.08305	43.392	12.042	22
23	3.8197	0.2618	0.02128	0.08128	46.996	12.303	23
24	4.0489	0.2470	0.01968	0.07968	50.816	12.550	24
25	4.2919	0.2330	0.01823	0.07823	54.865	12.783	25
26	4.5494	0.2198	0.01690	0.07690	59.156	13.003	26
27	4.8223	0.2074	0.01570	0.07570	63.706	13.211	27
28	5.1117	0.1956	0.01459	0.07459	68.528	13.406	28
29	5.4184	0.1846	0.01358	0.07358	73.640	13.591	29
30	5.7435	0.1741	0.01265	0.07265	79.058	13.765	30
31	6.0881	0.1643	0.01179	0.07179	84.802	13.929	31
32	6.4534	0.1550	0.01100	0.07100	90.890	14.084	32
33	6.8406	0.1462	0.01027	0.07027	97.343	14.230	33
34	7.2510	0.1379	0.00960	0.06960	104.184	14.368	34
35	7.6861	0.1301	0.00897	0.06897	111.435	14.498	35
36	8.1473	0.1227	0.00839	0.06839	119.121	14.621	36
37	8.6361	0.1158	0.00786	0.06786	127.268	14.737	37
38	9.1543	0.1092	0.00736	0.06736	135.904	14.846	38
39	9.7035	0.1031	0.00689	0.06689	145.058	14.949	39
40	10.2857	0.0972	0.00646	0.06646	154.762	15.046	40

Table A.1: **Interest Table for Life Cycle Costing—6% Discount Rate**

Factor:	SCA Single Compound Amount	PW Single Present Worth	USF Uniform Sinking Fund	PP Periodic Payment	UCA Uniform Compound Amount	PWA Present Worth Annuity	
Given:	P	F	F	P	A	A	
Find:	F	P	A	A	F	P	
n							n
1	1.0700	0.9346	1.00000	1.07000	1.000	0.935	1
2	1.1449	0.8734	0.48309	0.55309	2.070	1.808	2
3	1.2250	0.8163	0.31105	0.38105	3.215	2.624	3
4	1.3108	0.7629	0.22523	0.29523	4.440	3.387	4
5	1.4026	0.7130	0.17389	0.24389	5.751	4.100	5
6	1.5007	0.6663	0.13980	0.20980	7.153	4.767	6
7	1.6058	0.6227	0.11555	0.18555	8.654	5.389	7
8	1.7182	0.5820	0.09747	0.16747	10.260	5.971	8
9	1.8385	0.5439	0.08349	0.15349	11.978	6.515	9
10	1.9672	0.5083	0.07238	0.14238	13.816	7.024	10
11	2.1049	0.4751	0.06336	0.13336	15.784	7.499	11
12	2.2522	0.4440	0.05590	0.12590	17.888	7.943	12
13	2.4098	0.4150	0.04965	0.11965	20.141	8.358	13
14	2.5785	0.3878	0.04434	0.11434	22.550	8.745	14
15	2.7590	0.3624	0.03979	0.10979	25.129	9.108	15
16	2.9522	0.3387	0.03586	0.10586	27.888	9.447	16
17	3.1588	0.3166	0.03243	0.10243	30.840	9.763	17
18	3.3799	0.2959	0.02941	0.09941	33.999	10.059	18
19	3.6165	0.2765	0.02675	0.09675	37.379	10.336	19
20	3.8697	0.2584	0.02439	0.09439	40.995	10.594	20
21	4.1406	0.2415	0.02229	0.09229	44.865	10.836	21
22	4.4304	0.2257	0.02041	0.09041	49.006	11.061	22
23	4.7405	0.2109	0.01871	0.08871	53.436	11.272	23
24	5.0724	0.1971	0.01719	0.08719	58.177	11.469	24
25	5.4274	0.1842	0.01581	0.08581	63.249	11.654	25
26	5.8074	0.1722	0.01456	0.08456	68.676	11.826	26
27	6.2139	0.1609	0.01343	0.08343	74.484	11.987	27
28	6.6488	0.1504	0.01239	0.08239	80.698	12.137	28
29	7.1143	0.1406	0.01145	0.08145	87.347	12.278	29
30	7.6123	0.1314	0.01059	0.08059	94.461	12.409	30
31	8.1451	0.1228	0.00980	0.07980	102.073	12.532	31
32	8.7153	0.1147	0.00907	0.07907	110.218	12.647	32
33	9.3253	0.1072	0.00841	0.07841	118.933	12.754	33
34	9.9781	0.1002	0.00780	0.07780	128.259	12.854	34
35	10.6766	0.0937	0.00723	0.07723	138.237	12.948	35
36	11.4239	0.0875	0.00672	0.07672	148.913	13.035	36
37	12.2236	0.0818	0.00624	0.07624	160.337	13.117	37
38	13.0793	0.0765	0.00580	0.07580	172.561	13.193	38
39	13.9948	0.0715	0.00539	0.07539	185.640	13.265	39
40	14.9745	0.0668	0.00501	0.07501	199.635	13.332	40

Table A.2: **Interest Table for Life Cycle Costing—7% Discount Rate**

Factor:	SCA Single Compound Amount	PW Single Present Worth	USF Uniform Sinking Fund	PP Periodic Payment	UCA Uniform Compound Amount	PWA Present Worth Annuity	
Given:	P	F	F	P	A	A	
Find:	F	P	A	A	F	P	
n							n
1	1.0800	0.9259	1.00000	1.08000	1.000	0.926	1
2	1.1664	0.8573	0.48077	0.56077	2.080	1.783	2
3	1.2597	0.7938	0.30803	0.38803	3.246	2.577	3
4	1.3605	0.7350	0.22192	0.30192	4.506	3.312	4
5	1.4693	0.6806	0.17046	0.25046	5.867	3.993	5
6	1.5869	0.6302	0.13632	0.21632	7.336	4.623	6
7	1.7138	0.5835	0.11207	0.19207	8.923	5.206	7
8	1.8509	0.5403	0.09401	0.17401	10.637	5.747	8
9	1.9990	0.5002	0.08008	0.16008	12.488	6.247	9
10	2.1589	0.4632	0.06903	0.14903	14.487	6.710	10
11	2.3316	0.4289	0.06008	0.14008	16.645	7.139	11
12	2.5182	0.3971	0.05270	0.13270	18.977	7.536	12
13	2.7196	0.3677	0.04652	0.12652	21.495	7.904	13
14	2.9372	0.3405	0.04130	0.12130	24.215	8.244	14
15	3.1722	0.3152	0.03683	0.11683	27.152	8.559	15
16	3.4259	0.2919	0.03298	0.11298	30.324	8.851	16
17	3.7000	0.2703	0.02963	0.10963	33.750	9.122	17
18	3.9960	0.2502	0.02670	0.10670	37.450	9.372	18
19	4.3157	0.2317	0.02413	0.10413	41.446	9.604	19
20	4.6610	0.2145	0.02185	0.10185	45.762	9.818	20
21	5.0338	0.1987	0.01983	0.09983	50.423	10.017	21
22	5.4365	0.1839	0.01803	0.09803	55.457	10.201	22
23	5.8715	0.1703	0.01642	0.09642	60.893	10.371	23
24	6.3412	0.1577	0.01498	0.09498	66.765	10.529	24
25	6.8485	0.1460	0.01368	0.09368	73.106	10.675	25
26	7.3964	0.1352	0.01251	0.09251	79.954	10.810	26
27	7.9881	0.1252	0.01145	0.09145	87.351	10.935	27
28	8.6271	0.1159	0.01049	0.09049	95.339	11.051	28
29	9.3173	0.1073	0.00962	0.08962	103.966	11.158	29
30	10.0627	0.0994	0.00883	0.08883	113.283	11.258	30
31	10.8677	0.0920	0.00811	0.08811	123.346	11.350	31
32	11.7371	0.0852	0.00745	0.08745	134.214	11.435	32
33	12.6760	0.0789	0.00685	0.08685	145.951	11.514	33
34	13.6901	0.0730	0.00630	0.08630	158.627	11.587	34
35	14.7853	0.0676	0.00580	0.08580	172.317	11.655	35
36	15.9682	0.0626	0.00534	0.08534	187.102	11.717	36
37	17.2456	0.0580	0.00492	0.08492	203.070	11.775	37
38	18.6253	0.0537	0.00454	0.08454	220.316	11.829	38
39	20.1153	0.0497	0.00419	0.08419	238.941	11.879	39
40	21.7245	0.0460	0.00386	0.08386	259.057	11.925	40

Table A.3: **Interest Table for Life Cycle Costing—8% Discount Rate**

Factor:	SCA Single Compound Amount	PW Single Present Worth	USF Uniform Sinking Fund	PP Periodic Payment	UCA Uniform Compound Amount	PWA Present Worth Annuity	
Given:	P	F	F	P	A	A	
Find:	F	P	A	A	F	P	
n							n
1	1.1000	0.9091	1.00000	1.10000	1.000	0.909	1
2	1.2100	0.8264	0.47619	0.57619	2.100	1.736	2
3	1.3310	0.7513	0.30211	0.40211	3.310	2.487	3
4	1.4641	0.6830	0.21547	0.31547	4.641	3.170	4
5	1.6105	0.6209	0.16380	0.26380	6.105	3.791	5
6	1.7716	0.5645	0.12961	0.22961	7.716	4.355	6
7	1.9487	0.5132	0.10541	0.20541	9.487	4.868	7
8	2.1436	0.4665	0.08744	0.18744	11.436	5.335	8
9	2.3579	0.4241	0.07364	0.17364	13.579	5.759	9
10	2.5937	0.3855	0.06275	0.16275	15.937	6.145	10
11	2.8531	0.3505	0.05396	0.15396	18.531	6.495	11
12	3.1384	0.3186	0.04676	0.14676	21.384	6.814	12
13	3.4523	0.2897	0.04078	0.14078	24.523	7.103	13
14	3.7975	0.2633	0.03575	0.13575	27.975	7.367	14
15	4.1772	0.2394	0.03147	0.13147	31.772	7.606	15
16	4.5950	0.2176	0.02782	0.12782	35.950	7.824	16
17	5.0545	0.1978	0.02466	0.12466	40.545	8.022	17
18	5.5599	0.1799	0.02193	0.12193	45.599	8.201	18
19	6.1159	0.1635	0.01955	0.11955	51.159	8.365	19
20	6.7275	0.1486	0.01746	0.11746	57.275	8.514	20
21	7.4002	0.1351	0.01562	0.11562	64.002	8.649	21
22	8.1403	0.1228	0.01401	0.11401	71.403	8.772	22
23	8.9543	0.1117	0.01257	0.11257	79.543	8.883	23
24	9.8497	0.1015	0.01130	0.11130	88.497	8.985	24
25	10.8347	0.0923	0.01017	0.11017	98.347	9.077	25
26	11.9182	0.0839	0.00916	0.10916	109.182	9.161	26
27	13.1100	0.0763	0.00826	0.10826	121.100	9.237	27
28	14.4210	0.0693	0.00745	0.10745	134.210	9.307	28
29	15.8631	0.0630	0.00673	0.10673	148.631	9.370	29
30	17.4494	0.0573	0.00608	0.10608	164.494	9.427	30
31	19.1943	0.0521	0.00550	0.10550	181.943	9.479	31
32	21.1138	0.0474	0.00497	0.10497	201.138	9.526	32
33	23.2252	0.0431	0.00450	0.10450	222.252	9.569	33
34	25.5477	0.0391	0.00407	0.10407	245.477	9.609	34
35	28.1024	0.0356	0.00369	0.10369	271.024	9.644	35
36	30.9127	0.0323	0.00334	0.10334	299.127	9.677	36
37	34.0039	0.0294	0.00303	0.10303	330.039	9.706	37
38	37.4043	0.0267	0.00275	0.10275	364.043	9.733	38
39	41.1448	0.0243	0.00249	0.10249	401.448	9.757	39
40	45.2593	0.0221	0.00226	0.10226	442.593	9.779	40

Table A.4: **Interest Table for Life Cycle Costing—10% Discount Rate**

Factor:	SCA Single Compound Amount	PW Single Present Worth	USF Uniform Sinking Fund	PP Periodic Payment	UCA Uniform Compound Amount	PWA Present Worth Annuity	
Given:	P	F	F	P	A	A	
Find:	F	P	A	A	F	P	
n							n
1	1.1200	0.8929	1.00000	1.12000	1.000	0.893	1
2	1.2544	0.7972	0.47170	0.59170	2.120	1.690	2
3	1.4049	0.7118	0.29635	0.41635	3.374	2.402	3
4	1.5735	0.6355	0.20923	0.32923	4.779	3.037	4
5	1.7623	0.5674	0.15741	0.27741	6.353	3.605	5
6	1.9738	0.5066	0.12323	0.24323	8.115	4.111	6
7	2.2107	0.4523	0.09912	0.21912	10.089	4.564	7
8	2.4760	0.4039	0.08130	0.20130	12.300	4.968	8
9	2.7731	0.3606	0.06768	0.18768	14.776	5.328	9
10	3.1058	0.3220	0.05698	0.17698	17.549	5.650	10
11	3.4785	0.2875	0.04842	0.16842	20.655	5.938	11
12	3.8960	0.2567	0.04144	0.16144	24.133	6.194	12
13	4.3635	0.2292	0.03568	0.15568	28.029	6.424	13
14	4.8871	0.2046	0.03087	0.15087	32.393	6.628	14
15	5.4736	0.1827	0.02682	0.14682	37.280	6.811	15
16	6.1304	0.1631	0.02339	0.14339	42.753	6.974	16
17	6.8660	0.1456	0.02046	0.14046	48.884	7.120	17
18	7.6900	0.1300	0.01794	0.13794	55.750	7.250	18
19	8.6128	0.1161	0.01576	0.13576	63.440	7.366	19
20	9.6463	0.1037	0.01388	0.13388	72.052	7.469	20
21	10.8038	0.0926	0.01224	0.13224	81.699	7.562	21
22	12.1003	0.0826	0.01081	0.13081	92.503	7.645	22
23	13.5523	0.0738	0.00956	0.12956	104.603	7.718	23
24	15.1786	0.0659	0.00846	0.12846	118.155	7.784	24
25	17.0001	0.0588	0.00750	0.12750	133.334	7.843	25
26	19.0401	0.0525	0.00665	0.12665	150.334	7.896	26
27	21.3249	0.0469	0.00590	0.12590	169.374	7.943	27
28	23.8839	0.0419	0.00524	0.12524	190.699	7.984	28
29	26.7499	0.0374	0.00466	0.12466	214.583	8.022	29
30	29.9599	0.0334	0.00414	0.12414	241.333	8.055	30
31	33.5551	0.0298	0.00369	0.12369	271.293	8.085	31
32	37.5817	0.0266	0.00328	0.12328	304.848	8.112	32
33	42.0915	0.0238	0.00292	0.12292	342.429	8.135	33
34	47.1425	0.0212	0.00260	0.12260	384.521	8.157	34
35	52.7996	0.0189	0.00232	0.12232	431.663	8.176	35
36	59.1356	0.0169	0.00206	0.12206	484.463	8.192	36
37	66.2318	0.0151	0.00184	0.12184	543.599	8.208	37
38	74.1797	0.0135	0.00164	0.12164	609.831	8.221	38
39	83.0812	0.0120	0.00146	0.12146	684.010	8.233	39
40	93.0510	0.0107	0.00130	0.12130	767.091	8.244	40

Table A.5: **Interest Table for Life Cycle Costing—12% Discount Rate**

	Escalation Rate										
n	1.00%	2.00%	3.00%	4.00%	5.00%	6.00%	7.00%	8.00%	9.00%	10.00%	n
1	0.953	0.962	0.972	0.981	0.991	1.000	1.009	1.019	1.028	1.038	1
2	1.861	1.888	1.916	1.944	1.972	2.000	2.028	2.057	2.086	2.115	2
3	2.726	2.779	2.833	2.888	2.944	3.000	3.057	3.115	3.173	3.232	3
4	3.550	3.637	3.725	3.815	3.907	4.000	4.095	4.192	4.291	4.392	4
5	4.335	4.462	4.591	4.724	4.860	5.000	5.143	5.290	5.441	5.595	5
6	5.084	5.256	5.433	5.616	5.805	6.000	6.201	6.409	6.623	6.844	6
7	5.797	6.019	6.251	6.491	6.741	7.000	7.269	7.549	7.839	8.140	7
8	6.476	6.755	7.046	7.350	7.668	8.000	8.347	8.710	9.089	9.485	8
9	7.124	7.462	7.818	8.192	8.586	9.000	9.435	9.893	10.375	10.881	9
10	7.740	8.143	8.568	9.019	9.496	10.000	10.534	11.099	11.697	12.329	10
11	8.328	8.798	9.298	9.830	10.397	11.000	11.643	12.327	13.056	13.832	11
12	8.888	9.428	10.006	10.625	11.289	12.000	12.762	13.578	14.454	15.392	12
13	9.422	10.034	10.695	11.406	12.173	13.000	13.892	14.854	15.891	17.010	13
14	9.930	10.618	11.364	12.172	13.049	14.000	15.032	16.153	17.369	18.690	14
15	10.414	11.180	12.014	12.924	13.916	15.000	16.183	17.476	18.889	20.433	15
16	10.876	11.720	12.645	13.661	14.776	16.000	17.346	18.825	20.452	22.242	16
17	11.316	12.240	13.259	14.384	15.627	17.000	18.519	20.199	22.059	24.119	17
18	11.735	12.740	13.856	15.094	16.470	18.000	19.703	21.599	23.712	26.067	18
19	12.134	13.222	14.435	15.790	17.305	19.000	20.898	23.025	25.411	28.088	19
20	12.515	13.685	14.998	16.473	18.132	20.000	22.105	24.479	27.159	30.186	20
21	12.877	14.131	15.546	17.144	18.952	21.000	23.323	25.959	28.955	32.363	21
22	13.223	14.560	16.077	17.801	19.764	22.000	24.552	27.468	30.803	34.622	22
23	13.552	14.973	16.594	18.447	20.568	23.000	25.793	29.005	32.703	36.966	23
24	13.865	15.370	17.096	19.080	21.364	24.000	27.046	30.571	34.657	39.398	24
25	14.164	15.752	17.584	19.701	22.153	25.000	28.311	32.167	36.666	41.923	25
26	14.449	16.120	18.058	20.310	22.935	26.000	29.587	33.793	38.732	44.543	26
27	14.720	16.474	18.519	20.908	23.709	27.000	30.876	35.449	40.857	47.261	27
28	14.979	16.815	18.966	21.495	24.476	28.000	32.176	37.137	43.042	50.082	28
29	15.225	17.143	19.401	22.070	25.236	29.000	33.489	38.857	45.288	53.010	29
30	15.460	17.458	19.824	22.635	25.988	30.000	34.815	40.609	47.598	56.048	30
31	15.683	17.761	20.234	23.189	26.734	31.000	36.153	42.394	49.973	59.201	31
32	15.896	18.053	20.633	23.733	27.472	32.000	37.503	44.212	52.416	62.473	32
33	16.099	18.334	21.021	24.266	28.203	33.000	38.866	46.065	54.928	65.868	33
34	16.293	18.605	21.398	24.789	28.928	34.000	40.242	47.953	57.511	69.391	34
35	16.477	18.865	21.764	25.303	29.645	35.000	41.631	49.877	60.167	73.048	35
36	16.653	19.115	22.120	25.807	30.356	36.000	43.034	51.837	62.898	76.842	36
37	16.820	19.356	22.465	26.301	31.061	37.000	44.449	53.834	65.706	80.779	37
38	16.979	19.588	22.801	26.786	31.758	38.000	45.878	55.869	68.594	84.865	38
39	17.131	19.811	23.128	27.261	32.449	39.000	47.320	57.942	71.564	89.105	39
40	17.276	20.026	23.445	27.728	33.133	40.000	48.776	60.054	74.617	93.506	40

Table A.6: **Present Worth of an Escalating Annual Amount—6% Discount Rate**

					Escalation Rate						
n	1.00%	2.00%	3.00%	4.00%	5.00%	6.00%	7.00%	8.00%	9.00%	10.00%	n
1	0.944	0.953	0.963	0.972	0.981	0.991	1.000	1.009	1.019	1.028	1
2	1.835	1.862	1.889	1.917	1.944	1.972	2.000	2.028	2.056	2.085	2
3	2.676	2.728	2.781	2.835	2.889	2.944	3.000	3.056	3.114	3.171	3
4	3.470	3.554	3.640	3.727	3.817	3.907	4.000	4.094	4.190	4.288	4
5	4.219	4.341	4.466	4.595	4.727	4.862	5.000	5.142	5.287	5.437	5
6	4.927	5.092	5.262	5.438	5.619	5.807	6.000	6.199	6.405	6.617	6
7	5.594	5.807	6.028	6.257	6.496	6.743	7.000	7.267	7.543	7.831	7
8	6.224	6.489	6.765	7.054	7.356	7.671	8.000	8.344	8.703	9.078	8
9	6.819	7.139	7.475	7.828	8.199	8.590	9.000	9.431	9.884	10.361	9
10	7.381	7.759	8.158	8.581	9.028	9.500	10.000	10.529	11.088	11.679	10
11	7.911	8.349	8.816	9.312	9.840	10.402	11.000	11.636	12.314	13.035	11
12	8.411	8.912	9.449	10.023	10.637	11.295	12.000	12.755	13.563	14.428	12
13	8.883	9.449	10.058	10.714	11.420	12.181	13.000	13.883	14.835	15.861	13
14	9.329	9.961	10.645	11.385	12.188	13.057	14.000	15.022	16.131	17.334	14
15	9.750	10.449	11.210	12.038	12.941	13.926	15.000	16.172	17.451	18.848	15
16	10.147	10.914	11.753	12.673	13.681	14.786	16.000	17.332	18.796	20.404	16
17	10.522	11.357	12.276	13.289	14.406	15.639	17.000	18.504	20.166	22.004	17
18	10.876	11.780	12.780	13.889	15.118	16.483	18.000	19.686	21.562	23.649	18
19	11.210	12.182	13.265	14.471	15.817	17.320	19.000	20.879	22.983	25.340	19
20	11.525	12.566	13.732	15.037	16.503	18.149	20.000	22.084	24.432	27.079	20
21	11.823	12.933	14.181	15.588	17.176	18.970	21.000	23.300	25.907	28.866	21
22	12.104	13.281	14.613	16.123	17.836	19.783	22.000	24.527	27.410	30.703	22
23	12.369	13.614	15.030	16.643	18.484	20.589	23.000	25.765	28.941	32.592	23
24	12.620	13.931	15.430	17.148	19.120	21.387	24.000	27.015	30.501	34.534	24
25	12.856	14.233	15.816	17.639	19.744	22.178	25.000	28.277	32.089	36.530	25
26	13.079	14.522	16.188	18.117	20.356	22.961	26.000	29.551	33.708	38.583	26
27	13.289	14.796	16.545	18.581	20.957	23.737	27.000	30.836	35.357	40.693	27
28	13.488	15.058	16.889	19.032	21.546	24.506	28.000	32.134	37.036	42.861	28
29	13.676	15.308	17.220	19.470	22.125	25.268	29.000	33.444	38.747	45.091	29
30	13.853	15.546	17.539	19.896	22.693	26.022	30.000	34.765	40.490	47.384	30
31	14.020	15.773	17.846	20.310	23.250	26.770	31.000	36.100	42.266	49.740	31
32	14.178	15.989	18.142	20.713	23.796	27.510	32.000	37.446	44.074	52.163	32
33	14.327	16.195	18.426	21.104	24.333	28.244	33.000	38.806	45.917	54.653	33
34	14.467	16.391	18.700	21.484	24.859	28.971	34.000	40.178	47.794	57.214	34
35	14.600	16.579	18.963	21.854	25.376	29.690	35.000	41.563	49.706	59.846	35
36	14.725	16.757	19.217	22.213	25.883	30.404	36.000	42.960	51.654	62.552	36
37	14.843	16.928	19.461	22.562	26.381	31.110	37.000	44.371	53.638	65.334	37
38	14.955	17.090	19.696	22.902	26.869	31.810	38.000	45.795	55.659	68.193	38
39	15.060	17.244	19.923	23.231	27.348	32.503	39.000	47.233	57.718	71.133	39
40	15.160	17.392	20.141	23.552	27.818	33.190	40.000	48.683	59.816	74.156	40

Table A.7: **Present Worth of an Escalating Annual Amount—7% Discount Rate**

	Escalation Rate										
n	1.00%	2.00%	3.00%	4.00%	5.00%	6.00%	7.00%	8.00%	9.00%	10.00%	n
1	0.935	0.944	0.954	0.963	0.972	0.981	0.991	1.000	1.009	1.019	1
2	1.810	1.836	1.863	1.890	1.917	1.945	1.972	2.000	2.028	2.056	2
3	2.628	2.679	2.731	2.783	2.836	2.890	2.945	3.000	3.056	3.112	3
4	3.393	3.474	3.558	3.643	3.730	3.818	3.908	4.000	4.093	4.189	4
5	4.108	4.226	4.347	4.471	4.598	4.729	4.863	5.000	5.141	5.285	5
6	4.777	4.936	5.099	5.268	5.443	5.623	5.809	6.000	6.197	6.401	6
7	5.402	5.606	5.817	6.036	6.264	6.500	6.745	7.000	7.264	7.538	7
8	5.987	6.239	6.501	6.776	7.062	7.361	7.674	8.000	8.341	8.696	8
9	6.534	6.837	7.154	7.488	7.838	8.207	8.593	9.000	9.427	9.876	9
10	7.046	7.401	7.777	8.173	8.593	9.036	9.505	10.000	10.524	11.077	10
11	7.525	7.935	8.370	8.834	9.326	9.850	10.407	11.000	11.630	12.301	11
12	7.972	8.438	8.937	9.469	10.039	10.649	11.302	12.000	12.747	13.547	12
13	8.391	8.914	9.476	10.082	10.733	11.434	12.188	13.000	13.875	14.817	13
14	8.782	9.363	9.991	10.671	11.407	12.203	13.066	14.000	15.012	16.110	14
15	9.148	9.787	10.483	11.239	12.062	12.959	13.935	15.000	16.161	17.426	15
16	9.490	10.188	10.951	11.786	12.699	13.700	14.797	16.000	17.319	18.768	16
17	9.810	10.566	11.398	12.312	13.319	14.428	15.651	17.000	18.489	20.134	17
18	10.110	10.924	11.824	12.819	13.921	15.142	16.497	18.000	19.670	21.525	18
19	10.390	11.261	12.230	13.307	14.507	15.843	17.335	19.000	20.861	22.942	19
20	10.651	11.580	12.618	13.777	15.076	16.531	18.165	20.000	22.063	24.386	20
21	10.896	11.881	12.987	14.230	15.629	17.207	18.988	21.000	23.277	25.856	21
22	11.125	12.166	13.340	14.666	16.167	17.870	19.802	22.000	24.502	27.353	22
23	11.339	12.434	13.676	15.086	16.691	18.520	20.610	23.000	25.738	28.878	23
24	11.539	12.688	13.996	15.490	17.199	19.159	21.410	24.000	26.985	30.431	24
25	11.727	12.928	14.302	15.879	17.694	19.785	22.202	25.000	28.245	32.013	25
26	11.902	13.154	14.594	16.254	18.174	20.401	22.987	26.000	29.515	33.625	26
27	12.066	13.367	14.872	16.615	18.642	21.004	23.765	27.000	30.798	35.266	27
28	12.219	13.569	15.137	16.963	19.096	21.597	24.536	28.000	32.092	36.938	28
29	12.362	13.760	15.390	17.297	19.538	22.178	25.300	29.000	33.399	38.640	29
30	12.496	13.940	15.631	17.620	19.967	22.749	26.056	30.000	34.717	40.374	30
31	12.621	14.110	15.861	17.930	20.385	23.309	26.806	31.000	36.048	42.140	31
32	12.738	14.270	16.080	18.229	20.791	23.859	27.548	32.000	37.391	43.939	32
33	12.848	14.422	16.290	18.517	21.186	24.399	28.284	33.000	38.746	45.771	33
34	12.950	14.565	16.489	18.794	21.569	24.928	29.013	34.000	40.114	47.638	34
35	13.046	14.701	16.680	19.061	21.942	25.448	29.735	35.000	41.495	49.538	35
36	13.136	14.828	16.861	19.318	22.305	25.958	30.450	36.000	42.889	51.474	36
37	13.220	14.949	17.034	19.565	22.658	26.459	31.159	37.000	44.295	53.446	37
38	13.298	15.063	17.199	19.804	23.001	26.951	31.861	38.000	45.714	55.454	38
39	13.371	15.171	17.357	20.033	23.334	27.433	32.557	39.000	47.147	57.500	39
40	13.440	15.272	17.507	20.254	23.658	27.907	33.246	40.000	48.593	59.583	40

Table A.8: **Present Worth of an Escalating Annual Amount—8% Discount Rate**

	\multicolumn{10}{c}{Escalation Rate}	
n	1.00%	2.00%	3.00%	4.00%	5.00%	6.00%	7.00%	8.00%	9.00%	10.00%	n
1	0.918	0.927	0.936	0.945	0.955	0.964	0.973	0.982	0.991	1.000	1
2	1.761	1.787	1.813	1.839	1.866	1.892	1.919	1.946	1.973	2.000	2
3	2.535	2.584	2.634	2.684	2.735	2.787	2.839	2.892	2.946	3.000	3
4	3.246	3.324	3.403	3.483	3.566	3.649	3.735	3.821	3.910	4.000	4
5	3.899	4.009	4.123	4.239	4.358	4.480	4.605	4.734	4.865	5.000	5
6	4.498	4.645	4.797	4.953	5.115	5.281	5.453	5.630	5.812	6.000	6
7	5.048	5.234	5.428	5.628	5.837	6.053	6.277	6.509	6.750	7.000	7
8	5.553	5.781	6.019	6.267	6.526	6.796	7.078	7.372	7.680	8.000	8
9	6.017	6.288	6.572	6.871	7.184	7.513	7.858	8.220	8.601	9.000	9
10	6.443	6.758	7.090	7.441	7.812	8.203	8.616	9.053	9.513	10.000	10
11	6.834	7.194	7.575	7.981	8.411	8.868	9.354	9.870	10.418	11.000	11
12	7.193	7.598	8.030	8.491	8.983	9.510	10.072	10.672	11.314	12.000	12
13	7.523	7.972	8.455	8.973	9.530	10.127	10.770	11.460	12.202	13.000	13
14	7.825	8.320	8.853	9.429	10.051	10.723	11.449	12.233	13.082	14.000	14
15	8.103	8.642	9.226	9.860	10.549	11.297	12.109	12.993	13.954	15.000	15
16	8.358	8.941	9.576	10.268	11.024	11.849	12.752	13.738	14.818	16.000	16
17	8.593	9.218	9.903	10.653	11.477	12.382	13.377	14.470	15.674	17.000	17
18	8.808	9.475	10.209	11.018	11.910	12.895	13.985	15.189	16.523	18.000	18
19	9.005	9.713	10.496	11.362	12.323	13.390	14.576	15.895	17.363	19.000	19
20	9.187	9.934	10.764	11.688	12.718	13.867	15.151	16.588	18.196	20.000	20
21	9.353	10.139	11.015	11.996	13.094	14.326	15.711	17.268	19.022	21.000	21
22	9.506	10.329	11.251	12.287	13.454	14.769	16.255	17.936	19.840	22.000	22
23	9.647	10.505	11.471	12.562	13.797	15.196	16.784	18.591	20.650	23.000	23
24	9.776	10.668	11.678	12.822	14.124	15.607	17.299	19.235	21.454	24.000	24
25	9.894	10.819	11.871	13.069	14.437	16.003	17.800	19.867	22.250	25.000	25
26	10.003	10.960	12.052	13.301	14.735	16.384	18.287	20.488	23.038	26.000	26
27	10.102	11.090	12.221	13.521	15.020	16.752	18.761	21.097	23.820	27.000	27
28	10.194	11.211	12.380	13.729	15.291	17.107	19.222	21.695	24.594	28.000	28
29	10.278	11.323	12.528	13.926	15.551	17.448	19.671	22.283	25.361	29.000	29
30	10.355	11.426	12.667	14.112	15.799	17.777	20.107	22.859	26.122	30.000	30
31	10.426	11.523	12.798	14.287	16.035	18.095	20.532	23.426	26.875	31.000	31
32	10.491	11.612	12.920	14.453	16.261	18.400	20.944	23.982	27.622	32.000	32
33	10.551	11.695	13.034	14.610	16.476	18.695	21.346	24.527	28.362	33.000	33
34	10.606	11.771	13.141	14.759	16.682	18.979	21.736	25.063	29.095	34.000	34
35	10.657	11.843	13.241	14.899	16.878	19.252	22.116	25.589	29.821	35.000	35
36	10.703	11.909	13.335	15.032	17.065	19.516	22.486	26.106	30.541	36.000	36
37	10.745	11.970	13.423	15.158	17.244	19.770	22.845	26.613	31.254	37.000	37
38	10.784	12.027	13.505	15.276	17.415	20.014	23.195	27.111	31.961	38.000	38
39	10.820	12.079	13.582	15.389	17.578	20.250	23.535	27.600	32.661	39.000	39
40	10.853	12.128	13.654	15.495	17.733	20.478	23.866	28.080	33.355	40.000	40

Table A.9: **Present Worth of an Escalating Annual Amount—10% Discount Rate**

	\multicolumn{10}{c}{Escalation Rate}	
n	1.00%	2.00%	3.00%	4.00%	5.00%	6.00%	7.00%	8.00%	9.00%	10.00%	n
1	0.902	0.911	0.920	0.929	0.938	0.946	0.955	0.964	0.973	0.982	1
2	1.715	1.740	1.765	1.791	1.816	1.842	1.868	1.894	1.920	1.947	2
3	2.448	2.495	2.543	2.591	2.640	2.690	2.740	2.791	2.842	2.894	3
4	3.110	3.183	3.258	3.335	3.413	3.492	3.573	3.655	3.739	3.825	4
5	3.706	3.810	3.916	4.025	4.137	4.252	4.369	4.489	4.612	4.738	5
6	4.244	4.380	4.521	4.666	4.816	4.970	5.129	5.293	5.462	5.636	6
7	4.729	4.900	5.078	5.262	5.452	5.650	5.856	6.068	6.289	6.517	7
8	5.166	5.373	5.589	5.814	6.049	6.294	6.550	6.816	7.094	7.383	8
9	5.561	5.804	6.060	6.328	6.609	6.903	7.213	7.537	7.877	8.234	9
10	5.916	6.197	6.492	6.804	7.133	7.480	7.846	8.232	8.639	9.069	10
11	6.237	6.554	6.890	7.247	7.625	8.026	8.451	8.902	9.381	9.889	11
12	6.526	6.880	7.256	7.658	8.086	8.542	9.029	9.549	10.103	10.694	12
13	6.787	7.176	7.593	8.039	8.518	9.031	9.581	10.172	10.805	11.486	13
14	7.022	7.446	7.902	8.394	8.923	9.494	10.109	10.773	11.489	12.263	14
15	7.234	7.692	8.187	8.723	9.303	9.931	10.613	11.352	12.155	13.026	15
16	7.426	7.916	8.449	9.028	9.659	10.346	11.095	11.911	12.802	13.775	16
17	7.598	8.120	8.689	9.312	9.993	10.738	11.555	12.450	13.433	14.511	17
18	7.754	8.306	8.911	9.575	10.306	11.109	11.994	12.970	14.046	15.234	18
19	7.894	8.475	9.114	9.820	10.599	11.460	12.414	13.471	14.643	15.945	19
20	8.020	8.629	9.302	10.047	10.874	11.793	12.815	13.954	15.224	16.642	20
21	8.134	8.769	9.474	10.258	11.132	12.108	13.198	14.420	15.789	17.327	21
22	8.237	8.897	9.632	10.454	11.374	12.405	13.565	14.869	16.340	18.000	22
23	8.330	9.013	9.778	10.636	11.600	12.687	13.914	15.302	16.875	18.660	23
24	8.414	9.119	9.912	10.805	11.813	12.954	14.249	15.720	17.396	19.309	24
25	8.489	9.216	10.035	10.961	12.012	13.207	14.568	16.123	17.904	19.947	25
26	8.557	9.304	10.148	11.107	12.199	13.445	14.873	16.512	18.397	20.573	26
27	8.619	9.384	10.252	11.242	12.374	13.672	15.164	16.886	18.878	21.187	27
28	8.674	9.456	10.348	11.368	12.538	13.886	15.443	17.247	19.345	21.791	28
29	8.724	9.523	10.436	11.484	12.692	14.088	15.709	17.596	19.800	22.384	29
30	8.769	9.583	10.517	11.593	12.836	14.280	15.963	17.931	20.243	22.967	30
31	8.809	9.638	10.592	11.693	12.971	14.461	16.205	18.255	20.674	23.539	31
32	8.846	9.689	10.660	11.787	13.098	14.633	16.437	18.568	21.094	24.100	32
33	8.879	9.734	10.723	11.873	13.217	14.796	16.659	18.869	21.502	24.652	33
34	8.909	9.776	10.781	11.954	13.328	14.949	16.871	19.159	21.899	25.194	34
35	8.935	9.814	10.835	12.028	13.433	15.095	17.073	19.439	22.286	25.726	35
36	8.960	9.848	10.884	12.098	13.531	15.233	17.266	19.709	22.662	26.249	36
37	8.981	9.880	10.929	12.162	13.623	15.363	17.450	19.970	23.028	26.763	37
38	9.001	9.908	10.970	12.222	13.709	15.486	17.627	20.221	23.385	27.267	38
39	9.019	9.934	11.008	12.278	13.789	15.603	17.795	20.463	23.731	27.762	39
40	9.035	9.958	11.043	12.329	13.865	15.714	17.956	20.696	24.069	28.248	40

Table A.10: **Present Worth of an Escalating Annual Amount—12% Discount Rate**

Ratio of Initial Cost to First Year Savings	Fuel Price Escalation Rate					
	0%	2%	4%	6%	8%	10%
1	1.06	1.04	1.02	1.00	0.98	0.96
2	2.19	2.12	2.06	2.00	1.95	1.90
3	3.14	3.25	3.12	3.00	2.89	2.80
4	4.71	4.44	4.20	4.00	3.82	3.67
5	6.12	5.67	5.31	5.00	4.74	4.51
6	7.66	6.97	6.44	6.00	5.64	5.33
7	9.35	8.34	7.59	7.00	6.52	6.12
8	11.22	9.79	8.77	8.00	7.39	6.89
9	13.33	11.32	9.98	9.00	8.25	7.64
10	15.73	12.94	11.21	10.00	9.09	8.37
11	18.51	14.68	12.48	11.00	9.92	9.08
12	21.85	16.53	13.77	12.00	10.74	9.78
13	25.99	18.53	16.10	13.00	11.54	10.45
14	31.45	20.70	16.47	14.00	12.33	11.11
15	39.52	23.07	17.87	15.00	13.11	11.75
16	55.24	25.67	19.30	16.00	13.88	12.38
17	never	28.56	20.78	17.00	14.64	12.99
18	never	31.81	22.31	18.00	15.39	13.59
19	never	35.53	23.87	19.00	16.13	14.18
20	never	39.88	25.49	20.00	16.86	14.75

Table A.11: **Years to Payback at 6% Discount Rate**

Ratio of Initial Cost to Annual Savings	Fuel Price Escalation Rate					
	0	2	4	6	8	10
1	1.07	1.05	1.03	1.01	0.99	0.97
2	2.23	2.16	2.09	2.03	1.97	1.92
3	3.48	3.32	3.18	3.06	2.95	2.84
4	4.86	4.56	4.31	4.10	3.91	3.74
5	6.37	5.88	5.48	5.15	4.87	4.62
6	8.05	7.28	6.68	6.21	5.81	5.48
7	9.95	8.78	7.93	7.28	6.75	6.32
8	12.13	10.40	9.23	8.36	7.68	7.14
9	14.70	12.16	10.57	9.45	8.60	7.94
Example 10	17.79	14.08	11.97	10.55	9.52	8.72
11	21.72	16.19	13.42	11.67	10.43	9.49
12	27.09	18.54	14.94	12.80	11.33	10.24
13	35.59	21.19	16.53	13.93	12.22	10.97
14	57.82	24.22	18.19	15.09	13.10	11.70
15	never	27.77	19.93	16.25	13.98	12.40
16	never	32.05	21.77	17.43	14.85	13.10
17	never	37.44	23.70	18.62	15.71	13.78
18	never	44.72	25.75	19.82	16.57	14.44
19	never	55.98	27.93	21.04	17.42	15.10
20	never	82.16	30.25	22.27	18.26	15.74

Payback

0% Price Escalation: 17.8 years
10% Price Escalation: 8.7 years

Table A.12: **Years to Payback at 7% Discount Rate**

Ratio of Initial Cost to First Year Savings	Fuel Price Escalation Rate					
	0%	2%	4%	6%	8%	10%
1	1.08	1.06	1.04	1.02	1.00	0.98
2	2.27	2.19	2.12	2.06	2.00	1.95
3	3.57	3.40	3.25	3.12	3.00	2.89
4	5.01	4.69	4.43	4.20	4.00	3.83
5	6.64	6.09	5.66	5.30	5.00	4.74
6	8.50	7.62	6.95	6.43	6.00	5.64
7	10.67	9.28	8.31	7.58	7.00	6.53
8	13.27	11.13	9.74	8.75	8.00	7.40
9	16.54	13.19	11.26	9.96	9.00	8.26
10	20.91	15.52	12.86	11.19	10.00	9.10
11	27.55	18.22	14.57	12.44	11.00	9.94
12	41.82	21.41	16.40	13.73	12.00	10.76
13	never	25.31	18.37	15.06	13.00	11.56
14	never	30.35	20.49	16.41	14.00	12.36
15	never	37.44	22.79	17.80	15.00	13.14
16	never	49.57	25.32	19.23	16.00	13.92
17	never	never	28.11	20.69	17.00	14.68
18	never	never	31.23	22.20	18.00	15.43
19	never	never	34.77	23.75	19.00	16.17
20	never	never	38.85	25.35	20.00	16.90

Table A.13: **Years to Payback at 8% Discount Rate**

Ratio of Initial Cost to First Year Savings	Fuel Price Escalation Rate					
	0%	2%	4%	6%	8%	10%
1	1.11	1.08	1.06	1.04	1.02	1.00
2	2.34	2.26	2.19	2.12	2.06	2.00
3	3.74	3.55	3.39	3.24	3.12	3.00
4	5.36	4.99	4.68	4.42	4.19	4.00
5	7.27	6.59	6.07	5.64	5.30	5.00
6	9.61	8.42	7.58	6.93	6.42	6.00
7	12.63	10.55	9.22	8.28	7.57	7.00
8	16.89	13.08	11.04	9.70	8.74	8.00
9	24.16	16.21	13.06	11.20	9.94	9.00
10	never	20.31	15.34	12.79	11.16	10.00
11	never	26.30	17.95	14.48	12.41	11.00
12	never	37.52	21.01	16.28	13.70	12.00
13	never	never	24.72	18.21	15.01	13.00
14	never	never	29.39	20.29	16.36	14.00
15	never	never	35.75	22.54	17.74	15.00
16	never	never	45.73	24.99	19.15	16.00
17	never	never	70.45	27.69	20.60	17.00
18	never	never	never	30.70	22.10	18.00
19	never	never	never	34.08	23.63	19.00
20	never	never	never	37.94	25.21	20.00

Table A.14: **Years to Payback at 10% Discount Rate**

Ratio of Initial Cost to First Year Savings	Fuel Price Escalation Rate					
	0%	2%	4%	6%	8%	10%
1	1.13	1.10	1.08	1.06	1.04	1.02
2	2.42	2.33	2.25	2.18	2.12	2.06
3	3.94	3.72	3.54	3.38	3.24	3.11
4	5.77	5.32	4.96	4.66	4.41	4.19
5	8.09	7.20	6.55	6.04	5.63	5.29
6	11.23	9.49	8.35	7.54	6.91	6.41
7	16.17	12.39	10.43	9.16	8.25	7.56
8	28.40	16.40	12.89	10.95	9.66	8.72
9	never	22.88	15.90	12.93	11.15	9.92
10	never	42.04	19.79	15.16	12.72	11.14
11	never	never	25.26	17.70	14.39	12.38
12	never	never	34.61	20.65	16.16	13.66
13	never	never	never	24.18	18.06	14.97
14	never	never	never	28.56	20.10	16.30
15	never	never	never	34.34	22.30	17.67
16	never	never	never	42.88	24.69	19.08
17	never	never	never	59.52	27.31	20.52
18	never	never	never	never	30.21	22.00
19	never	never	never	never	33.45	23.52
20	never	never	never	never	37.12	25.08

Table A.15: **Years to Payback at 12% Discount Rate**

Appendix B Energy-Estimating Data

Energy-Estimating Data

This appendix assists the reader in budgeting and developing preliminary energy estimates for various types of facilities. These tables, when coupled with engineering judgment, allow the design professional to estimate the energy consumption of various design alternatives. Chapter 5, "Estimating Life Cycle Costs," provides a discussion of the use of the tables.

Numerous energy consumption methods are available to the reader for similar but not much more detailed analysis. A brief discussion of these methods is contained in Chapter 5. The bibliography also contains sources for more detailed information.

Table 1 – Building Type	Conditions	Qualifications	Loss Factor*
Factories & Industrial Plants General Office Areas at 70°F	One Story	Skylight in Roof	6.2
		No Skylight in Roof	5.7
	Multiple Story	Two Story	4.6
		Three Story	4.3
		Four Story	4.1
		Five Story	3.9
		Six Story	3.6
	All Walls Exposed	Flat Roof	6.9
		Heated Space Above	5.2
	One Long Warm Common Wall	Flat Roof	6.3
		Heated Space Above	4.7
	Warm Common Walls on Both Long Sides	Flat Roof	5.8
		Heated Space Above	4.1
Warehouses at 60°F	All Walls Exposed	Skylights in Roof	5.5
		No Skylight in Roof	5.1
		Heated Space Above	4.0
	One Long Warm Common Wall	Skylight in Roof	5.0
		No Skylight in Roof	4.9
		Heated Space Above	3.4
	Warm Common Walls on Both Long Sides	Skylight in Roof	4.7
		No Skylight in Roof	4.4
		Heated Space Above	3.0

*Note: This table tends to be conservative particularly for new buildings designed for minimum energy consumption.

Table 2 – Outside Design Temperature Correction Factor (for Degrees Fahrenheit)									
Outside Design Temperature	50	40	30	20	10	0	-10	-20	-30
Correction Factor	.29	.43	.57	.72	.86	1.00	1.14	1.28	1.43

Source: 2003 *Means Mechanical Cost Data*

Table B.1a: **Factors for Determining Heat Loss for Various Types of Buildings**

$$E = \left(\frac{H_L \times D \times 24}{\Delta t \times \eta \times V}\right) (C_D)(C_F)$$

where
- E = Fuel or energy consumption for the estimate period.
- H_L = Design heat loss, including infiltration, Btu per hour.
- D = Number of 65 F degree days for the estimate period.
- Δt = Design temperature difference, Fahrenheit.
- η = Rated full load efficiency, decimal.
- V = Heating value of fuel, consistent with H_L and E.
- C_D = Interim correction factor for heating effect vs. degree days.
- C_F = Interim part-load correction factor for fueled systems only; equals 1.0 for electric resistance heating.

Heat Loss vs. Degree Days Interim Factor C_D

Outdoor Design Temp, F	-20	-10	0	+10	+20
Factor C_D	0.57	0.64	0.71	0.79	0.89

The multipliers in the above Table are high for mild climates and low for cold regions, are not in error as might appear. For equivalent buildings, those in warm climates have a greater portion of their heating requirements on days when the mean temperature is close to 65°F, and thus the actual heat loss is not reflected.

Part-Load Correction Factor for Fuel-Fired Equipment

Percent oversizing	0	20	40	60	80
Factor C_F	1.36	1.56	1.79	2.04	2.32

Because equipment performance at extremely low loads is highly variable, it is strongly recommended that the values in the above Table not be extrapolated.

Table B.1b: **Heating Energy Calculation, Modified Degree Day Formula**

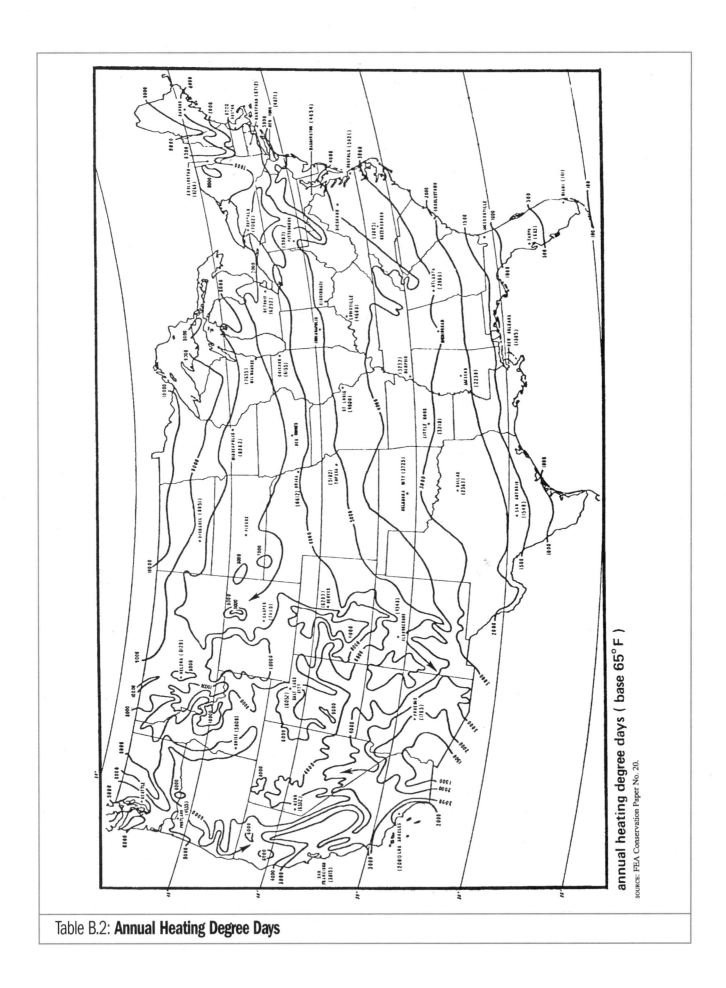

Table B.2: **Annual Heating Degree Days**

Air-Conditioning Basics

Type of Building	Tons of A/C per S.F.	Ductwork # / SF of Building Area			
		Rooftop Unit Single Zone	Rooftop Unit Multizone	Self-contained Air- or Water-cooled	Split system Air- or Water-cooled
Apartments - Individual	0.0022	0.267	0.533	0.240	0.227
Corridors	0.0018	0.218	0.436	0.196	0.185
Auditoriums and Theaters	0.0033	0.400	0.800	0.360	0.340
Banks	0.0042	0.500	1.000	0.450	0.425
Barber Shops	0.0040	0.480	0.960	0.432	0.408
Bars and Taverns	0.0111	1.333	2.667	1.200	1.133
Beauty Parlors	0.0056	0.667	1.333	0.600	0.567
Bowling Alleys	0.0057	0.686	1.371	0.617	0.583
Churches	0.0030	0.364	0.727	0.327	0.309
Cocktail Lounges	0.0057	0.686	1.371	0.617	0.583
Computer Rooms	0.0118	1.412	2.824	1.271	1.200
Dental Offices	0.0043	0.522	1.043	0.470	0.443
Department Store, Basement	0.0029	0.343	0.686	0.309	0.291
Main Floor	0.0033	0.400	0.800	0.360	0.340
Upper Floor	0.0025	0.300	0.600	0.270	0.255
Dormitory Rooms	0.0033	0.400	0.800	0.360	0.340
Corridors	0.0025	0.300	0.600	0.270	0.255
Dress Shops	0.0036	0.429	0.857	0.386	0.364
Drug Stores	0.0067	0.800	1.600	0.720	0.680
Factories	0.0033	0.400	0.800	0.360	0.340
High Rise Office, Ext. Rooms	0.0038	0.456	0.913	0.411	0.388
Int. Rooms	0.0031	0.369	0.738	0.332	0.314
Hospitals, Core	0.0036	0.429	0.857	0.386	0.364
Perimeter	0.0038	0.462	0.923	0.415	0.392
Hotel, Guest Rooms	0.0036	0.436	0.873	0.393	0.371
Corridors	0.0025	0.300	0.600	0.270	0.255
Public Spaces	0.0045	0.545	1.091	0.491	0.464
Industrial Plants, Offices	0.0031	0.375	0.750	0.338	0.319
General Offices	0.0029	0.343	0.686	0.309	0.291
Plant Areas	0.0033	0.400	0.800	0.360	0.340
Libraries	0.0042	0.500	1.000	0.450	0.425
Low Rise Office, Exterior	0.0031	0.375	0.750	0.338	0.319
Interior	0.0028	0.333	0.667	0.300	0.283
Medical Centers	0.0024	0.282	0.565	0.254	0.240
Motels	0.0024	0.282	0.565	0.254	0.240
Office (small suite)	0.0036	0.429	0.857	0.386	0.364
Post Office, Individual Office	0.0035	0.421	0.842	0.379	0.358
Central Area	0.0038	0.462	0.923	0.415	0.392
Residences	0.0017	0.200	0.400	0.180	0.170
Restaurants	0.0050	0.600	1.200	0.540	0.510
Schools and Colleges	0.0038	0.462	0.923	0.415	0.392
Shoe Stores	0.0045	0.545	1.091	0.491	0.464
Shopping Centers, Supermarkets	0.0029	0.343	0.686	0.309	0.291
Retail Stores	0.0040	0.480	0.960	0.432	0.408
Specialty	0.0050	0.600	1.200	0.540	0.510

Note: In addition to the ductwork an allowance should be made for diffusers, registers, insulation and accessories such as turning vanes, volume dampers, fire dampers and access doors.

Source: 2003 *Means Estimating Handbook*

Central Air Conditioning Watts per S.F., BTU's per Hour per S.F. of Floor Area and S.F. per Ton of Air Conditioning

Type Building	Watts per S.F.	BTUH per S.F.	S.F. per Ton	Type Building	Watts per S.F.	BTUH per S.F.	S.F. per Ton	Type Building	Watts per S.F.	BTUH per S.F.	S.F. per Ton
Apartments, Individual	3	26	450	Dormitory, Rooms	4.5	40	300	Libraries	5.7	50	240
Corridors	2.5	22	550	Corridors	3.4	30	400	Low Rise Office, Ext.	4.3	38	320
Auditoriums & Theaters	3.3	40	300/18*	Dress Shops	4.9	43	280	Interior	3.8	33	360
Banks	5.7	50	240	Drug Stores	9	80	150	Medical Centers	3.2	28	425
Barber Shops	5.5	48	250	Factories	4.5	40	300	Motels	3.2	28	425
Bars & Taverns	15	133	90	High Rise Off.-Ext. Rms.	5.2	46	263	Office (small suite)	4.9	43	280
Beauty Parlors	7.6	66	180	Interior Rooms	4.2	37	325	Post Office, Int. Office	4.9	42	285
Bowling Alleys	7.8	68	175	Hospitals, Core	4.9	43	280	Central Area	5.3	46	260
Churches	3.3	36	330/20*	Perimeter	5.3	46	260	Residences	2.3	20	600
Cocktail Lounges	7.8	68	175	Hotels, Guest Rooms	5	44	275	Restaurants	6.8	60	200
Computer Rooms	16	141	85	Public Spaces	6.2	55	220	Schools & Colleges	5.3	46	260
Dental Offices	6	52	230	Corridors	3.4	30	400	Shoe Stores	6.2	55	220
Dept. Stores, Basement	4	34	350	Industrial Plants, Offices	4.3	38	320	Shop'g. Ctrs., Sup. Mkts.	4	34	350
Main Floor	4.5	40	300	General Offices	4	34	350	Retail Stores	5.5	48	250
Upper Floor	3.4	30	400	Plant Areas	4.5	40	300	Specialty Shops	6.8	60	200

*Persons per ton

Source: 2003 *Means Electrical Cost Data*

12,000 BTUH = 1 ton of air conditioning

Table B.3: Air Conditioning Data

Table B.4: Equivalent Full-Load Hours of Operation Per Year

	Atlanta	Balti-more	Boston	Chicago	Dallas	Denver	Detroit	Los Angeles	Miami	Milwau-kee	Minne-apolis	New Orleans	Okla. City	Phila-delphia	Phoenix	Portland Ore.	San Fran.	Saint Louis	Wash., D.C.	New York
Restaurants	1750	1620	1050	1250	2240	1050	1250	1150	2020	1050	1050	2020	2240	1480	2240	1050	450	2020	1820	1430
Drug Stores	1700	1580	1030	1220	2170	1030	1220	1120	1850	1030	1030	1950	2170	1440	2170	1030	400	1950	1580	1400
Cafeterias	1370	1270	825	990	1750	825	990	910	1580	825	825	1580	1750	1160	1750	825	350	1580	1270	1120
Jewelry Stores	1020	950	620	750	1300	620	750	700	1170	620	620	1170	1300	875	1300	620	250	1170	950	850
Barber Shops	1020	950	620	750	1300	620	750	700	1170	620	620	1170	1300	875	1300	620	250	1170	950	850
Night Clubs	1010	940	610	730	1280	610	730	675	1150	610	610	1150	1280	860	1280	610	240	1150	940	840
Theaters	650–1000	600–1000	400–650	500–800	850–1400	400–650	500–800	475–750	800–1300	400–650	400–650	800–1300	850–1400	550–920	850–1400	400–650	200–400	800–1300	600–1000	550–900
Dress Shops	940	870	565	675	1200	565	675	630	1080	565	565	1080	1200	800	1200	565	225	1080	870	780
Large Offices	915	850	550	660	1180	550	660	610	1060	550	550	1060	1180	775	1180	550	200	1060	850	750
Department Stores	850	790	515	650	1100	515	650	600	1000	515	515	1000	1100	725	1100	515	175	1000	790	700
Specialty Shops (5 & 10)	840	780	510	640	1080	510	640	590	975	510	510	975	1080	710	1080	510	175	975	780	700
Residences	810	750	490	600	1050	490	600	550	950	490	490	950	1050	690	1050	490	170	950	750	670
Shoe Stores	650	600	400	500	850	400	500	475	775	400	400	775	850	575	850	400	150	775	600	550
Beauty Shops	625	580	380	450	800	380	450	425	750	380	380	750	800	540	800	380	150	750	530	525
Small Offices	540	500	425	450	700	425	450	410	650	425	425	650	700	490	700	425	125	650	500	475
Recreation Spaces	520	480	450	400	675	450	400	380	650	450	450	650	675	470	675	450	125	650	480	450
Funeral Parlors	460	425	350	375	600	350	375	350	575	350	350	575	600	410	600	350	100	575	425	400

SOURCE: Trane Air Conditioning Manual.

Ventilation Volume

$$V = \frac{CS \times S.F.}{PO}$$

Where
- V = CFM air supplied
- CS = Average CFM/SF of outside air (0.1 to 0.3)
- S.F. = Total Square Footage
- PO = Percent Outside Air (0.1 to 0.25)

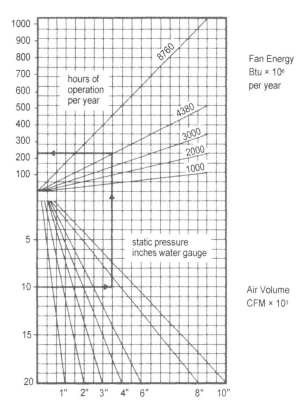

(Normal design loss = 0.1" per 100' of duct)

Source: *Federal Energy Administration*

Table B.5: **Ventilation Energy**

IESNA recommended illumination levels of foot-candles

Commercial Buildings

Type	Description	Foot-candles
Bank	Lobby	50
	Customer Areas	70
	Teller Stations	150
Offices	Routine Work	100
	Accounting	150
	Drafting	200
	Corridors, Halls, Washrooms	30
Schools	Reading or Writing	70
	Drafting, Labs, Shops	100
	Libraries	70
	Auditoriums, Assembly	15
	Auditoriums, Exhibition	30
Stores	Circulation Areas	30
	Stock Rooms	30
	Merchandise Areas, Service	100
	Self-Service Areas	200

Industrial

Type	Description	Foot-candles
Assembly	Rough bench & machine work	50
	Medium bench & machine work	100
	Fine bench & machine work	500
Inspection Areas	Ordinary	50
	Difficult	100
	Highly Difficult	200
Material Handling	Loading	20
	Stock Picking	30
	Packing, Wrapping	50
Stairways	Service Areas	20
Washrooms	Service Areas	30
Storage Areas	Inactive	5
	Active, Rough, Bulky	10
	Active, Medium	20
	Active, Fine	50
Garages	Active Traffic Areas	20
	Service & Repair	100

Lighting Energy Calculations

$$E = \frac{W \times S.F. \times Hr}{1000},$$

E = annual lighting Energy (kWh)
W = Approximate Watts/S.F. for are considered
S.F. = Square Footage of area served
Hr. = Annual Use Hours

Source: RSMeans

Table B.6: IESNA Recommended Illumination Levels of Foot-candles

Lamp Type	Smaller Sizes	Middle Sizes	Larger Sizes
Low Pressure Sodium	90	120	150
High Pressure Sodium	84	105	126
Metal Halide	67	75	93
Fluorescent	66	74	70
Mercury	44	51	57
Incandescent	17	22	24
Source: NEMA			

Table B.7: **Approximate Initial Lumens per Watt**

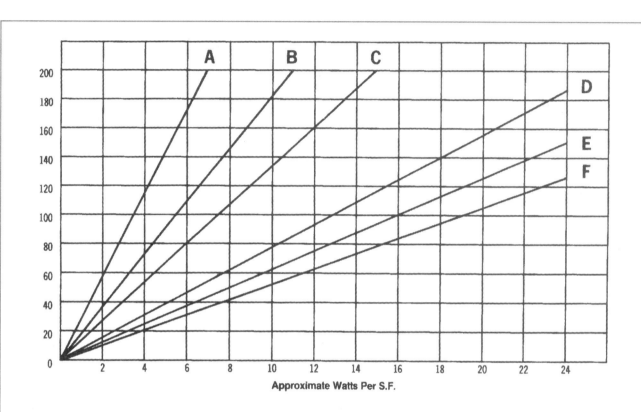

Due to the many variables involved, use for preliminary estimating only:

A. Fluorescent – industrial
B. Fluorescent – lens unit, Fixture types B & C
C. Fluorescent – louvered unit
D. Incandescent – open reflector, Type D
E. Incandescent – lens unit, Type A
F. Incandescent – down light, Type B

Source: 2003 *Means Electrical Cost Data*

Table B.8: **Approximate Watts Per Square Foot for Popular Fixture Types**

Appendix C: Elevator and Escalator LCC Considerations

By Edward Robinson

Elevators

Background Information

Definitions

The Safety Code for Elevators and Escalators (ASME A-17.1) defines an elevator as a hoisting and lowering mechanism equipped with a car or platform that moves in guide rails, serves two or more landings, and is classified by the following types: Freight, Gravity, Inclined, Multi-deck, Observation, Passenger, Power, Electric, Hydraulic, Direct Plunger Hydraulic, Electro-Hydraulic, Roped Hydraulic, Rack and Pinion, Screw Column, Private Residence, Rooftop, Shipboard, Sidewalk, and Construction.

Elevator Types and Limitations

As the definitions imply, there are numerous types of elevators, although most of them can be broken down into two classifications, and further into two types:

Electric elevators will be classified as either a *passenger* or a *freight* elevator. This is determined by the construction, as well as the landing door operation. Elevators might be referred to as a "hospital" elevator or a "building service" elevator, but in reality these are passenger elevators with a narrow and deep car, and wider doors to accommodate loading.

Electric elevators are further broken down into *geared* or *gearless traction* elevators. Geared elevators are used for slow to medium speed floors, 50 to 350 feet per minute (FPM). Although the normal break in machines is 350 FPM, some manufacturers can provide up to 450 FPM. Gearless are used for high-speed applications, normally above 14 floors of travel where the elevator can run at high speed, 500–1400 feet per minute.

Passenger elevators are rated by the inside net area of the platform which is listed in the A17.1 Code. Normally, this is not a focal point unless the project requires an odd size platform. It is best to use standard size elevators whenever possible because of cost impact. *Note: One should be careful about referencing manufacturers' literature (as in sources such as the Sweet Catalog) because it is possible to unknowingly specify a duty, capacity, hoistway size, etc., that eliminates competitive bidding.*

Freight elevators are sized and rated for the type of load they will be required to carry. They can be over-rated for a given size platform, but must not be under-rated. An example would be a Class A—Freight elevator of 4,000 lbs. capacity with a platform size of 80 square feet inside area. This would be the minimum capacity of this size platform, although this size platform could be designed to be rated for 10,000 lbs. or more.

The minimum size platform for any given capacity is listed in the A17.1 Code of Freight elevators are classified in different categories: Class A, B & C. Class A is used for general freight, Class B for carrying automobiles, and Class C for heavy, concentrated loads. Class C is also broken down into three additional types, Class C1, C2, and C3.

Hydraulic elevators are rated and sized the same as electric elevators and are used for low-rise buildings where speed is not important. They are substantially lower in initial costs and annual maintenance. Also, they do not require a penthouse above the roof. It should be pointed out that hydraulic elevators have larger electrical requirements. Although there is no limitation to height of travel, strong consideration should be given to limiting the travel to less than 45 feet with a maximum of 4 stops. An exception would be an extremely low usage freight elevator or one for building users with disabilities. Normal operating speed is 50–150 FPM. Although 225 FPM is obtainable, it is not recommended because of the increased cost, electrical requirements and operating cost. When 225 FPM is required, an electric elevator might be more suitable.

Elevator Layout Within Building Space

Although it places some restraints on the designer, the way to achieve the least number of elevators is to locate them in a center location. A rough rule of thumb is that the maximum number of people served by any group or bank of elevators should not exceed 1,600. An example would be an 8-story building, which, because of the building population, would require 8 elevators to adequately handle the 5-minute handling capacity. Rather than provide an 8-car group to theoretically have an elevator for every floor, two 4-car groups might be provided. Another consideration in the maximum distance to walk to the elevators.

The ADA requirements also have an impact on the arrangements of elevators, by requiring a minimum of time each door remains open. This is determined by the distance from the landing push-button to the furthest landing door of the group. There is a formula for this condition.

Estimated Elevator Peak Passenger Traffic

Following is a sample assessment:

    Full-load passenger traffic

    5-Story office building

Building located in a major city

Maximum 6 hours of peak-load elevator passenger traffic

Average 4 hours of peak-load elevator passenger traffic

(Note: Elevator consultants can provide an objective assessment, beyond the presentation of elevator sales people. They should be experienced and knowledgeable in elevator applications, having worked for elevator contractors.)

On most projects, it is possible to determine the number and capacity by calling one of the elevator companies. Remember that during the schematics phase, engineering data obtained from an elevator contractor should be increased approximately 10%, as it will vary with each manufacturer.

Rough Rules of Thumb for Elevator Capacity

During the preliminary design stage, a rough rule of thumb is one elevator for every two floors. This rule normally applies unless the building footprint is spread out over a larger than normal area. Four elevators normally will handle 8 floors and sometimes up to 10 floors. Six elevators will handle up to 14 floors. Remember, this is a rule of thumb only and should be used only in the very preliminary schematic design phase.

Another rule of thumb is that any core or bank of elevators should be limited to serve a maximum of 1,600 people.

When a building is contemplated above 4 floors, consideration should be given for obtaining an elevator consultant. Above 6 or more floors, the requirements for elevators differ with each and every building. Therefore, an elevator consultant and transportation traffic study is strongly recommended. The following information will be required by the consultant prior to performing a traffic study:

- Location of building
- Type of building—office/hospital/warehouse, mercantile, etc.
- Single or multi-tenant and employee working hours
- Building population
- Floor heights
- Cafeterias or auditoriums
- Any high concentration of people
- Means of employee travel to the building
- Attached or remote parking areas
- Means of egress from the building

The number of elevators required is determined by the average interval and the number of people they can handle in a five-minute period. For a moderately priced building, the design might be for 11% of the population with a 35-second interval, whereas a first-class, single-tenant, office building may require 16% of the population with a 25-second interval. These requirements not only change with the tenant, but also with the location of the building. An example would be a building located in a small suburb in Michigan versus one in Manhattan.

Special attention should be directed toward hospital projects, as they require material handling elevators as well as passenger elevators. Additional systems within the hospital might reduce the elevator requirements. These include: trash and linen systems, automatic load and unload car lift systems, dumbwaiters, and automatic box conveyers (more commonly referred to as *electric track vehicle systems*).

For existing buildings, a survey of the existing elevators will be required. An elevator consultant is usually also required, as it is against most state elevator codes for anyone to enter a hoistway without a General Inspector's, Contractor's, or Elevator Journey-person's license. It is sometimes possible to contact the company currently maintaining the equipment and obtain their recommendations. This will normally be adequate unless a full-blown modernization is planned, in which case a consultant will definitely be required. Note that the majority of existing elevators do not meet ADA requirements.

Discipline Requirements:

Each discipline requires different information at different times within the contract development process.

Design:
- Number and approximate size
- Penthouse requirements

Architectural:
- Number of elevators (consider whether you need a consultant)
- Hoistway size and location
- Maximum spacing of guide rail supports
- Hoistway ventilation requirements
- Run-by clearances
- Width of elevator corridor
- Entrance and cab design requirements
- Special specification requirements

Structural:
- Hoistway size
- Pit depth
- Waterproofing requirements
- Pit sump requirements
- Openings in pit floor
- Structural loadings
- Maximum beam sizes under machine room slab
- Divider beam locations and maximum sizes

Electrical:
- Horsepower requirements, add 10%
- Disconnect locations; locate next to door strike and within sight of machines
- Pit and machine room lighting requirements, panel to be located in machine room
- Communication requirements
- Cathodic protection requirements, if any
- Emergency power requirements

Mechanical: Heating and ventilating requirements

Estimating: Budget costs

Elevator Installation Costs

The cost of elevators is similar to that of cars. It depends on the type. For example, a 2-stop hydraulic type can be purchased for as little as $28,000, while a high-rise freight elevator might cost up to $325,000.

The following are approximate budget prices based on the East Coast:

3-stop, 10,000 lbs. hydraulic freight with Class A loading and power operated doors: $170,000.

Add $15,000 for each additional landing served.

Add 20% if the class of loading is increased to C1 or C2 loading.

5-stop, 3,000 lbs. geared, electric passenger @ 350 FPM: $95,000.

Add $9,000 for each additional landing served.

5-stop, 4,500 lbs. geared, electric passenger @ 350 FPM: $105,000.

Add $9,000 for each additional landing served.

5-stop, 4,500 lbs. gearless, electric passenger @ 700 FPM: $255,000.

The code allows what is referred to as *Special Elevating Devices* to be installed in lieu of full-blown elevators. These units are normally installed specifically for handicapped building users, and there are specific limitations associated with their use: size of car, vertical travel (15 ft.), number of landings served (2) and square feet of the area served (15,000 S.F.).

It would be very difficult to try to itemize a budget price for each and every application. Prices of specific units vary from type, quantity, capacity, speed, manufacturer's standard applications, cab, entrances and fixture design, location of the project, as well as time of year.

These prices are just rough guidelines. It will be necessary to obtain budget prices from a local elevator contractor or consultant for each project. Also, remember that the bid price for government projects and projects in most large cities can be 25–30% higher than what would normally be expected.

Elevator Codes

The location of the project determines which elevator code must be followed. Some states, and even cities, have written their own codes. Some have copied the A-17.1 with minor modifications, and some with major modifications, thereby creating the potential need for an addendum or bulletin for those who are not knowledgeable of the code in force. When starting a project, it is important to have the current code for that city or state.

A few items that will change with the different states are the number of elevators in any one hoistway, hoistway smoke vent requirements, firefighter's service, communication requirements, location of elevator machine room, etc. Review of applicable codes is a must. A local elevator contractor or consultant should also be able to point out any major differences.

When laying out elevator arrangements, the *Vertical Transportation Standards* publication should be followed. Each manufacturer differs slightly from their competition and will request changes and modifications so they can bid their standards. CAUTION: Research every request in detail prior to agreeing to accommodate requested changes. Determine whether the change can be made without penalizing the competitor, whether it will require changes to the drawings and specifications, and whether it will add cost.

Guide Rail Attachments

The vertical distance between guide rail attachments to the building should not exceed 14 feet. It is possible to span up to 18 feet by providing structural channel rail backing behind the guide rail, but the dimensions on each side of the platform will need to be increased by one inch on each side. When the distance exceeds 18 feet, it will be necessary to provide a structural rail ladder unless the hoistway is constructed of concrete, providing an opportunity for an attachment at the required spacing.

Do not attempt to locate an attachment into a masonry block wall unless seven or eight courses of block have been filled solid and doled together. The reason is not that the insert pulls out of the masonry, but rather the vibration (over time) causes the mortar to break, and the only thing holding the insert is the weight of the wall.

Heating and Ventilation

The A-17.1 code requires the elevator machine room to be heated and ventilated. With the development of total solid state equipment, some manufacturers require the temperature in the machine room not to exceed 95°F. Depending on the project location, this basically states that all machine rooms will require conditioned air. If it becomes impractical to air-condition the room, the temperature can be increased to 105°F, but the specifications will need to state that the proposed equipment shall be designed to operate within this environment.

Hydraulic elevator machine room access doors should be located within 25 feet of an elevator entrance. Careful consideration should be given during preliminary design. This will be caught when applying for a building permit.

The code requires that the machine room door swing out, be a minimum height and width, be self-closing and self-locking.

The electrical disconnect should be located near the door and within sight of the machine. If this is not possible, a second means of disconnecting will have to be provided. If the machine room is sprinklered, shunt trips will be required. The room lighting switch should be adjacent to the door, on the strike-side of the door.

A separate lighting panel should be located next to the electrical disconnect with separate circuits for the pit lighting, cab lighting, machine room lighting, etc.

Access requirements to the machine room are clearly spelled out in the code. Basically it states that you can enter a mechanical equipment room to get to the elevator machine room but you cannot enter an elevator machine room to get to a mechanical room. It also states there is to be nothing in the machine room that does not apply to the elevator or will require access by someone other than an elevator mechanic. If there is a difference in floor level (8 inches or more) from the top access area (roof, etc.) to the machine room floor, a metal ship's ladder, not exceeding 60 degrees shall be provided with the top landing edge at least 24 inches away from the swing line of the door. It is recommended that you review the A-17.1 Code on this item.

The size of machine rooms can vary significantly, but keep in mind that if a hydraulic machine room requires a 7' × 9' machine room (63 S.F.), you cannot provide a room 4' × 16' long. Again, follow the Standards book.

The location of a hydraulic machine room can also become critical because of the inherent noise created by the hydraulic pump unit. Normally, the machine room is located on the lowest level, although it could be located on any floor. Remember, the only criteria for the location is that it be within 25 feet of an entrance.

Hoistway smoke vents are normally required when an elevator travels more than 25 feet or serves more than 3 landings. The A-17.1 requires that a smoke vent be provided in accordance with the local building code. The requirements of the vent are normally 3.5% or a minimum of 3 S.F. (whichever is greater) of the area of the hoistway through a non-combustible duct to the outside area with 1/3 being open and provisions to open the other two-thirds when needed. Remember, if you use a louver it is only 50% efficient, and as a result, the size of the louver and the opening into the hoistway will be double. Some locations will grant a variance to this requirement for emergency conditions. This area is one to remember, as it is very often overlooked.

Another of the most overlooked design elements is the door operation. This is due to the fact that 75–80% of elevator shut-downs are caused by the door operating system; therefore, use center opening doors in lieu of two-speed doors whenever possible. Center opening doors are faster, more reasonable, and reliable. However, they normally require a larger hoistway width.

Escalators and Moving Ramps

Escalators and moving walks are in a class by themselves. Escalators and walks are provided in airports, shopping centers, and arenas where it is necessary to handle a large number of passengers in a short period of time. Escalators can handle some 5,000 people/hour. Sometimes they are provided for convenience or to handle high traffic between special functions within a building. Normally, they do very little to reduce the traffic on the elevators.

Certain manufacturers make three different sized units, while two other manufacturers make four sizes of units. The three common sizes are 24", 32", and 48". The other size, 40", would only be used under special conditions. The size refers to the width of the balustrade.

Escalators are driven by a single or two-speed AC motor, operating at a speed of 90 FPM or 120 FPM. One manufacturer provides an escalator with 125 FPM (the maximum speed allowed by code). The value of providing the two-speed units is questionable. If the units are installed in a project where the same number of people will use the escalators every day, then specify 120 FPM. In a shopping center or an airport, specify 90 FPM. If a manufacturer offers 125 FPM in lieu of 120 FPM, accept it at no additional cost to the contract.

Escalators are designed to operate at a 30-degree angle. Therefore, the distance between work points is the same regardless of the manufacturer. The major difference is the horizontal distance from each work point to the support members. It is recommended that the location and pit sizes be designed for the manufacturer requiring the largest pit. It's best if the bottom pit in a given location has provisions to alter the upper support location to accommodate the other manufacturers.

Approximate Budget Prices for Escalator Installation

The cost ranges between $80,000 and $90,000 for a 12' rise. The cost increases approximately $2,000 for each additional foot in rise. This price varies with the time of year. Budget prices should be obtained from at least two manufacturers.

Note: See Part 2, Section 7 for maintenance costs for elevators and escalators.

Appendix D: Sample LCC Scope of Services

Consultant Proposal to Perform LCC Services

This appendix provides design professionals and owners with appropriate material to be used in defining and negotiating life cycle costing services. Phases of work to be accomplished as well as hours to complete the work are illustrated by assuming a typical building and size.

The consultant proposed to furnish comprehensive, life cycle costing (LCC) analysis consulting services on behalf of the owner. The consultant will prepare the necessary economic studies on the selected project in order to meet the LCC objectives as defined by the owner. Normally 5% to 10% savings are targeted when LCC is applied early in the design process.

The goal of this effort will be to minimize life cycle costs without sacrificing essential functions, by using multi-disciplinary teams at various stages of design and following up with a post occupancy evaluation. Each team will analyze the project's systems using a formal methodology outlined in the book by Stephen Kirk and Alphonse Dell'Isola, *Life Cycle Costing for Facilities*. LCC input should occur early in the schematic and design development phases. The LCC application will be devoted to actions feasible during these phases, focusing on alternative design selections encompassing the areas of greatest savings potential.

Personnel
All work will be conducted under the direction of personnel trained in LCC methodology. The LCC tasks will be performed using a multi-disciplined approach to optimize the total cost of ownership.

Scope of Work
Phase I: Data Collection and Analysis
The consultant will collect all information relative to the project from the designer and owner. The consultant team members will then review the project documents and familiarize themselves with the program and design parameters to identify required functions. A life cycle cost model will be

prepared using information supplied by the designer and owner. This effort can occur concurrently with programming efforts or during the early stages of schematic design.

If desired, the consultant will conduct a one-day LCC executive seminar in a location selected by the client. Attendance is expected from design personnel involved with this project and interested client representatives, particularly those to be assigned to maintain and operate the proposed facility. The seminar will consist of a half-day of formal presentation and a half-day of project familiarization and information collection. Each participant will be presented with a printed copy of the briefing. The seminar objectives include a summary of LCC concepts and a discussion of procedures during the schematic and design development phases of the project.

Phase II: Schematic and Design Development Review and Interim Report

The LCC team will again be assembled for a three- to five-day study to prepare in-depth LCC studies. Areas for LCC will be isolated. Related ratios, e.g., building area net/gross, dollar/ton of air conditioning, maintenance cost per square foot per year, and energy usage as Btu per square foot per year, will also be evaluated. An idea listing of potential LCC savings will be generated by brainstorming techniques. The owner's and designer's input will be solicited to assist in selected areas for in-depth study. The in-depth study format will follow the methodology outlined in the text, *Life Cycle Costing for Facilities*. Each alternative will have life cycle costs estimated. The emphasis in this phase will be on component and system decisions. The team will focus on areas where implementation is feasible and where it can be accomplished quickly to realize optimum LCC cost potential. Subsequently, preliminary findings of the study will be presented to the owner for timely consideration. An interim report will be prepared by the consultant on completion of the workshop for review with the designer and owner.

It should be noted that, as the size or complexity of the project increases, additional reviews should be performed, i.e., at the programming, concept, schematic, and design development stages of the design process. Personnel requirements would increase proportionately.

Phase III: Post Occupancy Evaluation and Final Report

After the project has been bid and occupied, the owner will be asked to collect energy, maintenance, and other LCC data. The consultant will review the results of his/her input and compare with the actual building data. The consultant will attempt to reconcile the data with the design team and make recommendations for future studies and corrective (if any) actions. The consultant will assemble the result of each of the phases of the study into a final report. Six copies of the final report will be presented to designer-owner for review and comment. The contract will be considered complete at the conclusion of this phase.

Phase IV: Computer Simulation Energy Profile Systems (Optional)

The consultant will prepare an energy simulation of the proposed facility design using one of the nationally recognized computer models recommended by the American Society of Heating, Refrigeration, and Air Conditioning Engineers (ASHRAE). This input will be used by the consultant to augment data used in decision-making and for post occupancy evaluation.

Personnel Requirements: Typical Project, $6 to $10 Million Range

Phase I: Data Collection and Analysis

LCC team coordinator	2 days
LCC teams	4 days
Typist	1 day
Total	7 days

Phase II: Schematic and Design Development Review and Interim Report

LCC team coordinator	8 days
LCC team:	
Architectural	2 days
Structural	1 day
Mechanical, energy	3 days
Electrical	2 days
Estimator	2 days
Subtotal	18 days
Interim report	
LCC team coordinator	3 days
LCC team	2 days
Typist	3 days
Subtotal	8 days
Total for each review	26 days
Two reviews (2 × 26)	52 days (schematic and design development)

Phase III: Post Occupancy Evaluation and Final Report

LCC team coordinator	6 days
LCC team	10 days
Architectural	2 days
Total	18 days

Phase IV: Computer Simulation Energy Profile Systems (Optional)

Lump sum approximation ($0.05 to $0.10 per square foot) as required.

LCC Services, Grand Total: 77 days (not including Phase IV)

Note: To this total must be added other costs including travel, per diem, and report reproduction.

Appendix E: Life Cycle Costing Forms

This appendix provides the design professional with recommended formats to perform a life cycle cost analysis. The forms are designed to be used in conjunction with the approach discussed in Chapter 4, "Life Cycle Cost Analysis." Both Chapter 4 and Chapter 7 provide a discussion in the use of these forms. Chapter 8, "Case Studies," illustrates by examples how these sheets can be utilized. Included here are the following LCC forms:

E-1	Life Cycle Costing Summary
E-2	Idea Comparison, Advantages and Disadvantages
E-3	Weighted Evaluation
E-4	Life Cycle Cost Analysis—Annualized Costs
E-5	Life Cycle Cost Analysis—Present Worth Costs*
E-6	Payback Period, Simple and Discounted
E-7	Confidence Index Computation
E-8	Sensitivity Break-Even Analysis
E-9	Life Cycle Cost Summary, Total Project
E-10	Estimated Costs, Back-up Calculations
E-11	Life Cycle Data Sheet
E-12	Initial Cost Model
E-13	Energy Model
E-14	Life Cycle Cost Model

*Note: Also consult the book's website, **www.rsmeans.com/supplement/67341.asp**, *for useful Excel spreadsheets.*

Life Cycle Costing Summary

Project _____   Item _____

Team Members _____   Team _____

Team Members _____   Date _____

Summary of Change (Study Area, Original & Proposed)

Estimated Savings Cost Summary:

	Arch.	Struct.	Mech.	Elec.	Site	Total
Initial Cost Savings						
Life Cycle Cost Savings (PW)						
Annual Energy Savings (EU)						

Percent Savings Initial _____

Percent Savings Life Cycle _____

Percent Savings Energy _____

Figure E.1: **Life Cycle Costing Summary**

Idea Comparison

Idea	Advantages	Disadvantages	Estimated Potential Savings	
			Initial	Life Cycle

Figure E.2: **Idea Comparison, Advantages and Disadvantages**

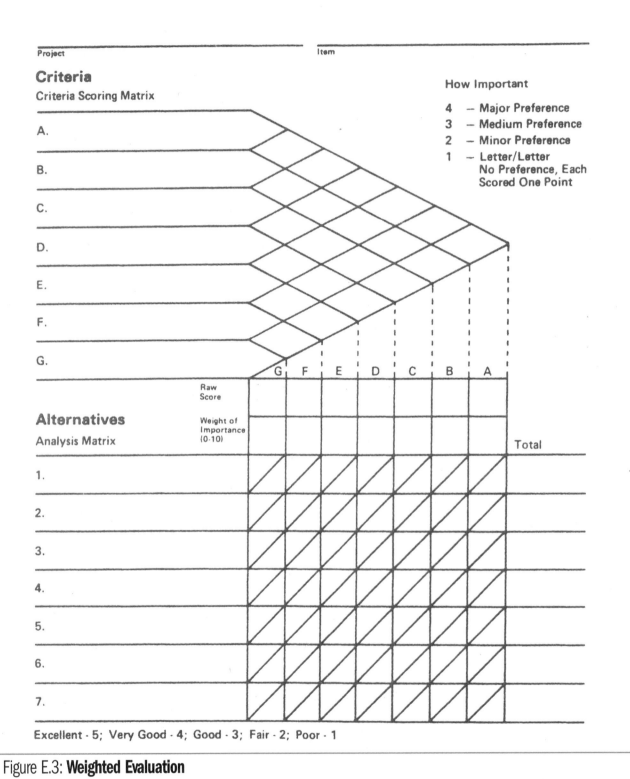

Figure E.3: **Weighted Evaluation**

Life Cycle Cost Analysis
Using Annualized Costs

Item _____   Date _____

		Original	Alt. No. 1	Alt. No. 2
Collateral & Instant Contract Costs	**Initial Costs**			
	Base Cost			
	Interface Costs			
	a. _____			
	b. _____			
	c. _____			
	Other Initial Costs			
	a. _____			
	b. _____			
	c. _____			
	Total Initial Cost Impact (IC)			
	Initial Cost Savings			
Salvage & Replacement Costs	Single Expenditures @ _____ Interest			
	Present Worth			
	1. Year _____ Amount			
	PW = Amount x (PW Factor ____) =			
	2. Year _____ Amount			
	Amount x (PW Factor ____) =			
	3. Year _____ Amount			
	Amount x (PW Factor ____) =			
	4. Year _____ Amount			
	Amount x (PW Factor ____) =			
	5. Year _____ Amount			
	Amount x (PW Factor ____) =			
	Salvage Amount x (PW Factor ____) =			
Life Cycle Costs (Annualized)	Annual Owning & Operating Costs			
	1. Capital IC x (PP _____) =			
	Recovery _____ Years @ _____ %			
	Replacement Cost: PP x PW			
	a. Year _____			
	b. Year _____			
	c. Year _____			
	d. Year _____			
	e. Year _____			
	Salvage:			
	2. Annual Costs			
	a. Maintenance			
	b. Operations			
	c. _____			
	d. _____			
	e. _____			
	3. Total Annual Costs			
	Annual Difference (AD)			
	4. Present Worth of Annual Difference			
	(PWA Factor _____) x AD			

Input Data ↕ Output ↕

PP - Periodic Payment to pay off loan of $1.
PWA - Present Worth of Annuity (What $1 payable periodically is worth today).
PW - Present Worth (What $1 due in future is worth today).

▓ Future Costs
☐ Present Costs

Figure E.4: **Life Cycle Cost Analysis—Annualized Costs**

Life Cycle Cost Analysis
General Purpose Work Sheet

Study Title: _____
Discount Rate: _____ Date: _____
Life Cycle (Years): _____ Present Time: _____

Initial/Collateral Costs
- A. _____
- B. _____
- C. _____
- D. _____
- E. _____
- F. _____

Total Initial/Collateral Costs
Initial Cost PW Difference

Replacement/Salvage (Single Expenditure) — Year | PW Factor
- A. _____
- B. _____
- C. _____
- D. _____
- E. _____
- F. _____

Total Replacement/Salvage Costs

Annual Costs — Diff. Escal. Rate | PWA w/Escal.
- A. _____
- B. _____
- C. _____
- D. _____
- E. _____
- F. _____

Total Annual Costs

LCC
- Total Life Cycle Costs (Present Worth)
- Life Cycle Cost PW Difference
- Discounted Payback — PP Factor
- Total Life Cycle Costs (Annualized)

PW=Present Worth PWA=Present Worth of Annuity PP=Periodic Payment

Columns (for each): Alternative 1 Describe: / Alternative 2 Describe: / Alternative 3 Describe: / Alternative 4 Describe: — each with Estimated Costs | Present Worth

(Discounted Payback: Years; Total Life Cycle Costs (Annualized): Per Year)

Figure E.5: **Life Cycle Cost Analysis—General Purpose Worksheet**

Life Cycle Cost Analysis
Using Payback Period

Life Cycle Costing Estimate Payback Period Work Sheet Study Title: _____ Discount Rate: _____ Economic Life: _____ Annual Savings Differential Escal. Rate: _____		Current Situation	Alternative 1 Describe:	Alternative 2 Describe:	Alternative 3 Describe:	Alternative 4 Describe:
Initial Investment Costs	Initial Investment Costs A. _____ B. _____ C. _____ D. _____ E. _____ F. _____ G. _____ Total Initial Investment Costs	▓▓▓				
Annual Savings	Annual Costs A. _____ B. _____ C. _____ D. _____ E. _____ F. _____ G. _____ Total Annual Costs					
	Annual Savings (Current — Alternative)	▓▓▓				
SPP	Simple Payback Period = $\dfrac{\text{Investment Cost}}{\text{Annual Savings}}$	▓▓▓				
DPP	Discounted Payback Period (Table DPP)	▓▓▓				

Figure E.6: **Payback Period, Simple and Discounted**

LIFE CYCLE COST ANALYSIS (Present Worth Method)
CONFIDENCE INDEX (CI) COMPUTATION

Project/Location:

Subject:
Description:

Project Life Cycle = _____ YEARS
Discount Rate = _____ %

COST ITEMS:	ESTIMATES RANGE			DIFFERENCES IN ESTIMATES				PRESENT WORTH		
	LOW	HIGH	BEST	LOW SIDE	HIGH SIDE	DELTA %	ok	BEST ESTIMATE	DELTA	DELTA^2

ALTERNATIVE L(ow)

Totals

ALTERNATIVE H(igh)

Totals

Note = If high and low 90% estimates > 25%, then use sensitivity analysis

Sum

(Sum)^1/2

Confidence Index = $\dfrac{\text{PW(High)} - \text{PW(Low)}}{(\text{PW Diff(High)}^{\wedge}2 + \text{PW Diff(Low)}^{\wedge}2)^{\wedge}1/2}$ = Difference =

Confidence Assignment:
Low: CI < 0.15
Medium: 0.15 < CI < 0.25
High: CI > 0.25

Figure E.7: **Confidence Index Computation**

Life Cycle Cost Analysis
Sensitivity Break-Even Analysis

Project Title ____________________

Installation & Location ____________________

Design Feature ____________________

Alt.No. __________ Title ____________________

☐ $ x 10^3

☐ $ x 10^6

	Parameter Studied	Representative Values	Alternative ____ Parameter Value	Alternative ____ Net PW	Alternative ____ Parameter Value	Alternative ____ Net PW	Alternative ____ Parameter Value	Alternative ____ Net PW	Alternative ____ Parameter Value	Alternative ____ Net PW
Sensitivity Test 1		Low								
		Intermediate								
		Best Estimate								
		Intermediate								
		High								
Notes										
Sensitivity Test 2		Low								
		Intermediate								
		Best Estimate								
		Intermediate								
		High								
Notes										

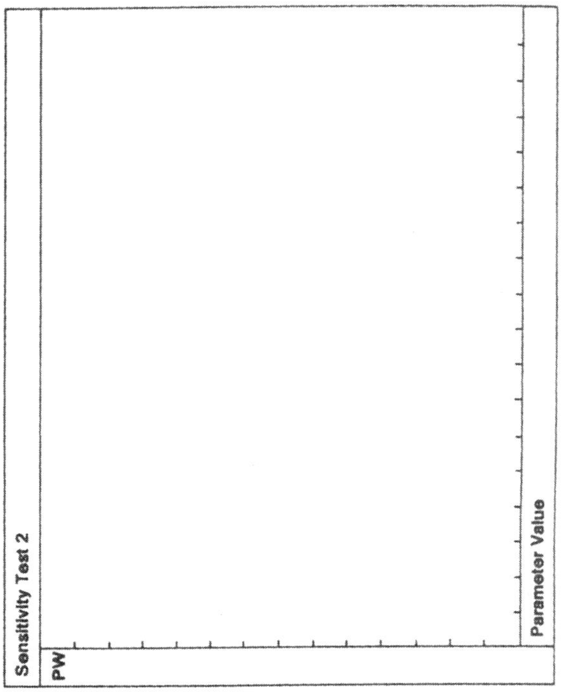

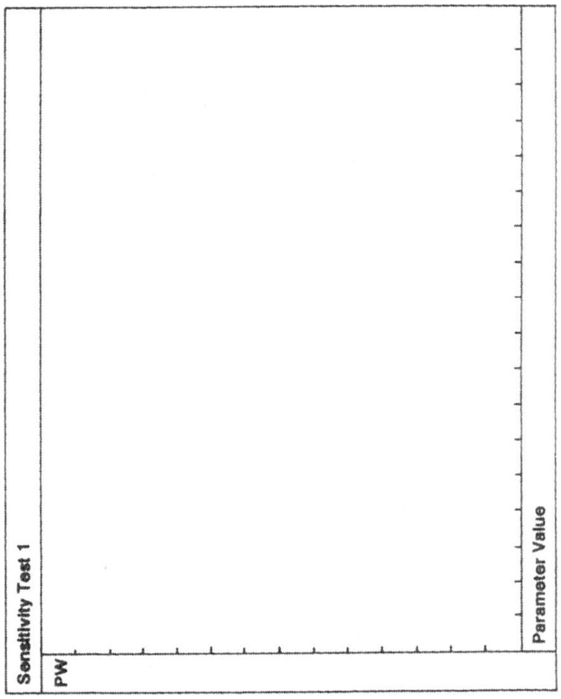

Sheet ________ Of ________

Figure E.8: **Sensitivity Break-Even Analysis**

Life Cycle Cost Summary

Project/Location _____ Date _____

Economic Assumptions: Discount Rate _____ Life Cycle _____ Escalation Rate: Energy _____

Method of Analysis: ☐ Constant Dollars ☐ Current Dollars   Staffing _____ Other _____

	Item:	Quantity/UM	Cost/UM	Cost/Year	Econ. Factor	PW Cost
Initial	Year				*	PW
	Year				*	PW
	Year				*	PW
	Year				*	PW
	Total Initial Cost (Facility, Land, Design & Management, Equipment)					
Energy					PWA	
					PWA	
	Total Energy Cost (Electrical, Natural Gas, Oil, etc.)					
Maintenance and Custodial					PWA	
					PWA	
					PWA	
					PWA	
	Total Maintenance and Custodial Cost (Routine, Non-Scheduled, etc.)					
Repairs and Replacements	Year				*	PW
	Year				*	PW
	Year				*	PW
	Year				*	PW
	Year				*	PW
	Year				*	PW
	Total Repair & Replacement Cost (Items, year replaced, etc.)					
Alterations					PWA	
					PWA	
					PWA	
	Total Alteration Cost (Due to functional need modifications)					
Staffing					PWA	
					PWA	
	Total Staffing Cost (Professional, Non-Professional)					
Associated					PWA	
					PWA	
					PWA	
					PWA	
	Total Associated Cost (Water, Chemicals, Property Taxes, Insurance, etc.)					
LCC	Total Present Worth Cost				$	
	Total Equivalent Uniform Annual Cost				PP	

Notes:

PW Present Worth   PWA Present Worth Annuity   PP Periodic Payment   *Cost @ Year

Figure E.9: **Life Cycle Cost Summary, Total Project**

Estimated Costs

Item	Quantity	Unit Meas.	Unit Cost	Total Cost

Figure E.10: **Estimated Costs, Back-up Calculations**

LIFE CYCLE DATA

ITEM DESCRIPTION	UNIT OF MEASURE	MAINTENANCE DESCRIPTION	MAINTENANCE ANNUAL COST			ENERGY DEMAND (EU)	REPLACE-MENT LIFE (YRS)	PERCENT REPLACED
			LABOR	MATERIAL	EQUIPMENT			

Figure E.11: **Life Cycle Data Sheet**

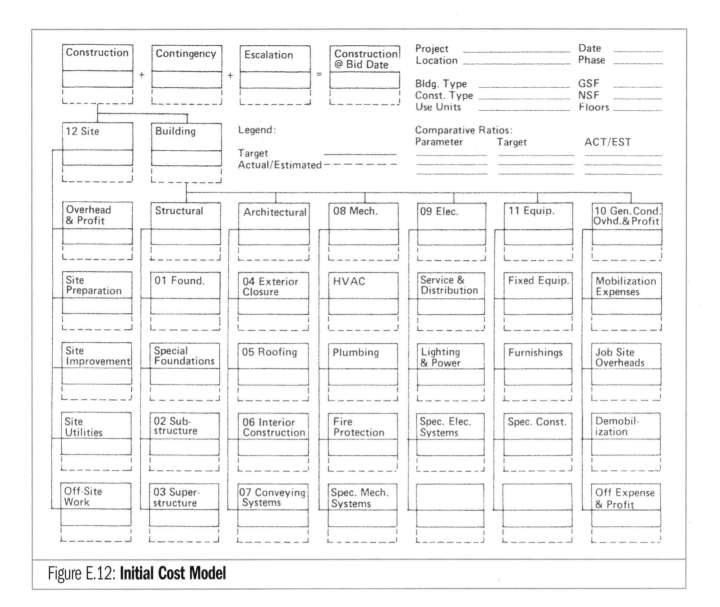

Figure E.12: **Initial Cost Model**

Energy Model Worksheet

Project: _____ **Date:** _____
Location: _____ **Phase:** _____
Bldg. Type: _____
G.S.F. _____
N.S.F. _____   **ENERGY MODEL**
Floors: _____
Energy Units (E.U.) _____ **Target:** _____
Use Units _____ **Actual:** _____

$$\text{Facility Energy (EU/YR)} = \text{Energy SF/YR (EU/YR GSF)}$$

Top level:
- Facility Energy EU/YR
- Exterior Energy EU/YR
- Interior Energy EU/YR

Categories (columns): Site | Offices | Area A | Area B | Area C | Services | Support | (Auxiliaries)

Site (EU/YR)	Offices (EU/SF × S.F. × Oper.Hrs. = EU/YR)	Area A (EU/SF × S.F. × Oper.Hrs. = EU/YR)	Area B (EU/SF × S.F. × Oper.Hrs. = EU/YR)	Area C (EU/SF × S.F. × Oper.Hrs. = EU/YR)	Services (EU/SF × G.S.F × Oper.Hrs. = EU/YR)	Support (EU/SF × G.S.F × Oper.Hrs. = EU/YR)	(EU/SF/YR × Oper.Hrs. = EU/YR)
Site Lighting EU/YR	EU/SF	EU/SF	EU/SF	EU/SF		EU/SF	EU/SF/YR
Parking EU/YR	Lighting EU/SF	Lighting EU/SF	Lighting EU/SF	Lighting EU/SF	Elevators EU/SF	Emerg. Light EU/SF	Cooling EU/SF/YR
Other	Power EU/SF	Power EU/SF	Power EU/SF	Power EU/SF	Fire Protect EU/SF	Security EU/SF	Heating EU/SF/YR
Emerg. Power EU	Equipment EU/SF	Comp. Equip. EU/SF	Cooking Equip. EU/SF	Print Equip. EU/SF	Other	Other	Auxiliaries EU/SF/YR
Other	HVAC: Fans EU/SF	HVAC: Fans EU/SF	HVAC: Fans EU/SF	HVAC: Fans EU/SF			Other

Figure E.13: **Energy Model**

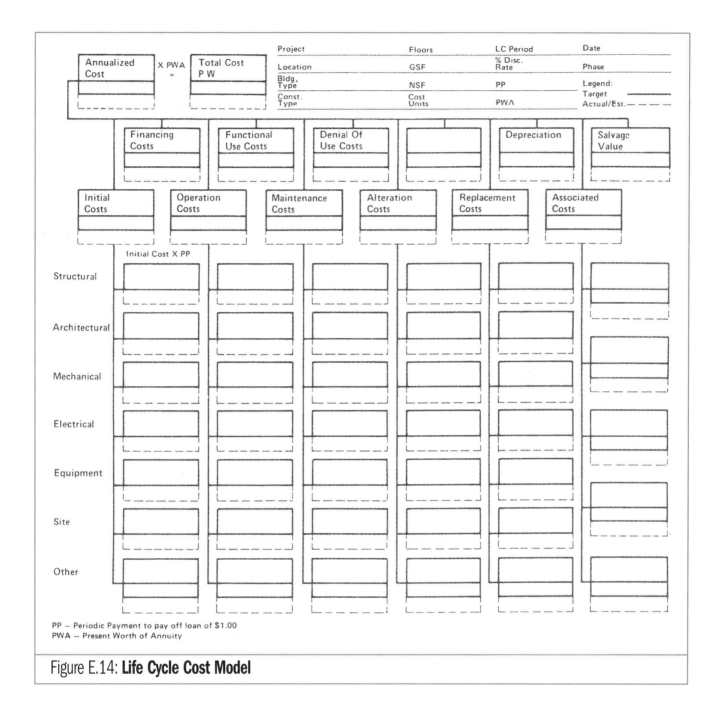

Figure E.14: **Life Cycle Cost Model**

Appendix F: Selected Government Requirements for LCC

Life Cycle Costing should be legislated for all state-financed construction inasmuch as it is capable of producing large savings in tax dollars and energy resources.

Judith C. Toth
Member of House of Delegates, Maryland Legislature, paper delivered to AIE/AACE Joint Conference, Washington, D.C., October 5-6, 1977

This appendix lists *excerpts* from selected federal and state government requirements that provide the reader with some indication of the broad terminology and scope of services necessary to satisfy agency requirements.

Example Life Cycle Life Cost Law, State of Alaska

The state of Alaska established in July 3, 1975 a law that requires procurement of public facilities to be analyzed based on life cycle costs. Chapter 216 of the laws of Alaska were amended as follows:

Section I. As 35.10 is amended by adding new section to read:

Article 5. Public Facility Procurement Policy

Section 35.10.160 Findings and Purpose

The legislature finds that since the needs of the state for physical facilities of all kinds are diverse, the planning, design, and construction of public facilities should be executed in accordance with facility procurement policies developed by the department and reviewed annually by the legislature.

AS 35.10.170. **Duties of Department**

In addition to other duties prescribed by statute, the department shall:

(1) develop facility procurement policies for the planning, design, construction, maintenance, and operation of public facilities of the state;

(2) develop and maintain an inventory of physical facilities currently owned or occupied by the state;

(3) make projections of future public facility needs of the state, analyze facilities needed, and establish methodology for program planning and facilities project planning, design, and construction, based upon:

> (A) a justification of the level of service anticipated by the program agency, utilizing population projections and estimates approved by the governor;
>
> (B) consideration of the geographical area to be served by the facility and relevant data concerning the agency's existing public facilities in that area;
>
> (C) the date by which the services are to be provided;
>
> (D) alternative program methods for providing the services; and
>
> (E) pertinent data requested by the department in accordance with procedures developed under AS 35.10.180.

(4) engage in experimental projects as necessary relating to any available or future method of facility procurement, design, or construction and any method of improving existing design, planning, and construction techniques;

(5) develop life cycle costs of public facilities of the state;

(6) develop life cycle costing methodologies for the following special purposes:

> (A) budget forecasting to support facility program planning and analysis;
>
> (B) systematic cost estimating to forecast planning, design, and construction;
>
> (C) budget forecasting to support development of annual maintenance and operating strategies and life cycle cost plans;
>
> (D) alternative methods of space acquisition and space equalization which will maximize the effectiveness of public funds;

(7) apply for and accept, on behalf of the state, grants from the federal government or an agency of it or from another state foundation, corporation, association, or individual for any of the functions or purposes of the department and may expend any of the money received under this section for any of the functions or purposes.

Section AS 35.10.180. Physical Facility Procurement and Planning Policies

(a) The department shall develop and keep current by periodic revision physical facility procurement and planning policies for rural schools, public buildings, and other state facilities, and shall develop regulations and guidelines for the implementation of these policies.

(b) In developing and revising these policies the department shall seek public review and evaluation by any reasonable means and shall

> (1) consult and cooperate with officials of the federal government, local governments, other political subdivisions of the state and other interested persons regarding physical facility procurement planning;

(2) request and receive from an agency or other unit of the state government the assistance and data needed to carry out the requirements of this section.

(c) The commissioner shall submit copies of proposed policies and plans annually, within 10 days after the legislature convenes, to the legislature. The legislature may approve, reject, or modify the policies and plans by concurrent resolution. If the legislature fails to act during the legislative session, the policies and plans are approved.

Section 35.10.190. Coordination by Department

(a) The department shall coordinate the procurement of physical facilities for the state to ensure the greatest cost savings of planning, design, and contractual techniques.

(b) When the state or an agency of the state determines that a public facility is to be constructed or renovated, it shall, unless exempted by regulations of the department, submit to the department an application for a certificate that the proposed facility complies with adopted facility procurement policies. The department may reject the application but if it does so it shall state in writing the reasons for the rejection. If a written statement that the application is rejected does not issue within 30 days after receipt of the application by the department, unless the department and the applicant have agreed upon an extension of time for consideration, the certificate of compliance is not required. Except as provided otherwise by regulation or by this section, a public facility of the state may not be constructed or renovated by the state unless a certificate that the facility complies with adopted facility procurement policies has been issued.

AS 35.10.200. Definitions

(1) "life cycle costs" means analytic techniques which provide data to describe the first cost of procurement of public facilities and the maintenance cost, operation cost, and occupancy cost of the facilities;

(2) "policies" includes but is not limited to budget accounting and cost planning techniques, facility design techniques, and contractual techniques for the procurement of labor, materials, and contractual services.

OMB (Office of Management and Budget) Circular No. A-94 (revised; October 29, 1992) Life Cycle Costing Guidelines for the Federal Government

To the Heads of the Executive Departments and Establishments

Subject: Guidelines and Discount Rates for Benefit—Cost Analysis of Federal Programs

Following are extracts taken from A-94 which relates to life cycle costing:

1. *Purpose.* The goal of this Circular is to promote efficient resource allocation through well-informed decision-making by the Federal Government. It provides general guidance for conducting benefit-cost and cost-effectiveness analyses. It also provides specific guidance on the discount rates to be used in evaluating Federal programs whose benefits and costs are distributed over time. The general guidance will serve as a checklist of whether an agency has considered and properly dealt with all the elements for sound benefit-cost and cost-effectiveness analyses.

4. *Scope.* This Circular does not supersede agency practices which are prescribed by or pursuant to law, Executive Order, or other relevant circulars. The Circular's guidelines are suggested for use in the internal planning of Executive Branch agencies. The guidelines must be followed in all analyses submitted to OMB in support of legislative and budget-programs in compliance with OMB Circulars No. A-11, "Preparation and Submission of Annual Budget Estimates," and No. A-19, "Legislative Coordination and Clearance." These guidelines must also be followed in providing estimates submitted to OMB in compliance with Executive Order No. 12291, "Federal Regulation," and the President's April 29, 1992 memorandum requiring benefit-cost analysis for certain legislative proposals.

a. Aside from the exceptions listed below, the guidelines in this Circular apply to any analysis used to support Government decisions to initiate, renew, or expand programs or projects which would result in a series of measurable benefits or costs extending for three or more years into the future. The Circular applies specifically to:

1. Benefit-cost or cost-effectiveness analysis of Federal programs or policies.
2. Regulatory impact analysis.
3. Analysis of decisions whether to lease or purchase.
4. Asset valuation and sale analysis.

b. Specifically exempted from the scope of this Circular are decisions concerning:

1. Water resource projects (guidance for which is the approved *Economic and Environmental Principles and Guidelines for Water and Related Land Resources Implementation Studies*).
2. The acquisition of commercial-type services by Government or contractor operation (guidance for which is OMB Circular No. A-76).
3. Federal energy management programs (guidance for which can be found in the *Federal Register* of January 25, 1990, and November 20, 1990).
5. **General Principles.** *Benefit-cost analysis* is recommended as the technique to use in a formal economic analysis of government programs or projects. *Cost-effectiveness analysis* is a less comprehensive technique, but it can be appropriate when the benefits from competing alternatives are the same or where a policy decision has been made that the benefits must be provided.

a. **Net Present Value and Related Outcome Measures.** The standard criterion for deciding whether a government program can be justified on economic principles is *net present value*—the discounted monetized value of expected net benefits (i.e., benefits minus costs). Net present value is computed by assigning monetary values to benefits and costs, discounting future benefits and costs using an appropriate discount rate, and subtracting the sum total of discounted costs from the sum total of discounted benefits. Discounting benefits and costs transforms gains and losses occurring in different time periods to a common unit of measurement. Programs with positive net present value increase social resources and are generally preferred. Programs with negative net present value should generally be avoided. (Section 8 considers discounting issues in more detail.)

Although net present value is not always computable (and it does not usually reflect effects on income distribution), efforts to measure it can produce useful insights even when the monetary values of some benefits or costs cannot be determined. In these cases:

1. A *comprehensive enumeration* of the different types of benefits and costs, monetized or not, can be helpful in identifying the full range of program effects.
2. *Quantifying* benefits and costs is worthwhile, even when it is not feasible to assign monetary values; *physical measurements* may be possible and useful.

Other **summary effectiveness measures** can provide useful supplementary information to net present value, and analysts are encouraged to report them also. Examples include the number of injuries prevented per dollar of cost (both measured in present value terms) or a project's internal rate of return.

b. **Cost-Effectiveness Analysis.** A program is cost-effective if, on the basis of *life cycle cost* analysis of competing alternatives, it is determined to have the lowest costs expressed in present value terms for a given amount of benefits. Cost effectiveness analysis is appropriate whenever it is unnecessary or impractical to consider the dollar value of the benefits provided by the alternatives under consideration. This is the case whenever (i) each alternative has the same annual benefits expressed in monetary terms; or (ii) each alternative has the same annual affects, but dollar values cannot be assigned to their benefits. Analysis of alternative defense systems often falls in this category.

Cost-effectiveness analysis can also be used to compare programs with identical costs but differing benefits. In this case, the decision criterion is the discounted present value of benefits. The alternative program with the largest benefits would normally be favored.

c. **Elements of Benefit-Cost or Cost-Effectiveness Analysis.**

1. **Policy Rationale.** The rationale for the Government program being examined should be clearly stated in the analysis. Programs may be justified on efficiency grounds where they address market failure, such as public goods and externalities. They may also be justified where they improve the efficiency of the Government's internal operations, such as cost-saving investments.

2. **Explicit Assumptions.** Analyses should be explicit about the underlying assumptions used to arrive at estimates of future benefits and costs. In the case of public health programs, for example, it may be necessary to make assumptions about the number of future

beneficiaries, the intensity of service, and the rate of increase in medical prices. The analysis should include a statement of the assumptions, the rationale behind them, and a review of their strengths and weaknesses. Key data and results, such as year-by-year estimates of benefits and costs, should be reported to promote independent analysis and review.

3. **Evaluation of Alternatives.** Analyses should also consider alternative means of achieving program objectives by examining different program *scales*, different *methods* of provision, and different degrees of government *involvement*. For example, in evaluating a decision to acquire a capital asset, the analysis should generally consider: (i) doing nothing; (ii) direct purchase; (iii) upgrading, renovating, sharing, or converting existing government property; or (iv) leasing or contracting for services.

4. **Verification.** Retrospective studies to determine whether anticipated benefits and costs have been realized are potentially valuable. Such studies can be used to determine necessary corrections in existing programs, and to improve future estimates of benefits and costs in these programs or related ones.

Agencies should have a plan for periodic, results-oriented evaluation of program effectiveness. They should also discuss the results of relevant evaluation studies when proposing reauthorizations or increased program funding.

6. **Identifying and Measuring Benefits and Costs.** Analyses should include comprehensive estimates of the expected benefits and costs to *society* based on established definitions and practices for program and policy evaluation. Social net benefits, and not the benefits and costs to the Federal Government, should be the basis for evaluating government programs or policies that have effects on private citizens or other levels of government. Social benefits and costs can differ from private benefits and costs as measured in the marketplace because of imperfections arising from: (i) *external economies or diseconomies* where actions by one party impose benefits or costs on other groups that are not compensated in the market place; (ii) monopoly power that distorts the relationship between marginal costs and market prices; and (iii) taxes or subsidies.

a. **Identifying Benefits and Costs.** Both intangible and tangible benefits and costs should be recognized. The relevant cost concept is broader than private-sector production and compliance costs or government cash expenditures. Costs should reflect the opportunity cost of any resources used, measured by the return to those resources in their most productive application elsewhere. Below are some guidelines to consider when identifying benefits and costs.

b. **Measuring Benefits and Costs.** The principle of willingness-to-pay provides an aggregate measure of what individuals are willing to forego to obtain a given benefit. Market prices provide an invaluable starting point for measuring willingness-to-pay, but prices sometimes do not adequately reflect the true value of a good to society. Externalities, monopoly power, and taxes or subsidies can distort market prices.

7. **Treatment of Inflation.** Future inflation is highly uncertain. Analysts should avoid having to make an assumption about the general rate of inflation whenever possible.

　a. **Real or Nominal Values.** Economic analyses are often most readily accomplished using *real* or *constant-dollar* values, i.e., by measuring benefits and costs in units of stable purchasing power. (Such estimates may reflect expected future changes in relative prices, however, where there is a reasonable basis for estimating such changes.) Where future benefits and costs are given in *nominal* terms, i.e., in terms of the future purchasing power of the dollar, the analysis should use these values rather than convert them to constant dollars as, for example, in the case of lease-purchase analysis.

Nominal and real values must not be combined in the same analysis. Logical consistency requires that analysis be conducted either in constant dollars or in terms of nominal values. This may require converting some nominal values to real values, or vice versa.

　b. **Recommended Inflation Assumption.** When a general inflation assumption is needed, the rate of increase in the Gross Domestic Product deflator from the Administration's economic assumptions for the period of the analysis is recommended. For projects or programs that extend beyond the six-year budget horizon, the inflation assumption can be extended by using the inflation rate for the sixth year of the budget forecast. The Administration's economic forecast is updated twice annually, at the time the budget is published in January or February and at the time of the Mid-Session Review of the Budget in July. Alternative inflation estimates, based on credible private sector forecasts, may be used for sensitivity analysis.

8. **Discount Rate Policy.** In order to compute net present value, it is necessary to discount future benefits and costs. This discounting reflects the time value of money. Benefits and costs are worth more if they are experienced sooner. All future benefits and costs, including nonmonetized benefits and costs, should be discounted. The higher the discount rate, the lower is the present value of future cash flows. For typical investments, with costs concentrated in early periods and benefits following in later periods, raising the discount rate tends to reduce the net present value. (Technical guidance on discounting and a table of *discount factors* are provided in Appendix B.)

　a. **Real versus Nominal Discount Rates.** The proper discount rate to use depends on whether the benefits and costs are measured in real or nominal terms.

　　(1) A real discount rate that has been adjusted to eliminate the effect of expected inflation should be used to discount constant-dollar or real benefits and costs. A real discount rate can be approximated by subtracting expected inflation from a nominal interest rate.

(2) A nominal discount rate that reflects expected inflation should be used to discount nominal benefits and costs. Market interest rates are nominal interest rates in this sense.

b. **Public Investment and Regulatory Analyses.** The guidance in this section applies to benefit-cost analyses of public investments and regulatory programs that provide benefits and costs to the general public. Guidance related to cost-effectiveness analysis of internal planning decisions of the Federal Government is provided in Section 8.c.

In general, public investments and regulations displace both private investment and consumption. To account for this displacement and to promote efficient investment and regulatory policies, the following guidance should be observed.

(1) **Base-Case Analysis.** Constant-dollar benefit-cost analyses of proposed investments and regulations should report net present value and other outcomes determined using a real discount rate of 7%. This rate approximates the marginal pretax rate of return on an average investment in the private sector in recent years. Significant changes in this rate will be reflected in future updates of this Circular.

(2) **Other Discount Rates.** Analyses should show the sensitivity of the discounted net present value and other outcomes to variations in the discount rate. The importance of these alternative calculations will depend on the specific economic characteristics of the program under analysis. For example, in analyzing a regulatory proposal whose main cost is to reduce business investment, net present value should also be calculated using a higher discount rate than 7%.

Analyses may include among the reported outcomes the *internal rate of return* implied by the stream of benefits and costs. The internal rate of return is the discount rate that sets the net present value of the program or project to zero. While the internal rate of return does not generally provide an acceptable decision criterion, it does provide useful information, particularly when budgets are constrained or there is uncertainty about the appropriate discount rate.

(3) Using the *shadow price of capital* to value benefits and costs is the analytically preferred means of capturing the effects of government projects on resource allocation in the private sector. To use this method accurately, the analyst must be able to compute how the benefits
and costs of a program or project affect the allocation of private consumption and investment. OMB concurrence is required if this method is used in place of the base case discount rate.

c. **Cost-Effectiveness, Lease-Purchase, Internal Government Investment, and Asset Sales Analyses.** The Treasury's borrowing rates should be used as discount rates in the following cases:

(1) **Cost-Effectiveness Analysis.** Analyses that involve constant-dollar costs should use the real Treasury borrowing rate on marketable securities of comparable maturity to the period of analysis. This rate is

computed using the Administration's economic assumptions for the budget, which are published in January of each year. A table of discount rates based on the expected interest rates for the first year of the budget forecast is presented in Appendix C of this Circular. Appendix C is updated annually and is available upon request from OMB. Real Treasury rates are obtained by removing expected inflation over the period of analysis from nominal Treasury interest rates. (Analyses that involve nominal costs should use nominal Treasury rates for discounting, as described in the following paragraph.)

(2) **Lease-Purchase Analysis.** Analyses of nominal lease payments should use the nominal Treasury borrowing rate on marketable securities of comparable maturity to the period of analysis. Nominal Treasury borrowing rates should be taken from the economic assumptions for the budget. A table of discount rates based on these assumptions is presented in Appendix C of this Circular, which is updated annually. (Constant dollar lease-purchase analyses should use the real Treasury borrowing rate, described in the preceding paragraph.)

(3) **Internal Government Investments.** Some Federal investments provide "internal" benefits which take the form of increased Federal revenues or decreased Federal costs. An example would be an investment in an energy-efficient building system that reduces Federal operating costs. Unlike the case of a Federally funded highway (which provides "external" benefits to society as a whole), it is appropriate to calculate such a project's net present value using a comparable-maturity Treasury rate as a discount rate. The rate used may be either nominal or real, depending on how benefits and costs are measured.

Some Federal activities provide a mix of both Federal cost savings and external social benefits. For example, Federal investments in information technology can produce Federal savings in the form of lower administrative costs and external social benefits in the form of faster claims processing. The net present value of such investments should be evaluated with the 7% real discount rate discussed in Section 8.b. unless the analysis is able to allocate the investment's costs between provision of Federal cost savings and external social benefits. Where such an allocation is possible, Federal cost savings and their associated investment costs may be discounted at the Treasury rate, while the external social benefits and their associated investment costs should be discounted at the 7% real rate.

9. **Treatment of Uncertainty.** Estimates of benefits and costs are typically uncertain because of imprecision in both underlying data and modeling assumptions. Because such uncertainty is basic to many analyses, its effects should be analyzed and reported. Useful information in such a report would include the key sources of uncertainty; expected value estimates of outcomes; the sensitivity of results to important sources of uncertainty; and where possible, the probability distributions of benefits, costs, and net benefits.

a. **Characterizing Uncertainty.** Analyses should attempt to characterize the sources and nature of uncertainty. Ideally, probability distributions of potential benefits, costs, and net benefits should be presented. It should be recognized that many phenomena that are treated as deterministic or certain are, in fact, uncertain. In analyzing uncertain data, objective estimates of probabilities should be used whenever possible. Market data, such as private insurance payments or interest rate differentials, may be useful in identifying and estimating relevant risks. Stochastic simulation methods can be useful for analyzing such phenomena and developing insights into the relevant probability distributions. In any case, the basis for the probability distribution assumptions should be reported. Any limitations of the analysis because of uncertainty or biases surrounding data or assumptions should be discussed.

b. **Expected Values**. The expected values of the distributions of benefits, costs and net benefits can be obtained by weighting each outcome by its probability of occurrence, and then summing across all potential outcomes. If estimated benefits, costs and net benefits are characterized by point estimates rather than as probability distributions, the expected value (an unbiased estimate) is the appropriate estimate for use.

Estimates that differ from expected values (such as worst-case estimates) may be provided in addition to expected values, but the rationale for such estimates must be clearly presented. For any such estimate, the analysis should identify the nature and magnitude of any bias. For example, studies of past activities have documented tendencies for cost growth beyond initial expectations; analyses should consider whether past experience suggests that initial estimates of benefits or costs are optimistic.

c. **Sensitivity Analysis**. Major assumptions should be varied and net present value and other outcomes recomputed to determine how sensitive outcomes are to changes in the assumptions. The assumptions that deserve the most attention will depend on the dominant benefit and cost elements and the areas of greatest uncertainty of the program being analyzed. For example, in analyzing a retirement program, one would consider changes in the number of beneficiaries, future wage growth, inflation, and the discount rate. In general, sensitivity analysis should be considered for estimates of: (i) benefits and costs; (ii) the discount rate; (iii) the general inflation rate; and (iv) distributional assumptions. Models used in the analysis should be well documented and, where possible, available to facilitate independent review.

d. **Other Adjustments for Uncertainty**. The absolute variability of a risky outcome can be much less significant than its correlation with other significant determinants of social welfare, such as real national income. In general, variations in the discount rate are not the appropriate method of adjusting net present value for the special risks of particular

OMB (Office of Management and Budget) Circular No. A-31 (Revised; May 21, 1993) Value Engineering for the Federal Government

projects. In some cases, it may be possible to estimate *certainty-equivalents* which involve adjusting uncertain expected values to account for risk.

To the Heads of Executive Departments and Establishments

Subject: Value Engineering

1. **Purpose.** This Circular requires Federal Departments and Agencies to use value engineering (VE) as a management tool, where appropriate, to reduce program and acquisition costs.

4. **Background.** For the purposes of this Circular, value analysis, value management, and value control are considered synonymous with VE. VE is an effective technique for reducing costs, increasing productivity, and improving quality. It can be applied to hardware and software; development, production, and manufacturing; specifications, standards, contract requirements, and other acquisition program documentation; facilities design and construction. It may be successfully introduced at any point in the life-cycle of products, systems, or procedures. VE is a technique directed toward analyzing the functions of an item or process to determine "best value," or the best relationship between worth and cost. In other words, "best value" is represented by an item or process that consistently performs the required basic function and has the lowest total cost. In this context, the application of VE in facilities construction can yield a better value when construction is approached in a manner that incorporates environmentally-sound and energy-efficient practices and materials.

5. **Relationship to other management improvement processes.** VE is a management tool that can be used alone or with other management techniques and methodologies to improve operations and reduce costs. For example, the total quality management process can include VE and other cost cutting-techniques, such as life-cycle costing, concurrent engineering, and design-to-cost, approaches, by using these techniques as analytical tools in process and product improvement.

6. **Definitions.**

   b. **Life-cycle cost.** The total cost of a system, building, or other product, computed over its useful life. It includes all relevant costs involved in acquiring, owning, operating, maintaining, and disposing of the system or product over a specified period of time, including environmental and energy costs.

7. **Policy.** Federal agencies shall use VE as a management tool, where appropriate, to ensure realistic budgets, identify and remove nonessential capital and operating costs, and improve and maintain optimum quality of program and acquisition functions. Senior management will establish and maintain VE programs, procedures and processes to provide for the aggressive, systematic development and maintenance of the most effective, efficient, and economical and environmentally-sound arrangements for conducting the work of agencies, and to provide a sound basis for identifying and reporting accomplishments.

8. **Agency responsibilities.** To ensure that systemic VE improvements are achieved, agencies shall, at a minimum:

a. Designate a senior management official to monitor and coordinate agency VE efforts.

b. Develop criteria and guidelines for both in-house personnel and contractors to identify programs/projects with the most potential to yield savings from the application of VE techniques. The criteria and guidelines should recognize that the potential savings are greatest during the planning, design, and other early phases of project/program/system/product development. Agency guidelines will include:

1. Measuring the net life-cycle cost savings from value engineering. The net life-cycle cost savings from value engineering is determined by subtracting the Government's cost of performing the value engineering function over the life of the program from the value of the total saving generated by the value engineering function.
2. Dollar amount thresholds for projects/programs requiring the application of VE. The minimum threshold for agency projects and programs which require the application of VE is $1 million. Lower thresholds may be established at agency discretion for projects having a major impact on agency operations.

9. **Reports to OMB.** Each agency shall report the Fiscal Year results of using VE annually to OMB, except those agencies whose total budget is under $10 million or whose total procurement obligations do not exceed $10 million in a given fiscal year. The reports are due to OMB by December 31st of the calendar year, and should include the current name, address, and telephone number of the agency's VE manager.

Part I of the report asks for net life-cycle cost savings achieved through VE. In addition, it requires agencies to show the project/program dollar amount thresholds the agency has established for requiring the use of VE if greater than $1 million. If thresholds vary by category, show the thresholds for all categories. Savings resulting from VE proposals and VE change proposals should be included under the appropriate categories.

Part II asks for a description of the top 20 fiscal year VE projects (or all projects if there are fewer than 20). List the projects by title and show the net life-cycle cost savings and quality improvements achieved through application of VE.

Appendix G: Historical Development of LCC

Historical Development of LCC

Early Years, Engineering Economics

Written records pertinent to LCC are somewhat obscure. The first edition of *Principles of Engineering Economy* by Eugene L. Grant was published in 1930 and became the classic reference for all who dealt with the effects of engineering economics. One of the first government references to LCC was published in 1933 by the Comptroller General of the United States. With regard to tractor acquisitions, the General Accounting Office (GAO) supported acceptance of bids predicated on the total cost of the government after 8,000 hours of operation. Maintenance costs were included in the bid price. More than 25 additional rulings in the following years mandated LCC procurement for all types of equipment, culminating with the GAO report on LCC purchases for the Department of Defense, published in 1973. During the decade 1940 to 1950, when material and labor shortages resulting from the war effort caused extensive searches for substitute materials and production procedures, Lawrence D. Miles originated the idea of value engineering at the General Electric Company. As conceived, VE was much broader than life cycle cost analysis alone, incorporating the study of *functions* along with a *total cost* concept.

The earliest proponent of VE for the construction industry was Alphonse J. Dell' Isola who published an application guide, *Value Engineering in the Construction Industry,* in 1972. He was one of the first to observe and graphically display the relationship among decisions and their potential impact on total cost, pointing out that the initial planning and design choices have maximum effect on life cycle costs.

The years 1950 to 1960 were relatively static with regard to LCC development in the United States. During that period, however, P.A. Stone began his work in England at the Building Research Station, which, in the decade to follow, resulted in the publication of two major texts concerning cost-in-use.

The American Telephone and Telegraph Company (AT&T) published its first edition of *Engineering Economy* in 1952. Its purpose was to bring under one cover, for designers' use, a discussion of the basic principles of engineering economy and to outline the techniques for making comparative cost studies. This document states:

> It is the responsibility of the engineer to determine the plan which will meet the physical requirements in the most economical manner . . . In the telephone business (and other industries). . . the company has not only the desire, but the obligation to provide service. Therefore the only engineering question is: How can good service be provided at the lower over-all cost?

The importance of this statement is perhaps best described in the final sentence of the document: "The future success of a company rests largely on the quality of the engineering work done today."

Significance Recognized

In 1961, an important conference entitled, "Methods of Building Cost Analysis" was sponsored by the Building Research Institute in Washington, D.C. The papers presented procedures for developing life cycle cost analyses for buildings, their enclosures, lighting, and heating, ventilating, and air-conditioning systems. It was also during the early 1960s that an effort was undertaken by the Logistics Management Institute that led to publication in 1965 of a most influential report for the Assistant Secretary of Defense. The document concluded that, had total life cycle costs been considered, many Department of Defense contracts would have been awarded to other than low bidders at considerable overall savings to the government. As a consequence, in 1970 the Department of Defense published the first in a series of three guidelines for LCC procurement. The initial requirement was a consideration of replacement cost in the purchase of certain items (replacement siding on family housing, solid-state 15-megahertz oscilloscopes, computers, etc.). The policy was reinforced in 1971 by Department of Defense Directive 5000.1, which mandates LCC procurement for major systems acquisitions. The third instruction, *Life Cycle Costing Guide for System Acquisition,* was issued in 1973 and specified elementary-level life cycle cost analysis procedures. The first publication related to facilities was prepared by the Department of the Army, Office of Chief of Engineers, in 1971, but was never officially issued. It contained general and specific procedures for long-term economic analysis of barracks, warehouses, and 32 other facility types.

In November 1972, the comptroller general of the United States issued one of the most exhaustive and devastating reports to date on the life cycle costs for U.S. hospital facilities. The objective of the effort, conducted by Westinghouse researchers, was to "study the feasibility of reducing the cost of constructing hospital facilities. . ."[1] Existing hospitals throughout the United States were examined and tested against a variety of innovative design and construction solutions. The conclusion: "GAO believes that life cycle cost analysis is essential in the planning and design of all hospital construction projects." It was further demonstrated that operation and maintenance costs for such facilities could equal the initial investment within one to three years. This gave some additional credibility to similar values produced earlier by the National Bureau of Standards, now National Institute of Standards and Technology (NIST), which suggested that capital costs represent less than 2% of a building's total life cycle cost.

In 1975 two major pieces of legislation, the Energy Policy and Conservation Act and the Energy Conservation Act, reiterated the need for long-term analysis. Both have influenced procurement practices. Title III of the former provided $150 million to state governments in 1976 to develop statewide energy conservation plans. Building Energy Performance Standards (BEPS) became available in 1979 for implementation. The most recent NIST Handbook (NIST Handbook 135—Life Cycle Costing Manual for Federal Energy Management Program) provides detailed guidance.

Nearly every state has enacted energy legislation based on groundwork by the American Society of Heating, Refrigeration and Air Conditioning (ASHRAE) Standard 90-71, written in 1975, and the *ASHRAE Handbook HVAC Applications,* Chapter 35 "Owning and Operation Costs," written in 1999. This standard has been updated by ASHRAE.

Procedures Developed

During the same period, the U.S. General Services Administration began several fundamental projects to establish effective cost management and LCC. The UNIFORMAT framework for initial costs was published by GSA in 1975. It was an extension of a previous system developed for the American Institute of Architects and termed MasterFormat. The framework was extended in an LCC study prepared for the University of Alaska in 1974. The Alaskan system provides a coordinated code of accounts for operation and maintenance expenditures and general equations for executing long-term cost analyses.

Two facility LCC systems have been completed by GSA. The first, Life Cycle Planning and Budgeting Model, which began service in 1974, is a comprehensive, automated system designed to evaluate lease, construction, and renovation options to satisfy user-agency space requirements. It permits relatively non-technical personnel to provide input about floor areas and functional requirements (for instance, for workstations). Based on existing federal standards and series of quasi-engineering and other assumptions, it generates an anticipated building configuration. The configuration, transformed into UNIFORMAT categories, is then priced for capital, maintenance, and operation costs.

In 1974–1975, the second GSA system was completed. It consists of a unique bid package arrangement for building which incorporates the concept of LCC. The systems portions of three Social Security payment centers were put out for bid on a total cost basis. However, the designs have resulted in facilities with some of the lowest energy-consumption figures. Of course, many other systems procurement projects were executed during the late 1960s and early 1970s (School Construction

System Development, for example). However, the GSA buildings were the first major facilities purchased with total costs as a bid criterion.

With regard to equipment purchases, the U.S. Federal Supply Service (GSA) has adopted a life cycle procurement technique similar to that used by the Department of Defense. The entire procedure is described in a training manual published in 1977.[2] The methodology allows equipment and other purchases to be considered within a total cost context, which includes acquisition costs, initial logistics, and recurring costs.

Energy Highlighted

The energy and environmental concerns described previously have resulted in numerous federal and state laws to encourage or mandate conservation. These laws require the use of LCC to evaluate various design alternatives.

In 1974, Florida became the first state to formally adopt LCC and required consideration of initial, energy, operation, and maintenance costs as criteria for the design of buildings over 5,000 square feet. All the other states mentioned previously include LCC as part of energy legislation, with the exception of Alaska, which passed a law in 1975 requiring LCC procurement of all public facilities. The Florida bill further stated that the State Department of Transportation and Public Facilities would "develop life cycle costs of public facilities of the State and those public facilities of political subdivisions of the State. . ." and "develop life cycle methodologies for special purposes." The authors prepared a booklet, "Life Cycle Cost Analysis," designed for facilities procurement, for the state of Wyoming, and a similar document for the state of North Dakota in the early 1980s.

Since 1977, the state of Illinois Capital Development Board has required LCC analysis for the preparation of value engineering of conceptual design alternatives.[3]

In April 1977, the American Institute of Architects issued a formal set of guidelines for architects and engineering consultants electing to offer LCC as an additional service. The guide presents a method for computing the present worth and uniform annual cost for total building costs. It also recommends techniques for incorporating the results into the building decision process. See "The Architect's Handbook of Professional Practice" (2001) for typical life cycle and engineering provisions. The UNIFORMAT framework is offered as a vehicle for organizing both the required data and the analysis. The system is manual and requires a user to provide all cost and performance input data.

In addition, in conjunction with the American Consulting Engineers Council, the AIA has sponsored LCC seminars.

In the mechanical (energy) area in particular, a large number of energy analysis programs that incorporate long-term economic models are currently being utilized. The economic portion typically involves the computation of maintenance costs, interest rates, and depreciation, and performance of a cash-flow analysis and rate-of-return evaluation. One of the first, the TRACE program, developed by the Trane Company, will forecast consumption values for alternative fuels applied to various building systems and generate cost estimates for initial investment, utilities, and annual owning expenditures. The results may be evaluated on a present worth or uniform annual equivalent basis for heating, ventilating, and air-conditioning (HVAC) systems.

Public Law 95-619 established the National Energy Conservation Policy Act on November 9, 1978. This law mandates that all new federal buildings be *life cycle cost-effective* as determined by methods prescribed by the legislation. In the design of new federal buildings, cost evaluation shall be made on the basis of life cycle cost rather than initial cost. For existing buildings, retrofit measures are to be taken to improve their energy efficiency in general and to minimize their life cycle cost.

The Department of Health, Education, and Welfare (now Department of Health and Human Services), at the request of the GAO's November 1972 report, "Study of Health Facilities Construction Cost," sponsored a study, "Evaluation of the Health Facilities Building Process," dated March 31, 1973. The overall objectives of the study included the evaluation of existing processes of acquisition of those health facilities that are procured by federally assisted grant and loan programs. Special emphasis was placed on life cycle cost analysis and its impact on the procurement process.

Because of the interest generated by this document, another firm was commissioned to study LCC. The objective was to "bring the concept of life cycle budgeting and costing to bear on the process of planning, budgeting, designing, constructing, and operating health facilities." The result was a series of four volumes published from 1975-1976 entitled, *Life Cycle Budgeting and Costing as an Aid in Decision Making*.

While the activities described above were primarily the result of inflationary pressures and increasing technical sophistication of products and buildings (and thus, higher maintenance and operating costs), 1973 brought the Arab oil embargo and crystallized the energy crisis.

Public hearings were held in May and June of 1979 regarding a proposal by the Department of Energy (44 FR 25366, April 30, 1979) that LCC analysis be required in all new construction or retrofitting projects in federal buildings. The proposal would require LCC methods to be used in early design and planning to determine which energy-saving investment to use. For existing buildings the proposed change would require retrofit investments to be ranked by cost savings. The aim, according to the Department of Energy, was to reduce consumption of fossil fuels through use of solar and other renewable energy sources.

The Department of Energy issued a final rule to establish a methodology and procedures to conduct life cycle cost analyses on January 23, 1980. This methodology involves estimating and comparing the effects of replacing building systems with energy-saving alternatives in existing federal buildings and of selecting among alternative building designs containing different energy-using systems for a new federal building.

The last significant action occurred recently with the government emphasis on LEEDS. Emphasis was placed on the use of LCC methodology. The American Society for Testing and Materials (ASTM) during the 1980s developed a series of standards on building economics. The updated standards are listed below.

ASTM Standards for Building Economics

Guides

E 1185-02	Guide for Selecting Economic Methods for Evaluating Investments in Buildings and Building Systems
E 1185-02	Standard Guide for Selecting Techniques for Treating Uncertainty and Risk in the Economic Evaluation of Buildings and Building Systems

Practices

E 917-02	Standard Practice for Measuring Life-Cycle Costs of Buildings and Building Systems
E 964-02	Standard Practice for Measuring Benefit-to-Cost and Savings-to-Investment Ratios for Buildings and Building Systems
E 1057-99	Standard Practice for Measuring Internal Rate of Return and Adjusted Internal Rate of Return for Investments in Buildings and Building Systems
E 1074-93(e1)	Standard Practice for Measuring Net Benefits for Investments in Buildings and Building Systems
E 1121-02	Standard Practice for Measuring Payback for Investments in Buildings and Building Systems

Terminology

E 833-02a	Standard Terminology of Building Economics

Life Cycle Costing for Design Professionals, published by the authors in 1981, served as the foundation for these ASTM standards. The material contained in that publication is consistent with these ASTM standard practices. In 1985, recognizing the need for LCC data, the ASTM assisted in the preparation of a computerized database on building maintenance, repair, and replacement.

1. Study of Health Care Facilities Construction Costs, Report to the Congress, Comptroller General of the United States, November 1972.
2. *Life Cycle Costing Workbook: A Guide for the Implementation of Life Cycle Costing in the Federal Supply Services,* U.S. General Services Administration, Washington, 1977.
3. *Life Cycle Cost Analysis Manual,* Capital Development Board, State of Illinois, 401 South Spring Street, Springfield, IL 62706, Wayne Hucklebe.

Bibliography, Glossary & Index

Bibliography

Life Cycle Costing

Apartment Building Income/Expense Analysis. Chicago: The Institute of Real Estate Management.
   This report compiles both annual operating expenses and rental income information by building location, age, and type of ownership.

Blanchard, Benjamin S. *Design and Manage to Life Cycle Cost.* Portland, OR: M/A Press, 1978.

Brown, Robert J., and Rudolph R. Yanuck. *Life Cycle Costing: A Practical Guide for Energy Managers.* Atlanta: Fairmont Press, 1980.

Building Maintenance, Repair, and Replacement Database (BMDB) for Life-Cycle Cost Analysis. Philadelphia: American Society for Testing and Materials, 1992.

Dell'Isola, Alphonse J., and Stephen J. Kirk. *Life Cycle Cost Data.* New York: McGraw-Hill, 1983.

Dell'Isola, Alphonse J., and Stephen J. Kirk. *Life Cycle Costing for Design Professionals.* New York: McGraw-Hill, 1981 (Japanese Edition, 1985).

Dell'Isola, Michael, and Alphonse J. Dell'Isola. "Life Cycle Techniques for Design-Build-Finance." Paper presented at Design Build International Association Annual Conference, Boston, 2001.

Guidelines and Discount Rates for Benefit-Cost Analysis of Federal Programs. Washington, D.C.: Office of Management and Budget Circular A-94, 1992.

Evaluation of the Health Facilities Building Process, vol. IV, *Evaluation of LCC.* Washington, D.C.: U.S. Department of Health, Education, and Welfare, 1973.

"Federal Energy Management and Planning Programs: Life Cycle Cost Methodology and Procedures." *Federal Register*, vol. 55 (1990) and vol. 55, no. 224 (1990).

Feldman, Edwin B. *Building Design for Maintainability.* New York: McGraw-Hill, 1975.

Griffith, J.W. *Life Cycle Cost-Benefit Analysis: A Basic Course in Economic Decision Making*, Document PB-251 848/LK. Springfield, VA: National Technical Information Center, 1975.

Harper, G. Neil, ed. *Computer Applications in Architecture and Engineering.* New York: McGraw-Hill, 1968.

Haviland, David S. *Life Cycle Cost Analysis 2: Using it in Practice.* Washington, D.C.: The American Institute of Architects, 1978.

Iselin, Donald G., and Andrew C. Lemer, eds. *The Fourth Dimension in Building: Strategies for Minimizing Obsolescence.* Washington, D.C.: National Academy Press, 1993.

Kirk, Stephen J. "Economic Building Performance Model" (unpublished thesis). School of Architecture and Urban Design, University of Kansas, 1974.

Kirk, Stephen J. "Life Cycle Costing: Increasing Popular Route to Design Value." *Architectural Record*, 1979.

"Life Cycle Budgeting as an Aid to Decision Making, Building Information Circular." Washington, D.C.: U.S. Department of Health, Education, and Welfare, Office of Facilities Engineering and Property Management, 1977.

Life Cycle Cost Analysis, a Guide for Architects. Washington, D.C.: The American Institute of Architects, 1977.

Life Cycle Costing Guide. Washington, D.C.: Facilities Group, National Aeronautics and Space Administration, 1978.

Life Cycle Costing in the Public Buildings Service. Washington, D.C.: General Services Administration, 1977.

Life Cycle Costing—The Concept—Application in Conventional Design. Air Force Civil Engineering.

Life Cycle Costing Workbook: A Guide for the Implementation of Life Cycle Costing in the Federal Supply Service. Washington, D.C.: General Services Administration, 1977.

Life Cycle Management Model for Project Management of Buildings. Washington, D.C.: General Services Administration, Public Building Service, 1972.

LLC-1, Life Cycle Costing Procurement Guide. Washington, D.C.: U.S. Department of Defense, 1970.

LLC-2, Casebook—Life Cycle Costing in Equipment Procurement. Washington, D.C.: U.S. Department of Defense, 1970.

LLC-3, Life Cycle Costing Guide for System Acquisition. Washington, D.C.: U.S. Department of Defense, 1973.

National Energy Conservation Policy Act, Public Law 95-619, 1978.

Nielson, Kris R. "Tax Consideration in Building Design." *AIA Journal*, 1973.

Office Building Experience Exchange Report. Washington, D.C.: The Building Owners and Managers Association International.

> Report includes data and analyses of office building costs. Over 1,000 buildings are broken down by age, height, size, and location of buildings.

Operations and Maintenance. Alexandria: The Association of Higher Education Facilities Officers, 1992.

Peterson, Stephen R. *Energy Prices and Discount Factors for Life Cycle Cost Analysis.* Washington, D.C.: NIST, 1993.

Ruegg, Rosalie T. *Solar Heating and Cooling in Buildings: Methods of Economic Evaluation,* NBSIR 75-712. Washington, D.C.: National Bureau of Standards, 1975.

Sizemore, M.M., and H.O. Clark, and W.S. Ostrander. *Energy Planning for Buildings.* Washington, D.C.: The American Institute of Architects, 1979.

Tax Information on Depreciation. Internal Revenue Service Publication 534 (issued annually).

Terotechnology—An Introduction to the Management of Physical Resources. Surrey, UK: The National Terotechnology Centre, 1976.

UNIFORMAT: Automated Cost Control. Washington, D.C.: General Services Administration, 1975.

Williams, John E. "Life Cycle Costing: An Overview." Washington, D.C.: American Institute of Industrial Engineers and the American Association of Cost Engineers Joint Conference, 1977.

Economics

ASTM Standards on Building Economics, 3rd ed. Philadelphia: American Society for Testing and Materials, 1994.

Building Economics: Solving the Owner's Problems of the 80s. Morgantown, WV: American Association of Cost Engineers, 1981.

Economic Studies for Military Construction: Design Applications, Technical Manual 5-802-1. Washington, D.C.: U.S. Department of the Army, 1986

Economic Analysis and Program Evaluation for Resource Management, Army Regulation No. 11-28. Washington, D.C.: Headquarters, Department of the Army, 1975.

Economic Analysis Handbook (P-442). Washington, D.C.: Naval Facilities Engineering Command, 1980.

Engineering Economy, 2d ed. New York: American Telephone and Telegraph Company, Engineering Department, 1970.

Grant, Eugene L., and W. Grant Ireson. *Principles of Engineering Economy.* New York: The Ronald Press, 1970.

Kurt, Max. *Handbook of Engineering Economics.* New York: McGraw-Hill, 1984.

Jelen, F.C. (ed.). *Cost and Optimization Engineering*. New York: McGraw-Hill, 1970.

Johnson, Robert E. *The Economics of Building*. New York: John Wiley & Sons, 1990.

Ruegg, Roslie T., and Harold E. Marhsall. *Building Economics: Theory and Practice*. New York: Van Nostrand Reinhold, 1990.

Samuelson, Paul A. *Economics*, 9th ed. New York: McGraw-Hill, 1980.

Stone, P.A. *Building Design Evaluation: Costs in Use*. London: E. & F.N. Spon, 1975.

Turner, R. Gregory. *Construction Economics and Building Design, A Historical Approach*. New York: Van Nostrand Reinhold, 1986.

Value Engineering

Dell'Isola, A.J. *Value Engineering: Practical Applications. . . for Design, Construction, Maintenance & Operations*. Kingston, MA: R.S. Means, 1982.

Dell'Isola, A.J. *Value Engineering in the Construction Industry*, 3rd ed. New York: Van Nostrand Reinhold, 1982.

Fallon, Carlos. *Value Analysis to Improve Productivity*. New York: John Wiley & Sons, 1971.

Fallon, Carlos. *Value Analysis*, 2nd ed. Washington, D.C.: Lawrence D. Miles Value Foundation, 1980.

Fowler, Theodore C. *Value Analysis in Design*. New York: Van Nostrand Reinhold, 1990.

Kelly, John, and Steven Male. *Value Management in Design and Construction*. London: E. & F.N. Spon, 1993.

Kirk, Stephen J. *Improved Design Decision-Making Using Small Group Value Engineering Gaming/Simulation*, Dissertation. Ann Arbor: University of Michigan, 1992.

Kirk, Stephen. "Integrating Value Engineering into the Design Process." SAVE International Conference, May 1990.

Kirk, Stephen J. "Leadership in Design Team Innovation Using Value–Based Decision–Making Techniques." Cambridge: Harvard University Design School, Executive Education Seminars, 2002.

Kirk, Stephen. "Life Cycle Costing for Value Enhanced Healthcare." Chicago: SAVE International Conference, 1996.

Kirk, Stephen. "Post-Occupancy Value Engineering." SAVE International Conference, 1988.

Kirk, Stephen. "Quality Modeling: Defining Project Expectations." Phoenix: SAVE International Conference, 1994.

Kirk, Stephen. "Strategic Value Planning Using VENTURE Computer Simulation Modeling." Phoenix: SAVE International Conference, 1993.

Kirk, Stephen J. "Value Management Assistance in Design-Build." Washington, D.C.: SAVE International Conference, 1998.

Kirk, Stephen, and Robert Formisano. "Responding to New Market Opportunities: Program Value Management." Phoenix: SAVE International Conference, 1995.

Kirk, Stephen, and Kurt Gernerd, and Khaled Obeid. "Palau VM Study Paves Road to Success." Washington, D.C.: SAVE International Conference, 1998.

Kirk, Stephen, and Richard Hobbs, and Richard Turk. "Value-Based Team Decision-Making to 'Prove' Design Matters." San Diego: AIA National Conference, 2003.

Kirk, Stephen J., and Richard Hobbs, and Richard Turk. "Value-Based Team Design Decision-Making." Charlotte, NC: AIA National Conference, 2002, and Denver: SAVE International Conference, 2002.

Kirk, Stephen J., and Michael M. Paquette. "Scenario Learning for Value Master Planning." San Antonio: SAVE International Conference, 1999.

Kirk, Stephen J., and David Sherwood. "Conversations About Establishing a Value Management Program." Reno: SAVE International Conference, 2000.

Kirk, Stephen J., and Kent F. Spreckelmeyer. *Creative Design Decisions: A Systematic Approach to Problem Solving in Architecture.* New York: Van Nostrand Reinhold, 1988.

Kirk, Stephen J., and Kent F. Spreckelmeyer. *Enhancing Value in Design Decisions.* Lawrence, KS: KU Publishers, 1993.

Kirk, Stephen, and Robert Vrancken. "PIPP – Process Integration for Peak Performance, (Integrating Scenario Planning, Life Cycle Cost Analysis, and Value Management Techniques)." Cincinnati: AIA Conference, 1998.

Kirk, Stephen, and Jill Woller, and George Gish. "Value Enhanced Court Operations and Courthouse Masterplanning." Seattle: SAVE International Conference, 1997.

Miles, Lawrence D. *Techniques of Value Analysis and Engineering*, 2nd ed. New York: McGraw-Hill, 1972.

O'Brien, James. *Value Analysis in Design and Construction.* New York: McGraw-Hill, 1976.

Parker, Donald. *Value Engineering Theory.* Washington, D.C.: Lawrence D. Miles Value Foundation, 1977.

Parker, Donald E. *Management Application of Value Engineering.* Washington, D.C.: Lawrence D. Miles Value Foundation, 1977.

Value Engineering Circular No. A-131. Washington, D.C.: U.S. Office of Management and Budget, 1993.

Zimmerman, Larry W., and Glen D. Hart. *Value Engineering: A Practical Approach for Owners, Designers, and Contractors.* New York: Van Nostrant Reinhold, 1982.

Estimating

ASHRAE 2001 Handbook of Fundamentals. New York: American Society of Heating, Refrigerating, and Air-Conditioning Engineers, 1993.

ASHRAE Energy Standards for Buildings. New York: American Society of Heating, Refrigerating, and Air-Conditioning Engineers, 2001.

Automated Cost Control and Estimating System: Data Base. Washington, D.C.: General Services Administration, 1975.

Boeckh Building Valuation Manual. Milwaukee: Thomson Publishing, bimonthly.

Building Construction Cost Data. Kingston, MA: RSMeans, yearly.

Assemblies Cost Data. Kingston, MA: RSMeans, yearly.

Current Construction Costs. Walnut Creek, CA.: Lee Saylor, yearly.

Dell'Isola, Michael. *Architect's Essentials of Cost Management.* New York: John Wiley and Sons, Inc., 2002.

Design Cost & Data. Tampa: L.M. Rector, quarterly.

Dubin, Fred S., and Chalmers G. Long, Jr. *Energy Conservation Standards for Building Design, Construction, and Operation.* New York: McGraw-Hill, 1978.

General Construction Estimating Standards. Mesa, AZ: Richardson Engineering Services, Inc., yearly.

Liska, Roger W. *Means Facilities Maintenance Standards.* Kingston, MA: R.S. Means, 1988.

Miller, C. William. *Estimating and Cost Control in Electrical Construction Design.* New York: Van Nostrand Reinhold, 1978.

Ostwald, P.F. *Cost Estimating for Engineering and Management.* Englewood Cliffs, NJ: Prentice-Hall, 1974.

Ottaviano, Victor B. *National Mechanical Estimator.* Melville, NY: Ottaviano Technical Services, yearly.

Parker, Donald E., and Alphonse J. Dell'Isola. *Project Budgeting for Buildings.* New York: Van Nostrand Reinhold, 1991.

Trane Air Conditioning Manual. La Crosse, WI: Trane Co., 1977.

U.S. Air Force Manual 88-8, Engineering Weather Data. Washington, D.C.: U.S. Air Force, 1975.

Environmental Sustainability

LEED Reference Guide, Version 2.0. Washington, D.C.: U.S. Green Building Council, 2001.

Environmental Building News. BuildingGreen, Inc. www.buildinggreen.com

Environmental Design & Construction. www.eD.C.mag.com

Green Developments, Case Study Database. Rocky Mountain Institute. www.rmi.org

RSMeans. *Green Building: Project Planning & Cost Estimating.* Kingston, MA: RSMeans, 2002.

Glossary of Terms

Alteration Costs Those costs for anticipated modernization or changing of a facility or space to provide a new function.

Alternatives The different choices or methods by which functions may be attained.

Annualized Method Economic method that requires conversion of all present and future expenditures to a uniform annual cost.

Annuity A series of equal payments or receipts to be paid or received at the end of successive periods of equal time.

Associated Costs These costs may include functional use, denial of use, security and insurance, utilities (other than energy), waste disposal, start-up, etc.

Baseline Date The date to which all future and past benefits and costs are converted when a present worth method is used (usually the beginning of the study period) (synonym *base time*).*

Benefit/Cost Analysis A method of evaluating projects or investments by comparing the present worth or annual value of expected benefits to the present worth or annual value of expected costs.*

Benefit-to-Cost Ratio (BCR) Benefits divided by costs, where both are discounted to a present value or equivalent uniform annual value (synonym *benefit-cost ratio*).*

Brainstorming A widely used creativity technique for generating a large quantity and broad variety of ideas for alternative ways of solving a problem or making a decision. All judgment and evaluation are suspended during the free-wheeling generation of ideas.

Cash Flow The stream of monetary (dollar) values—costs and benefits—resulting from a project investment.

Constant Dollars Dollars of purchasing power in which actual prices are stated, including inflation or deflation. Constant dollars are tied to a reference year.*

Cost Growth An increase (or decrease) in the price of an individual item with or without a corresponding increase (or decrease) in value.

Current Dollars Dollars of purchasing power in which actual prices are stated, including inflation and deflation. In the absence of inflation or deflation, current dollars equal constant dollars.

Cost/Energy Model A diagram, graph, tabulation, or equation whose elements and sub-elements provide a breakdown of the actual, estimated, or target costs and energy units, or both, of an item and its sub-items.

Depreciation An accounting device that distributes the monetary value (less salvage value) of a tangible asset over the estimated years of productive or useful life. It is a process of allocation, not valuation.

Differential Price Escalation Rate The expected percent difference between the rate of increase assumed for a given item of cost (such as energy), and the general rate of inflation.

Discount Factor A multiplicative number (calculated from a discount formula for a given discount rate and interest period) that is used to convert costs and benefits occurring at different times to a common time.*

Discount Rate The rate of interest reflecting the investor's time value of money, used to determine discount factors for converting benefits and costs occurring at different times to a baseline data (synonym *interest rate*).*

Discounted Payback (DPB) Period The time required for the cumulative benefits from any investment to pay back the investment cost and other accrued costs considering the time value of money.*

Economic Life That period of time over which an investment is considered to be the least-cost alternative for meeting a particular objective. *

Energy Costs Costs of electricity and fuel required to operate the facility.

Engineering Economics A technique that allows the assessment of proposed engineering alternatives on the basis of considering their economic consequences over time.

Equivalent Dollars Dollars, both present and future, expressed in a common baseline reflecting the time value of money and inflation. (See *Present Worth Method* and *Annualized Method*.)

Equivalent Full-Load Hours Total energy consumption divided by the full-load energy input. This gives the number of hours a piece of equipment would need to operate at its full capacity to consume as much energy as it did operating at various part loads.

First Cost See *Initial Costs*.

Functional-Use Costs The costs of labor and services required to carry out the intended functions of the building.

Inflation A continuing rise in the general price levels, caused usually by an increase in the volume of money and credit relative to available goods. Inflation can also be described as a decline in the general purchasing power of a currency.

Initial Costs Costs associated with initial development of a facility, including project costs (fees, real estate, site, etc.) as well as construction costs (synonym *first cost*).

Interest Rate See *Discount Rate*.

Internal Rate of Return (IRR) The compound rate of interest that, when used to discount study period costs and benefits of a project, will make the two equal.

Investment Cost First cost and later expenditures which have substantial and enduring value (generally more than 1 year) for upgrading, expanding, or changing the functional use of a building or building subsystem.

Life Cycle See *Study Period*.

Life Cycle Costing An economic assessment of an item, system, or facility and competing design alternatives considering all significant costs of ownership over the economic life, expressed in terms of equivalent dollars.

Maintenance Costs The costs of regular custodial care and repair, annual maintenance contracts, and salaries of facility staff performing maintenance tasks.

Nonrecurring Cost Costs that are not incurred annually over the study period.

Obsolescence The condition of being antiquated, old-fashioned, or out of date, resulting when there is a change in the requirements or expectations regarding the shelter, comfort, profitability, or other dimension of performance that a building or building subsystem is expected to provide. Obsolescence may occur because of functional, economic, technical, or social and cultural change.

Operation Costs The expenses incurred during the normal operation of a building or a building system or component, including labor, materials, utilities, and other related costs.*

Payback Period The time it takes the savings resulting from a modification to pay back the costs involved. A simple payback period does not consider the time value of money. A discounted payback period does.

Present Worth Method Economic method that required conversion of costs and benefits by discounting future cash flows to a baseline date (synonym *present value method*).

Post Occupancy Evaluation (POE) Collection and analysis of information, particularly from users, to assess how well a facility's performance matches user needs and design intent.

Programming Activities that lead to determination of the specific scale, scope, and timing of facility construction; typically precedes consideration of spatial configurations, but when used in the architectural sense, includes determination of required floor areas and adjacencies of various uses of the facility.

Recurring Cost Costs that recur on a periodic basis throughout the life of a project.

Replacement Costs Building component replacement and related costs, included in the capital budget, that are expected to be incurred during the study period.*

Resale Value The monetary sum expected from the disposal of an asset at the end of its economic life, its useful life, or at the end of the study period.*

Return on Investment (ROI) See *Internal Rate of Return*.

Salvage Value The value of an asset, assigned for tax computation purposes, that is expected to remain at the end of the depreciation period.*

Savings-to-Investment Ratio (SIR) Either the ratio of present worth savings to present worth investment costs or the ratio of annualized savings to annualized investment cost.*

Sensitivity Analysis A test of the outcome of an analysis by altering one or more parameters from initially assumed values.*

Simple Payback (SPB) Period The time required for the cumulative benefits from an investment to pay back the investment costs and other accrued costs, not considering the time value of money.*

Study Period The length of time over which an investment is analyzed (synonyms *life cycle, time horizon*).

Sunk Cost Cost which has already been incurred and should not be considered in making a new investment decision.

Target (cost or energy) A goal, expectation, budget, or estimate of the minimum an item should cost (or consume energy) during one or more phases of its life cycle. There is often very poor correlation of the value of a cost (energy) target for an item relative to the needs and wants contained in the requirements, specifications, and conditions for the item. The value of the target is frequently based on how much the item is worth to the buyer and user of the item. The worth of an item, as used in value engineering, is the lowest possible target for the item.

Tax Elements Those assignable costs pertaining to taxes, credits, and depreciation.

Terminal Value The value at the end of the study period (not including investment cost) of net cash flows and their earnings from reinvestment.

Time Horizon See *Study Period*.

Time Value of Money The time-dependent value of money stemming both from changes in the purchasing power of money (that is, inflation or deflation), and from the real earning potential of alternative investments over time.

UNIFORMAT A framework of construction cost categories suggested by the AIA and developed by the U.S. General Services Administration to be used by professionals. Costs are organized by building systems, not by construction labor trades.

Useful Life The period of time over which an investment is considered to meet its original objective.*

Value Engineering (VE) A creative, function-oriented, multi-disciplined, team approach whose objective is to optimize the life cycle costs and performance of a facility.**

---

* ASTM E833-91a, *Standard Terminology of Building Economics*, May 1991.
** Alphonse J. Dell'Isola, *Value Engineering in the Construction Industry*, 3d ed. New York: Van Nostrand Reinhold, 1982.

Index

A

After-inflation discount rate, 43
AIA (American Institute of Architects), 5, 10, 372
Air conditioning data, 323
Alterability factor, 116
Alteration costs, 160, 383
Alternatives, 383
American Consulting Engineers Council, 372
American Institute of Architects, 5, 10, 372
American Society of Heating, Refrigeration and Air Conditioning Engineers (ASHRAE), 102, 159, 371
    ASHRAE Systems Handbook, 102
    Standard, 102
American Society for Testing and Materials (ASTM), 106, 107, 374
Americans with Disabilities Act (1990), 5
Analysis period, 47-49
    Arbitrary life, 48
    Component life, 47
        Common multiple, 47
    Costs, 49
    Facility life, 47
    Investment (mission) life, 47
    Organizational policy, 48
    Planning horizon, 48
    Present time, 49
    Recommended, 48
Annual heating degree days, 322
Annualized method, 383
    Calculations worksheet, 6
Annuity, 383
Artificial lighting, 192
ASHRAE (*see* American Society of Heating, Refrigeration and Air Conditioning Engineers)
AXCESS, 160

B

Baghouse, 195-196, 224-225
Bank, case study, 189, 211-214
Base year, 44
Baseline date, 383
Benefit/Cost Analysis, 383
Benefit-to-Cost Ratio (BCR), 383
BLAST, 102, 159
"BMDB (Building Maintenance, Repair and Replacement Database) for Life Cycle Cost Analysis," 106
Brainstorming, 156-158, 368
Building Construction Cost Data, 91
Building Cost Modifier, 91
Building Design for Maintainability, 104

Building exterior closure, sensitivity analysis example, 143-146
"Building Maintenance Repair and Replacement Database (BMDB) for Life Cycle Cost Analysis," 106
Building owning and operating labor trades, 114

C

Campus university planning, case study, 200-202, 230-232
Case studies, LCC 185-236
    Office to Museum Renovation, 186
    District Court Consolidation, 188
    Branch Bank Prototype Layout, 189
    Health Care Facility Layout, 189
    Daylighting, 192
    Glass/HVAC Replacement, 193
    Lighting System, 194
    Elevator Selection, 194
    Equipment Purchase Examples, 195
    Construction and Service Contracts, 197
    Chemical Stabilization: Plant– Dryers vs. Windows, 199
    Campus University Planning Using LCC and Choosing Advantages, 200
    High School Trade-Off Analysis of Initial vs. Staffing Costs, 203
    Highway for Regional Transportation Authority, 203
    Life Cycle Cost Assessment— HVAC System for a High School, 203
    Lease vs. Build LCC Analysis, 204
Cash flow, 383
Ceiling, coffered vs. flat, 192
Chemical stabilization plant: dryers vs. windows, 199-200
Chillers, case study, 198
Choosing by advantage (CBA), 201-202 (see Confidence index)
Collateral costs, 51
Computer simulation energy profile systems, 339
Confidence index, 126-127, 129-150
    Calculations, 131-133
    Computation form, 348
    Examples, 133-135
Constant dollars, 383
Construction and service contracts, 197-199
Consumer price index, 40
Contingency planning, 247-249
    documentation, 248-249
    planning approaches, 247-248
    role of, 247
Continuing costs, 52
Correctional facilities, staffing costs, 244
Cost analysis form, life cycle, 345-346
Cost/Energy Model, 384
Cost growth, 384
    Current dollar, 384
    Differential escalation, 44
Cost of ownership, 13, 42
Cost in use, 7
Costs types,
    Alteration and replacement, 50, 115
    Associated, 51
    Downtime, 51
    Common, 51
    Collateral, 51
    Continuing, 52
    Energy, 50
    Functional use, 51
    Initial project (investment), 50
    Operation and maintenance, 50
    Sunk, 52
    Terminal (salvage value), 51
    (See also Equivalent costs; Life cycle cost estimating)
Costing summary form, life cycle, 342, 350
Cost model form, life cycle, 353, 355
CSI (Construction Specifications Institute), 81, 82
Costs, estimated, back-up calculations form, 351
Court, case study, 188-189, 210
Current dollars, 384

Custodial costs (*see* Maintenance, repair and custodial costs), 160

D

Data processing center layout, 70, 71, 72
Data sheet, life cycle, 352,
Daylighting, case study, 192-193, 217-219
Demolition, 160
Dental clinic feasibility study worksheet, 73
Department of Health Education and Welfare, 7
Depreciation, 384
Design alternatives, 155-157
    brainstorming, 156-157
Design process
    application in, 17
    planning and feasibility, 21
Differential escalation, 384
DINDOW, 102
Discount factor, 384
Discount rate, 42
    Cost of borrowing money, 43
    Minimum attractive rate of return, 43
    Opportunity rate of return, 43
Discounted payback period, 62, 384
District court consolidation, 188
Documentation, LCC, 248-250
DOE-2, 159-160

E

Economic analysis (EA), 55-61
    Approaches, 61-64
    Benefits and costs, 60-61
        Non-monetary, 60
        Private sector treatment, 61
        Public sector treatment, 61
    Categories of cost, xix
    Cost-centered study, 56
    Initial-cost study, 56
    Office building example, 57
    Period, xix
    Present worth method, xxi, 37
    Procedures
        Payback period, 62
        Return on investment, 63
    Relationship between quality and cost, xxiii, 22
    Restricted study, 56
    Single-alternative study, 56
    Terminology, 55
    Types of
        Design, 59-60
        Investment, 59
        Primary, 59
        Secondary, 59
Economic criteria, checklist, 155
    identifying, 154
Economic life, 41, 384
Economic risk assessment, 125-151
    confidence index approach, 129-150
    range of probable values, 127-129
Economic study (*see* Economic analysis), 55
Economic tables, 305-318
    interest for life cycle costing, 306-310
    present worth of an escalating annual amount, 311-315
    years to payback, 316-318
Elevator LCC considerations, 329-336
    capacity, 331-332
    codes, 333-334
    definition, 329
    discipline requirements, 332-333
    guide rail attachments, 334
    heating and ventilation, 334-335
    layout within a building, 330
    installation costs, 333
    limitations, 329-330
    peak passenger traffic, 330-331
    types of, 239-330
Elevator selection, case study, 194-195, 223
Energy 10, 102
Energy conservation, 192
Energy Conservation Act, 371

Energy costs, 159-160, 239, 244, 245, 384
    Environmental (weather), 100
  Energy modeling
    Domestic hot water and equipment, 102
    General, 102
    Heating, cooling, and ventilation, 102
    Lighting, 102
    Power, 102
    Service center, 101
  Estimating methods
    Accuracy, 85
    Data, 99
    Degree days, 94
    Equivalent full-load hours, 94
    Hour by hour, 94
    Outside temperature bins, 95
  Impact, 93
  In life cycle cost analysis study, 55
  Performance
    Standards, 96
Energy-estimating data, 319-327
  air conditioning data, 323
  annual heating degree days, 322
  equivalent full-load hours of operation per year, 324
  factors for determining heat loss, 320
  heating energy calculation, 321
  IESNA recommended illumination levels, 326
  lumens per watt, 327
  ventilation energy, 325
  watts for popular fixture types, 327
Energy model form, 354
Engineering Economics, 384
Engineered Performance Standards (EPS), 106
Engineering economic study (analysis), 55
Engineering News-Record, 91
Environmental sustainability, 4
EPA (Environmental Protection Agency), 5
Equipment purchasing, case study, 195-197
Equivalent costs
  Annualized method, 39, 72
  Present worth method, 37
Equivalent dollars, 384
Equivalent full-load hours of operation per year, 324, 384
Escalator LCC considerations, 329, 335-336
Estimating life cycle costs, sources of information, 91
Evaluation,
  primary, 157-158
  weighted, 161-163
Exterior closure, sensitivity analysis example, 143-146

F

FAIA (Fellow of the American Institute of Architects), 10
Factories, factors for determining heat loss, 320
Factors for determining heat loss table, 320
Fenestration, 192
Final report, 338
Floor finishes, confidence index example, 134, 136
Forms
  Annualized costs in life cycle cost analysis, 345
  Confidence index computation, 348
  Energy model, 354
  Estimated costs, 351
  Idea comparison advantages and disadvantages, 343
  Initial cost model, 353
  Life cycle cost model, 355
  Life cycle costing summary, 342, 350
  Life cycle data sheet, 352
  Payback period, 347
  Sensitivity break-even analysis, 349
  Weighted evaluation, 344

Worksheet, general purpose, 346
FRESA, 102
Functional-use (staffing) costs, 384

G

Glass replacement, case study, 183
Glass, low-E, 193
Government requirements for LCC, 357-368
    Alaska, 357-359
    OMB Circular No. A-94 (Life Cycle Costing Guidelines for the Federal Government), 359-366
    OMB Circular No. A-131 (Value Engineering for the Federal Government), 367-368
    State of Alaska law, 357-359

H

Heating energy calculation, modified degree day formula, 321
Health care facility, 189-191, 215-216
    Glass/HVAC replacement, 193
    Layout, 189
High school trade-off analysis, case study, 203, 233
Highway for regional transportation authority, case study, 203, 234
Historical development of LCC, 369-374
Hot water system, sensitivity analysis example, 149-150
Hospital (*see* Health care facility), 11, 189
Hot2000, 102
HVAC, case study, 183, 203, 235
HVAC system, for health care facility, 190, 191, 220

I

Idea comparison form, 343
IESNA recommended illumination levels, 326

Inflation, 27, 39, 40, 43-47, 363, 384
    Constant dollar, 44
    Current dollar, 44
    Differential escalation, 44
    Nonrecurring costs, 45
    Recurring cost, 45
Initial costs, 385
Initial cost model form, 353
Installation costs, of elevators, 333
Interest formulas
    Periodic payment, 34
    Present worth of annuity, 34
    Single present worth, 33
    Uniform compound amount, 36
    Uniform sinking fund, 35
Interest tables, 35, 306-310
Internal rate of return (return on investment), 63, 385
Investment costs (*see* initial project costs), 86, 385

L

Lawrence Berkeley National Laboratory, 102
LCC (Life cycle costing)
    forms, 341-355
        confidence index computation, 348
        energy model, 345
        estimated costs, back-up calculations, 351
        idea comparison, 343
        initial cost model, 353
        life cycles cost analysis, 345-346, 350
        life cycle costing summary, 342
        life cycle cost model, 355
        life cycle data sheet, 352
        payback period, 347
        sensitivity break-even analysis, 349
        weighted evaluation, 344
    historical development, 369-374
    study areas, 242, 244, 245
    sources of data, 110

Lease vs. build LCC analysis, case study, 204, 236
LEEDS, 374
Life cycle, 41
Life cycle cost, 42, 385
Life cycle cost analysis (LCCA), 159-161
    approaches, 65
        present worth approach, 65-67
    (*see also* Case studies; Forms; Risk assessment)
    for design professionals, 7
    priorities, matrix, 246-247
    scope of services, sample, 337-340
Life cycle costing,
    strengths, 238-239
    weaknesses, 238-239
Life cycle cost analysis study
    Associated costs, 119-123
        Administrative costs, 120
        Denial-of-use costs, 121
        Interest (debt service), 120
        Miscellaneous costs, 122-123
        Staffing (functional-use) costs, 120
    Construction Specifications Institute, 81, 82
    Framework, 80-83
    Maintenance, repair, and custodial costs, 103-115
    Replacement costs, 108-109
    UNIFORMAT, 88, 91, 92, 104, 109
Life cycle costing (LCC)
    Assessment elements, 13
        Alteration and replacement costs, 14
        Associated costs, 14
        Denial-of-use costs, 14
        Financing costs, 13
        Functional-use costs, 14
        Logic, 15, 18
        Initial costs, 13
        Insurance, 14
        Maintenance costs, 13
        Salvage value, 15
    Definition, 12
    Historical development, 369-374
        Energy Highlighted, 372
        Procedures, 371-372
    Methodology
        Phase cost reduction, 17
        Present worth cost conversion, 15
    Program requirements, 24
    (*see also* Case studies; Forms; Government requirements)
Life Cycle Cost Data, 106
Life Cycle Cost Data, 255-301
    Introduction, 255
    Maintenance and replacement estimating data, 257-301
        Structural, 257-260
        Architectural, 260-268
        Mechanical, 268-289
        Electrical, 289-290
        Equipment, 297
        Sitework, 297
Life cycle cost modeling, 244-245
Life cycle costs and benefits, 158-162
Life cycle costs, initial, 192
Lighting system, 194
Lighting, artificial, 192
Lumens per watt, 327

M

Maintenance and repair costs, annual, 132, 134, 140, 147, 149
Maintenance costs, 160, 239, 244, 245
Maintenance, repair, and custodial costs, 385
    Estimating methods and data
        Labor trades, 112
        Labor wages, 112
Management considerations, 237-252
    contingency planning, 247-249
    level of effort, 239
    planning the overall study effort, 239-247
    strengths and weaknesses of life cycle costing, 238-239

Management briefing for life cycle costing, xv-xxiv
Management considerations, xv-xxiv
Manufacturing facilities, staffing costs, 244
Marshall and Swift Publication Company, 91
Marshall-Valuation service, 91
Maslow, Abraham, 9
MASTERCOST, 83
MasterFormat, 371
Mechanical Cost Data, 91
Microchip manufacturing facility, 245
Moving ramps, LCC considerations, 335-336
Museum renovation, 186-187, 205-209

N
National Bureau of Standards, 371
National Institute of Standards and Technology, 371
National Energy Conservation Policy Act (1978), 8
National Energy Conservation Policy, 373
Net present value, 360
Nonrecurring costs, 385
Nursing tower, 74, 75, 76

O
Obsolescence, facility, 385
Office building
    Initial costs, 83
    Life cycle cost distribution, 242
Office of Management and Budget (OMB)
    Circular A-94 (Life Cycle Costing Guidelines for the Federal Government), 359-366
    Circular A-131 (Value Engineering for the Federal Government), 357-359
Operation and maintenance costs, 385
Owner demands, 8

P
Payback period, 385
Payback period form, 347
Post Occupancy Evaluation (POE), 338, 385
Present worth (PW)
    Method, 385
    Tables of an escalating annual amount, 311-315
Probabilistic approach, to risk assessment, 125
Programming, 385
Project costs (*see* Initial costs), 83, 159
PW (Present worth) values, 126-129
PWA (Present worth of annuity), calculations, best estimate, 141

R
Recurring cost, 385
Refrigerator-freezer, 196, 225
Repair costs (*see* Maintenance, repair, and custodial costs), 160
Replacement costs, 160, 386
Resale Value, 386
RESFEN, 102
Retscreen, 102
Return on investment (internal rate of return), 386
Risk assessment, 125-151, 163, 171, 174
    confidence index approach, 129-150
    range of probable values, 127-129
Roofing maintenance, repair, and custodial costs, 107
Rooftop air-handling system, 66, 68, 69

S
Salvage value, 160, 386
Sash, fixed vs. operable, 192
Savings to investment ratio (SIR), 63, 386
Schematic and design development phases, 338

apportioning resources, 240-241
Scope of services, LCC, sample, 337-340
Scope of work, 337-338
Sensitivity analysis, 386
Sensitivity approach to risk assessment, 125, 127, 135-142, 143-151, 366
Sensitivity break-even analysis form, 349
Service contracts (*see* Construction and service contracts), 228
Simple payback period, 62, 386
Single compound amount (SCA)
Siting and layout of research laboratory, 134-135
Solar-assisted domestic hot water system risk assessment, 149-150
Staffing (functional-use) costs, 244
Study Period, 386
Sunk costs, 52, 386

T

Target (cost or energy), 386
Tax elements, 386
Terminal value, 386
THERM, 102
Time horizon, 386
Time value of money, 28-41, 386
    Cash flow diagrams, 28-31
    Interest formulas, 28, 31
        Single compound amount, 31, 32
        Single present worth, 31, 32
        Uniform compound amount, 31
    Interest tables, 35
        Periodic payment, 32
        Present worth of annuity, 33
        Single present worth, 33
        Uniform sinking fund, 33, 35
Total quality management (TQM), 4
TRACE, 160, 373
Trends, recent, 7
Trends, that support need for LCC, 3
    Environmental sustainability, 4
    Operational effectiveness (including re-engineering), 4
    Total quality management, 4

Value engineering, 4
Owners' Rising Expectations, 4

U

Uniform Building Component Format (UNIFORMAT), 83, 84-85, 104, 109, 371, 386
Unit life, 27
University Teaching Laboratory, Budget cost estimate, 89
    Cost model, 90
Useful life, 386
U.S. Department of Energy, 160, 373
U.S. Department of Health and Human Services, 373
U.S. General Services Administration, 371
U.S. Office of Management and Budget, 186

V

VAHBS, 6
Value engineering (VE), 4, 64, 387
    Interrelationship of LCCA and, 23
Value incentive clause (VIC), 226-227
Ventilation energy, 325
Verification of predominant uncertainty, 142-145

W

Wall, exterior study, 172-183
Warehouses, factors for determining heat loss, 320
Water-cooling system, sensitivity analysis example, 146-148
Watts for popular fixture types, 327
Weighted evaluation form, 344
Window washing, 192

Y

Years to payback tables, 316-318